THE ENLIGHTENMENT OF
JOSEPH PRIESTLEY

ROBERT E. SCHOFIELD

THE ENLIGHTENMENT OF
JOSEPH PRIESTLEY

*A Study of His Life
and Work from
1733 to 1773*

THE PENNSYLVANIA STATE UNIVERSITY PRESS
UNIVERSITY PARK, PENNSYLVANIA

Library of Congress Cataloging-in-Publication Data

Schofield, Robert E.
    The enlightenment of Joseph Priestley : a study of his life and
work from 1733 to 1773 / Robert E. Schofield.
        p.   cm.
    Includes bibliographical references and index.
    ISBN 0-271-01662-0 (alk. paper)
    1. Priestley, Joseph, 1733–1804.   2. Unitarian churches—Clergy—
Biography.   3. Chemists—Biography.   I. Title.
BX9869.P8S36   1997
540'.92—dc21
[B]   96-54484
                                                                      CIP

It is the policy of The Pennsylvania State University Press to use acid-free paper
for the first printing of all clothbound books. Publications on uncoated stock
satisfy the minimum requirements of American National Standard for Information
Sciences—Permanence of Paper for Printed Library Materials, ANSI Z39.48-
1992.

Frontispiece: The "Leeds Portrait" of Joseph Priestley, c. 1763–65. Artist
unknown. Painting in the possession of the Royal Society of London, reproduced
here by permission, Officers of the Royal Society.

Contents

List of Illustrations

Preface

This work was begun some thirty years ago as a labor of love and intense interest, when I first discovered there was no definitive biography of Joseph Priestley, one of the most distinguished polymaths of eighteenth-century England. It soon became painfully clear just why this was the case: Priestley was simply too many persons for any one easily to comprehend them all into a single study. Yet, in the long run, thoroughly to understand Priestley and his age, it must sometime be necessary to join those varied manifestations of his life into a study that might begin to integrate and illuminate the extraordinary career of this exemplar of British Enlightenment. Properly done, the subject of middle-class, Dissenting, intellectual achievements in eighteenth-century England would also be illuminated, as had not previously been done.

The task has been addressed with continued interest through various related research projects and publications, amid interruptions of teaching and changes of position. I now present the first fruit of an attempted solution to the Priestley manifold. It is only one of the two volumes I had planned for a complete story of Priestley's life. But volume 2 is not yet completed and retirement and two bypass operations have forced me to accept temporal realities. I can no longer choose to delay publication of half my work in anticipation of completing the whole.

A study of the first forty years of Priestley's life has, in any event, an intrinsic logic to justify its separate appearance. These were the years of preparation and trial, during which Priestley qualified for the achievements that were to make him famous. Even so, this is not even half of the definitive biography I had once hoped to write. A cursory scanning of the more than one hundred and fifty titles in Priestley's bibliography shows contributions in language study, English grammar, philosophy of education, rhetoric, politics, history, religion, and biblical criticism, as well as the science for which he is best known. He was not, perhaps, the premier contributor in any one of these, but he made significant contributions to each of them.

To appraise those contributions usefully required information that could only be supplied by the expertise of authorities in each appropriate field and, for too many areas, work relevant to Priestley had not been done.

This is not to say that Joseph Priestley himself has not been the subject of recent research and writing. The pages of the *Price-Priestley Newsletter* 1–4 (1977–80), which has become *Enlightenment and Dissent* (vol. 1, 1982), amply testify to a strong continued interest in the man. So also do many other publications, most notably those of Isaac Krammick on Priestley's politics, John McEvoy on the philosophy of his science, and Simon Schaffer in sociological history. I have incorporated some material from these studies, among others, as evidenced in my notes, but the major concern of most of them is for the work of the matured Priestley and must be dealt with in a second volume if I get to that. For this volume, I was primarily dependent (as inevitably I shall be in any future volume) on the writings of Priestley and the comments and criticisms of his contemporaries.

This, at least, is the first biography of Priestley's early years that attempts to include all of the many activities in which he was engaged and discuss all of the works which he wrote and published during the critical first forty years of his life. Though it is not definitive, it is, I think, complete. And having pointed out some dimensions of the problem and its importance; perhaps now all the various pieces can be studied individually and the whole then addressed by a biographer luckier and more skilled than I.

For all its inevitable inadequacies, I could not have achieved what I have without the assistance of many more persons and institutions than can possibly be thanked individually. Still, one must acknowledge the assistance of the history faculties and the administrations of the University of Kansas, Case Institute of Technology (later part of Case Western Reserve University), and Iowa State University. The financial assistance of the National Science Foundation, the American Philosophical Society, a Fullbright Fellowship, and the John Simon Guggenheim Foundation was essential, as were the hospitality and financial assistance of the Institute for Advanced Study, Princeton.

I have noted in the text my specific obligations to many of the individuals who aided my research. There were many others: librarians in general, of course (and what would the historian do without their help) and specifically those of the Royal Society of London, the Royal Society of Arts, the Dr. Williams's Library, and the American Philosophical Society. Generations of graduate students have helped with their comments—and occasional incomprehensions. I am also grateful for the advice of friends who have read and criticized portions of this work. They cannot be responsible for its blemishes, but they share the credit that there are not more. Among these friends are: my wife, Mary-Peale, of course; the Iowa State University history faculty colloquia, "The History Vigilantes"; and especially Dr. Achilles Avraamides, for their occasionally patient reading and acid comments.

And, finally, my thanks and acknowledgments are due the following institutions and persons for their permissions to quote, or cite extensively, from manuscripts or copyrighted materials:

The Bodleian Library, University of Oxford, for Thomas Bentley and Priestley letters to Samuel Pegge.

The British Library, Department of Manuscripts, for Priestley letters to Thomas Birch and to John Wilkes.

Burndy Library, Smithsonian Institution Libraries, Washington, D.C., for Priestley letter to Newcome Cappe.

Mr. G. A. Carter, for his edition of William Turner's "Warrington Academy."

Cheltham Society, Council for Herbert McLachlan, "Warrington Academy," *Remains Historical and Literary connected with the Palatine Counties of Lancaster and Chester*, n.s., 107 (1943).

John and David Clark, for the MS Diary of the Reverend Henry Crooke, in the Clark MSS Leeds District Archives.

Columbia University Press, for Mary P. Mack, *Jeremy Bentham: An Odyssey of Ideas* (New York, 1963).

Dickinson College, the Boyd Lee Spahr Library, Carlisle, Pennsylvania, for Priestley letter to John Seddon.

Duke University Press, for *The Correspondence of Richard Price*, vol. 1, ed. D. O. Thomas and Bernard Peach (Durham, N.C., 1983).

Edinburgh University, Library, for MS Minute Book, *Senatus Academicus*, vol. 1.

Enlightenment and Dissent, for "Joseph Priestley's Journal while at Daventry Academy," 13 (1994): 49–113.

Hakluyt Society, for *The Voyage of the Resolution and Adventure, 1771–1775*, ed. J. C. Beaglehole (Cambridge, 1961).

Heckmondwike, West Yorkshire, Upper United Reformed Church, for MS Minute-Book of the Independent Church, Heckmondwike, 1741–55.

Harris Manchester College, Oxford, Library, for Matthew Turner letter to John Seddon and MS Catalogue of Books in the Library of Warrington Academy.

The MIT Press, for *A Scientific Autobiography of Joseph Priestley*, ed. Robert E. Schofield (Cambridge, Mass., 1966).

The John Rylands University Library of Manchester, Director and University Librarian, for Priestley letters to John Seddon.

State Library of New South Wales, Mitchell Library, for *The Endeavour Journal of Joseph Banks, 1761–1771*, ed. John C. Beaglehole (Sidney, New South Wales, 1962).

The Royal Society of London, Council, for the "Leeds Portrait" of Joseph Priestley and copy, Priestley letter to Anna Laetitia Barbauld, Yates Priestleyana.

Southern Illinois University Press, for *A Course of Lectures on Oratory and Criticism by Joseph Priestley*, ed. Vincent M. Bevilacqua and Richard Murphy (Carbondale, Ill., 1965).

Dr. D. O. Thomas and Beryl Thomas for prior-to-publication transcripts of Richard Price shorthand draft letter to Priestley and Priestley's Daventry Journal.

The United Reformed Church in the United Kingdom, 87 Tavistock Place, London WC1H 9RT, Independent Press, for J. W. Ashley Smith, *The Birth of Modern Education: The Contribution of the Dissenting Academies, 1660–1800* (London, 1954, 1972).

Warrington Library, Cheshire County Council, Libraries and Archives, for MS Minutes of the Proceedings of the Trustees of Warrington Academy, and MS Minutes of the Circulating Library, Warrington, 1760–67.

The Dr. Williams's Library for Priestley letters to Caleb Rotheram and Theophilus Lindsey and photograph of lithograph of Priestley's birthplace.

I

BIRSTALL FIELDHEAD
AND HECKMONDWIKE,
1733–1752

In 1752, at age nineteen, Joseph Priestley was refused membership in the Independent Chapel of Heckmondwike, West Riding, Yorkshire. He had, the elders thought, unsound opinions respecting the sin of Adam![1] This event, traumatic enough in the life of any young eighteenth-century English Dissenter, was doubly so in that of the young Priestley. This was the church of his boyhood, center for the social and family activities of Nonconformists of the area. His aunt-guardian was a member of the congregation, as were his father, uncles and aunts, and several cousins. One of the latter, another Joseph Priestley, was elected deacon the year that young Joseph was rejected. John Kirkby, minister of the chapel for more than forty years, had been his teacher for a time and joined his aunt in urging the application for membership. And Priestley, aimed since childhood for the Dissenting ministry, was on the point of leaving home for further education.

1. Joseph Priestley, *Memoirs and Correspondence, 1733–1787*, vol. 1, part 1, p. 14; in *Theological and Miscellaneous Works of Joseph Priestley*, ed. John Towill Rutt (New York: Kraus Reprint, [1831] 1972; hereafter cited as W 1/1; page numbers follow, preceded by colon. MS Minute-Book of the Independent Church, Heckmondwike, 1741–1755, 36–38; a copy of these pages was sent me by the courtesy of the Reverend Arthur H. Tebbet, minister of Upper United Reformed Church, Heckmondwike, and is used by permission. Note the minute-book does not note rejection of Joseph's application.

Fig. 1. Joseph Priestley's birthplace, Birstal Fieldhead, Yorkshire, from a lithograph in the Dr. Williams's Library, London. Used by permission.

Writing about the rejection, some thirty-five years later, Priestley exhibited no animosity. Although the "rigor of the congregation" had "incommoded" him, he was grateful for the spiritual discipline they had instilled.[2] But time and a distinguished career can assuage ancient injuries. Rejection by the congregation marked a critical juncture in Joseph's life; it broke one of his few links to family and local mores and, limiting his options, directed his future activities along paths the church elders could scarcely have imagined. Their action was, however, a reaction to that vigorous independence of mind, custom, and authority that was to be characteristic of Joseph Priestley's entire life. That this independence should manifest itself so early suggests that it was, at least in part, the result of a personality and attitudes formed by an idiosyncratic childhood.

Joseph Priestley was born 13 March (old style) 1733, in his father's house in Fieldhead, Birstall Parish (Fig. 1), about six miles southwest of Leeds,

2. Priestley, W 1/1:15–16.

first child of Jonas Priestley and his wife Mary. Priestleys were (and are) a large family in Yorkshire and Lancashire; records back to the early sixteenth century attest their presence in the West Riding and Joseph's immediate family had been settled in Fieldhead for several generations. His grandfather Joseph and at least one uncle, John Priestley, were living there when he was born. Nevertheless, when he was taken from home, his "mother having children so fast," it was to live with his mother's father, Joseph Swift, a farmer and maltster of Shafton, near Wakefield and about fifteen miles from Fieldhead. He was probably about a year old when first taken to live at Shafton (brother Timothy was born in June 1734), and there he was to stay, as Priestley writes, "with little interruption till my mother's death."[3]

His mother was to bear five more children, three sons and two daughters, dying on 28 December 1739, shortly after the birth of his youngest brother. Joseph was not kept at Shafton all the five intervening years; he records that his mother taught him the Shorter Westminster Catechism "the little time that I was at home," and once required that he return a pin he had found while playing at his cousins'. He was not, however, to live at his family home until the death of his mother, when some of his brothers were sent to Shafton to take his place.

Birstall Parish, in which Fieldhead was located, is in the heart of the woolcloth-making district of the West Riding. Information is scant as to the size of Fieldhead as Priestley knew it. Birstall Village, about three miles away, was not a township and figures from the 1801 census included the towns of Greater and Lesser Gomersall, Heckmondwike, Birstall village, and Fieldhead, in a total population of 4,303. The parish was already studded with so many grey-stone hamlets as to make it nearly one continuous village, but there were also green fields and country lanes in which Priestley might walk on his way to and from the "school in the neighborhood" to which he was sent while living at home.[4] He was not to stay at home long;

---

3. Timothy Priestley, *A Funeral Sermon occasioned by the Death of the late Rev. Joseph Priestley. . . . To which is added, a True Statement of many important Circumstances relative to those Differences of Opinion which existed between the Two Brothers* (London: for Alex. Hogg and Co., 1805), 35, writes, "When about four years and a half old, Joseph and I went to abide with our grandfather. . . ." Joseph says nothing about his brother being with him when he first went to Shafton and it may be supposed that Timothy remembers a subsequent joint removal.

4. It is unlikely that the school was Birstall Grammar School, as this was poorly endowed, badly run, and in the nomination of the parish vicar, notoriously anti-Dissenter. Priestley probably attended a dame school in the vicinity where he would be taught reading, writing, and sums.

his father remarried in 1741 and Joseph was sent that year to live with his father's older sister and her husband, Sarah and John Keighley, at Heckmondwike.[5]

The Keighleys, childless after seventeen years of marriage, lived at Old Hall, an Elizabethan manor house some three miles from Fieldhead. In minor respects, Joseph's life was little changed. He had already lived mostly away from his family and, at Heckmondwike, was near enough to see them with some frequency. The chapel of the Independent congregation, of which the family were members, was nearby and they dined at Old Hall between Sunday services. Timothy Priestley retained memories of his older brother in casual meetings during this period. But in one important way, his situation was greatly changed. For John Keighley was a man of considerable property; when he died in 1745 leaving, in Joseph's words, "the greater part of his fortune to my aunt for life and much of it at her disposal after her death," Joseph, at age twelve, became her presumptive heir.

There is little doubt that his widowed aunt spoiled young Joseph. When she discovered how bright he was, she determined that he become a minister rather than follow the clothmaking and finishing trade of his grandfather, father, uncles, and cousins. Did his family resent these changes and act upon their resentments? This would help explain some of their later dealings; it might also contribute to the near-total absence of family influences or references appearing in his later life.

One cannot assume much from the absence of records, particularly for the early part of Priestley's life before fame prompted the saving of his letters and papers. The one extant substantial record of that period, Priestley's journal or diary for the period 24 April to 31 December 1754, reports letters written to father, sisters, and brother "Timmy"; a visit home during vacation from school, with calls to and from father, uncles, cousins, and brothers Timothy and Joshua and even a couple of days of communal labor mowing a field and making hay. Even so, it is a curiously impersonal record, with family members no more in evidence than the local ministers or host

---

5. Priestley dates his father's remarriage at 1745 (W 1/1:4), but his half-sister Ann, who died at twenty in 1763, must have been born in 1742, putting the second marriage at least in that year and more likely the previous one; see Samuel Parkes, "Mr. Parkes's Account of a Visit to Birstall, Dr. Priestley's Native Place," *Monthly Repository* 11 (1816): 274–76. Ronald A. M. Dixon of Thearne, "James Priestley and Cumberland University, U.S.A.," *Transactions of the Unitarian Historical Society* 4 (1927–30): 412–16, gives the date of the remarriage as 15 January 1741.

of friends also visited and written. And the later record is barren.[6] His published *Memoirs*, written to leave "some account of my friends and bene-factors"; the autobiographical remarks, in preface or text, in the more than one hundred and fifty books and pamphlets he had published during his lifetime; the nearly five hundred extant letters—all fail to provide significant information about his family.

Nothing but their names and occupations is recorded of his grandfathers, Joseph Swift and Joseph Priestley, though he lived with the former for nearly five years and the latter lived in Fieldhead and did not die until 1745.[7] His mother he scarcely knew. His stepmother, "a woman of good sense, as well as of religion," he knew little better; she had died by 1752. His father is identified by name and occupation in the *Memoirs* and there is reference to his "strong sense of religion," manifest in a later adoption of Whitfield's Calvinist Methodism. He died in 1779, without any other notice in Joseph's correspondence or publications. Of Joseph's three brothers, two sisters, and three half-sisters, Timothy is named in the *Memoirs* and appears in occasional letters or published references; the others might not exist so far as Priestley gives information about them. He had family and relations in Fieldhead, Leeds, and London, but served as minister in Leeds for six years without leaving any indication of visits to Fieldhead, nor to relatives in Leeds, and he frequently visited London, seeing friends (but not, appar-ently, his relatives) there.

Yet he was not a cold man, nor resentful or vindictive. Though his aunt disappointed his expectations and then disowned him, his references to her in the *Memoirs* are grateful and admiring. "She was truly a parent to me," a "truly pious and excellent woman, who knew no . . . use of wealth, or of talents . . . than to do good." In 1777, remonstrating upon Calvinism, he wrote: "I shall ever reflect with gratitude, that the person to whom . . . I have been under the greatest obligation, was at the same time a strict Calvinist, and in all respects as perfect a human . . . as I have yet been acquainted with. I had the fairest opportunity of observing . . . and I now frequently reflect . . . with satisfaction and improvement. . . . All who

6. For the journal, see Chapter II; some letters to sister Martha, during the crisis of 1791 and from the United States are extant, while brother Joshua recalled that Joseph provided for his permanent support before leaving England in 1794; see Samuel Parkes, "Mr. Parkes's Account," 275.

7. In his *Familiar Letters addressed to the Inhabitants of Birmingham* (Birmingham: for J. Johnson, 1790), 17, Priestley claims that his grandfather was a churchman, but does not identify which one; other sources suggest it was grandfather Priestley.

knew *her* will know that I do not exaggerate."[8] As late as 1803, his "good aunt" was recalled as among his "chief friends and benefactors." But no letters to her have been found; there are no references to her in extant correspondence prior to 1803, only two suggestions of visits to her after he left home for the academy and her death in 1764 passed without notice in the extant contemporary record.

No doubt there were visits home that went unrecorded, letters written that have since disappeared. Timothy Priestley recalls some such visits from school or from early church positions at Needham Market or Nantwich. But Timothy also suggests that Joseph was already theologically suspect and considered as outside the family circle by the time of these visits.[9] The combination of a childhood and youth lived independently of his family and a related emancipation from that family's religious opinions had, in fact as in appearance, cut him off from his family once he left home for a higher education.

For Dissenting Calvinists of the West Riding took their religion seriously. In 1715 there were more than sixty Presbyterian and Independent chapels licensed for services in Yorkshire and in most of these there would be members who remembered vividly the persecution of dissent prior to the Toleration Act of 1689. The Independent Chapel at Heckmondwike was a substantial reminder of those days and of the vigorous bigotry of the vicars of Birstall, in its withdrawn location, thatched-roof, village-cottage anonymity. By Priestley's youth, the process had well begun that was to separate Independents from English Presbyterians in matters of doctrine as well as of church government. Even in the West Riding the Presbyterians were showing signs of the drift toward unorthodoxy that Priestley was later to represent so well in his theological career. At the same time that other, parallel, religious movement of the century, the rise of the evangelical reaction, was also in evidence. When the Archbishop of York, Thomas Herring, sent out questionnaires respecting his diocese on assumption of office in 1743, there were reported twenty-two meeting-houses of Moravian pietists

---

8. Joseph Priestley, *The Doctrine of Philosophical Necessity Illustrated, being an Appendix to the Disquisitions relating to Matter and Spirit* (Birmingham: for J. Johnson, 1782), 2d ed., 199–200.

9. Timothy Priestley, *Funeral Sermon*, 39–41. Timothy's memories must be regarded with caution. His chronology is sometimes patently incorrect as well as in disagreement with the *Memoirs*. The *Sermon* is a curious document; Timothy's pride in his brother's accomplishments vies with complacency in their religious differences, self-advertisement, and the stretching of facts to "establish" some theological "truth."

and Methodists, and nearly all were in the district around Birstall.[10] Priestley grew up in this atmosphere of militant Dissent, in the form of evangelical Calvinism.

At the age of four, Joseph could repeat all of the one hundred and seven questions and answers of the Shorter Westminster Catechism "without missing a word."[11] Is this the period also in which the ideas of darkness and of malignant spirits and apparitions were so closely linked that some forty years later Priestley still associated them in an uneasiness about nightfall?[12] And is this the period in which he first began the stutter, "inherited" from his family, that plagued him all his life? During his first year at home, after his mother's death, he would "now and then" ask brother Timothy to kneel down with him while he prayed. He commenced then going with his family to the chapel at Heckmondwike, where he was catechized regularly and publicly till he was grown. There also he heard sermons twice each Sunday and, as he grew older, would recollect them at home and commit them to writing.

At Fieldhead, family prayers were said morning and evening; these were continued at his aunt's where he led the prayers after he reached seventeen. As a child at Heckmondwike, he attended the monthly prayer meetings of churchwomen at his aunt's. As he grew older he went, once a week, to a meeting of young men for spiritual conversation and prayer. During this period he also read "most of Mr. Bunyan's works and other authors on religion." Thus, as Priestley says, was he "brought up with sentiments of piety . . . and having, from my earliest years, given much attention to the subject of religion, I was as much confirmed as I well could be in the Principle of Calvinism."

When did his sentiments begin to change? About 1749–50 he had a serious illness, thought to be consumption, during which he had reason to believe that he would not live long. Already deeply concerned about religious matters, he became more so and, believing with the evangelical Calvinists that a new birth, a religious experience "produced by the immediate agency of the Spirit of God" was necessary to his salvation, he was in great distress, for he could not satisfy himself that he had had such an experience. He appears, in fact, to have lived some months in a state of terror, compar-

10. Ernest Axon, "Yorkshire Nonconformity in 1743," *Transactions of the Unitarian Historical Society* 5 (1931–34): 244–61.

11. Timothy Priestley, *Funeral Sermon*, 35.

12. *Examination of Dr. Reid's Inquiry into the Human Mind . . . Dr. Beattie's Essay . . . and Dr. Oswald's Appeal* (London: J. Johnson, [1774] 1775), 68.

ing himself to "Francis Spira, to whom . . . repentance and salvation were denied" and to Bunyan's "man in an iron cage."[13] Priestley's minister, John Kirkby, can have been of little help during his period of spiritual agony. Although Kirkby had studied at the Academy of John Chorlton, who has been described as a rational latitudinist, a man "of moderate, healing Principles," there is no indication that Kirkby himself publicly abated the rigors of the Calvinist position of election and original sin.[14] And Kirkby was seventy-two in 1750, ill, infirm, and unable adequately to serve his congregation.

During the next two years several ministers were brought to Heckmondwike as candidates to take Kirkby's place, candidates who generally stayed at Mrs. Keighley's home, where they were questioned by her ward. One of these, a Mr. Walker of Ashton-upon-Line, was a Baxterian, and Baxterians did not maintain the strict Calvinist position. The Reverend George Haggerstone, of nearby Hopton, with whom Priestley was to study mathematics and natural philosophy, was also a Baxterian and among the Dissenting ministers of the neighborhood, who frequented his aunt's home, there were "heretics" such as Mr. William Graham of Halifax who denied the doctrine of atonement, and Mr. Thomas Walker of Leeds, who was certainly an Arminian and may even then have been an Arian.[15]

Too honest to convince himself that he had experienced "re-birth," unwilling to believe that the venial sins of childhood eternally damned him or

13. W 1/1:12; the Francis Spira referred to was the, probably fictional, "hero" of a book, *The Horrible End of Francis Spiro*, inspirational reading for serious-minded Calvinists in the eighteenth century.

14. For Kirkby, see Alexander Gordon, *Cheshire Classic Minutes, 1691–1745* (London: Chiswick Press, for the Provincial Assembly of Lancashire and Cheshire, 1919), 185–86, and Frank Peel, *Nonconformity in Spen Valley* (Heckmondwike: Senior and Co., 1891), 94, with accounts of Kirkby's debating Arminian theology with the Methodist, John Nelson of Birstall, in 1745. For Chorlton and his academy, see J. W. Ashley Smith, *The Birth of Modern Education: The Contribution of the Dissenting Academies, 1660–1800* (London: Independent Press, 1954; The United Reformed Church in the United Kingdom, 86 Tavistock Place, London WC1H 9RT, 1972), 105–6; citations and quotations by permission. Some of Chorlton's students, using the books of Chetham's Library, read the works of the Dutch Arminian bishop Episcopius and even of the arch-heretics Socinus and Crellius, but there is no evidence that Kirkby had done so.

15. The names Independent (or Congregational) and Presbyterian, Baxterian, and Arminian, even Arian, though capable of precise definition, were so loosely used during the eighteenth century that each needs to be defined for specific occasions. For the moment it is sufficient to say that Baxterians held a midposition between Calvinist predestination and Arminian justification, believing some people attained by faith and works the salvation others had by election. Arians denied part of the orthodox formula for the Trinity.

that Adam's disobedience had done so, Priestley was greatly relieved to discover that there were respectable persons who did not hold that the "sin of Adam" alone made all mankind liable to the wrath of God and the pains of hell for ever. Priestley knew that this was the formal position of the congregation to which he applied for membership, for he had recited the catechism for many years and more recently seen Mr. Walker of Ashton-upon-Line rejected for his Baxterianism.[16] Yet he did not conceal his new opinions when questioned by the elders.

Did he believe that he would still be accepted? Perhaps not. All of his life he was to speak his mind without regard for the immediate practical consequences of his honesty. Yet it is just possible that the congregation, a few years earlier or later, might have found his minor unorthodoxy admissible, for his personal connections among them were formidable. During those years, however, in which Priestley was moving to Arminianism, the congregation, lacking an active minister, had set apart six of their number, "the gifted brethren," to conduct worship services. Already, by 1752, these brethren so dominated church activities that the efforts of others, especially Deacon Joseph Priestley, to obtain a new minister were handicapped; and these were the elders who sat in judgment on Priestley.

Events of the next few years show that some of the congregation, led by the "gifted brethren," had moved toward a primitive church fundamentalism of the Sandemanian type and finally, in 1760, they were expelled.[17] By that time it was too late, so far as young Joseph Priestley was concerned. Theological decision had led to educational decision and that led from the Arminianism with which he had left home to an Arianism that barred his ever returning.

For the congregation at Heckmondwike, under its new minister, remained Calvinist. More than that, the Reverend James Scott, brought to Heckmondwike through the negotiations of Deacon Joseph Priestley and his brother-in-law, the Reverend Edward Hitchin of London, became a leader of North England's resistance to the Arminian and Arian drift of Nonconformity. At the urging of Deacon Priestley and Edward Hitchin,

16. Answer to question 19, of the shorter catechism, *The Grounds and Principles of Religion contained in a Shorter Catechism according to the Advice of the Assembly of Divines at Westminster* (London, for the Lord Wharton Trustees, 1907), ed. William Carruthers, 6; also answers to questions 15–18, p. 5.

17. See Frank Peel, *Nonconformity*, 109–16, 147–49, for a discussion of this episode. The Sandemanians, named after Robert Sandeman, were "primitive" Christians, whose churches were led by elders without ministers.

and of the London "Northern Education Society" formed for the purpose, an academy was established at Heckmondwike in 1757, with Scott at its head, "by which many congregations might be blessed with godly preachers, sound in the faith and exemplary in their lives."[18] Timothy Priestley was one of the first students at Heckmondwike Academy and, though Joseph thought his education "imperfect," Timothy became an orthodox Calvinist minister in Lady Huntingdon's Connection.

By "imperfect" education, Joseph Priestley intended something more than a comment on the academy at Heckmondwike, though Scott *was* known more for the soundness of his divinity than of his learning. Timothy went, for a period, to the same grammar school that Joseph attended, but he then "for some time" followed his (and Joseph's) father's trade of cloth-dressing before studying under Scott. At the same age, Joseph was studying, on his own, at school, and with tutors, until he probably had more learning as he went off to his academy than Timothy was to acquire by graduation from Heckmondwike. The *Memoirs*, so thin in personal and family information, are detailed in discussing the nature and progress of Joseph's education. For what *is* recorded in the *Memoirs* and celebrated in correspondence and in dedications of published writings is the memory of the men, books, and institutions that nurtured the learning by which Priestley substituted a theological and intellectual background for that of family and local custom from which his upbringing and self-conscious revolt had separated him.

His aunt, having early decided that Joseph was to be a minister, sent him "to several schools in the neighbourhood, especially to a large free school, under the care of a clergyman Mr. Hague." These "several schools" cannot now be identified, but the "large free school" was probably the Grammar School at Batley, founded in 1612, with the unusual provision that the master was to teach reading and writing in English as well as prepare scholars in Latin and Greek for university entrance.[19] The requirements of the

18. T. G. Crippen, ed., "Early Nonconformist Academies: Heckmondwike and Northowram," *Transactions of the Congregational Historical Society* 6 (1913–15): 291–96.

19. Records of enrollment at Batley do not exist for 1744–46, during which time Joseph Priestley would have attended, nor was Joseph Hague master then, that position being held by Thomas Rhodes, vicar of Batley. But Rhodes, master since 1723, was exceedingly active as vicar and employed an usher from 1736 to 1744 and again in 1747. Hague, who graduated from Batley before going to Trinity College, Cambridge, was ordained deacon and licensed curate of Birstall in September 1744. It is probable that Hague was usher at Batley from that date to the late summer of 1746, when he became curate of Thornton, Bradford. These are the dates Priestley was attending the "large free school" in the neighborhood and no other school fits the conditions as well. The tradition that Priestley attended Batley is best treated in N. L. Frazer, "The Claim of Batley Grammar School to be the Alma Mater of the Rev.

school's endowment stipulated that boys of the parish of Batley, able to read the New Testament in English, were to be admitted free of charge. But Heckmondwike is not in Batley Parish. Some of the Keighley property in Heckmondwike was, however, in the manor of Batley and this may have entitled its owner to send a child to the school there.[20] Otherwise Mrs. Keighley paid the fees that the master was entitled to charge for students enrolled from outside the parish.

In any case, it seems that it was in this "large free school" (of about fifty students) that young Joseph "at the age of twelve or thirteen, first began to make any progress in the Latin tongue, and acquired the elements of Greek." The texts used were probably "Lilly's Latin Grammar," and the Eton Greek grammar.[21] Curiously, Priestley also learned there the shorthand system of Peter Annet. He was to use this system (with his own modifications) all of his life. He was so impressed that he wrote a set of verses commending the system that he sent to Annet, who published them as introductory to a new edition of his book describing the method. This was Priestley's first publication and it appears to have inaugurated a correspondence with Annet on the subject of determinism.[22] This is the period, at the age of sixteen to nineteen, during which Priestley was struggling with the Calvinist doctrines of the Independent Church and it is revealing to find that, at that time, he "maintained very strenuously the doctrine of liberty" against Annet's support of philosophical necessity (i.e., determinism).

During the holidays from Batley, Priestley studied Hebrew under John Kirkby and, when Joseph Hague left in 1746, Kirkby commenced a small

Joseph Priestley, F.R.S.," reprinted in *Transactions of the Unitarian Historical Society* 5 (1931–34): 133–44 and summarized in Derek N. R. Lester, *The History of Batley Grammar School, 1612–1962* (Batley: J. S. Newsome and Sons, [1962]), 15–17, 48–53.

20. Michael Sheard, *Records of the Parish of Batley in the County of York, Historical, Topographical, Ecclesiastical, Testamentary, and Genealogical* (Workshop: Robert White, 1894), 237–38.

21. These were the texts in use at Batley when Nicholas Carlisle collected his information for *A Concise Description of the Endowed Grammar Schools in England and Wales* (London: Baldwin, Cradock and Joy, 1818), 2:780–81. As the Latin grammar by William Lily and John Colet, first published in 1509 and reedited and republished into the mid-nineteenth century, was widely used at major grammar schools through the eighteenth century, there is little reason to suppose a change had been made in text from Priestley's time to that reported by Carlisle.

22. Peter Annet, *Expeditious Penmanship: or, Shorthand Improved, &c.* (London: for the Author, and sold by R. Baldwin, n.d. [British Library suggests 1750]); the poem appears on page 4; see Appendix to this chapter. The correspondence with Annet has disappeared. Annet wished to publish the letters, but Priestley "had the prudence not to consent," though he was "at that time very young, not having entered upon a course of academical learning." See the

school at which Priestley studied until 1749. Then the combination of
Kirkby's infirmities and Priestley's illness ended Joseph's formal education
for a time. Nothing is known of what was studied in Kirkby's school,
except the inference that lessons in Hebrew were continued.[23] Although he
soon recovered from the most dangerous aspect of his illness, it was not
thought that Priestley would be well enough to continue his preparations
for the ministry. His family determined to employ a customary (and usually
futile) eighteenth-century anodyne for consumption by sending him to the
warmer climate of Lisbon. A post was found for him in a countinghouse
there and to prepare for this new occupation Priestley taught himself enough
French, Italian, and "high Dutch" (i.e., German) to translate letters for a
merchant uncle.[24] But the recovery unexpectedly continuing, and mercantile
endeavors being unattractive to an inveterate scholar such as Priestley, he
began to seek an alternative.

A letter of advice, written by the Reverend Thomas Haynes, Dissenting
minister of Sheffield, in December 1750, notes, "I hear you intend some
other learned profession [than the ministerial]," but professions were few
in eighteenth-century England and if the intent was really a profession and
not just some quasi-learned occupation such as teaching or pharmacy, the
only ethical alternative open to a fervent Dissenter would be medicine.
Timothy declares that his brother seriously considered "physic," but there
is no confirming evidence from Haynes's letter or the course of self-
administered studies that Joseph commenced.[25]

preface to Priestley's *Doctrine of Philosophical Necessity Illustrated*, xxviii. Despite his dis-
claimer, the correspondence continued at least into 1754, but this is still well before Annet,
deistical freethinker, published his *Free Enquirer* (1761), for which he was convicted of
blasphemy.

23. The assumption by Ivan Poldhauf, *On the History of Some Problems of English Grammar before 1800* (Prague; Philosophical Faculty of Charles University, Prague Studies in English, 1948), 235 n, 137, that Priestley was inspired to study languages by Kirkby's publication
of an English school grammar in 1746 is incorrect. The notice in *Gentleman's Magazine* 16
(1746): 112, of the publication of this *New English Grammar; or, Guide to the English tongue,
with notes; together with a brief Latin Grammar* (Manby, 1746) cites its author as John Kirkby,
M. A. Priestley's Kirkby was not an M.A. while a Kirkby (1705–54), sometime tutor to
Edward Gibbon, is described in the *Dictionary of National Biography* as M.A., Cambridge,
1745, author of a Latin and an English grammar of which Gibbon spoke highly.

24. J. T. Rutt identifies this uncle as a "Mr. Priestley of Red House, Leeds," W 1/1:8n.
It seems more likely that he was John Priestley of Fieldhead, member of the mercantile firm
of Messrs. John Priestley and Sons (and including John Priestley Jr. and a Joseph Priestley of
Great St. Helen's, London) which was bankrupt in 1777; see G. D. Lamb and J. B. Place,
eds., "Extracts from the Leeds Intelligencer and the Leeds Mercury, 1777–1782," *Publications
of the Thoresby Society* 40 (1955): 1–247; ibid., 30: "Leeds Mercury 550, Tues. Aug. 1777."

25. The Haynes letter is published in W 1/1:9, and Timothy's statement in *Funeral Sermon*,
37. The tone of Haynes's letter suggests that he was a family friend, or even relative, but the
connection seems to have been his being minister to the church attended by Priestley cousins.

What is clear is that Priestley returned to more academic pursuits. He continued his study of languages; he instructed a "Mr. Thomas [John Tommas], a Baptist minister . . . of Gildersome" in Hebrew, thus becoming proficient in it himself, and taught himself "Chaldee and Syriac, and . . . began to read Arabic."[26] That mid-eighteenth-century Heckmondwike had available texts in these Mideastern languages is unlikely. The particular selection of languages suggests that, in fact, Priestley was learning them from a polyglot bible such as that left him years later by the Reverend William Graham, the "heretical" minister of Halifax, whose example and encouragement led Priestley "at my entrance on theological inquiries . . . to think for myself on subjects of the greatest importance."[27]

If he followed the advice of Haynes's letter, Joseph also continued his study of Latin and Greek, comparing his own readings with published translations of repute. Haynes particularly recommended Thomas Gordon's *Tacitus*. If Priestley obtained that work he would also have seen the annexed "Political Discourses" and been introduced to the anticlerical radical political theories of the "Commonwealth" tradition. Given his particular circumstances, no doubt he took special note of Gordon's arguments against any attempt whatever to dictate individuals' religious opinions.[28]

Priestley's education was further broadened under the direction of the Reverend George Haggerstone, of Hopton, who became his tutor in mathematics. It is difficult to determine how or why Haggerstone was selected for this position. There is some suggestion that the church at Heckmondwike was, in a sense, "mother church" to smaller Calvinist congregations in the area. This was clearly true for the church at Osset, to which Haggerstone removed in 1758, and it may also have been true for that at Hopton, in which case its minister and his curricular abilities could have been known to Kirkby and Mrs. Keighley. Perhaps the recent posthumous publication of Colin Maclaurin's *Account of Sir Isaac Newton's Philosophical Discoveries* (1748) reflected credit onto all graduates of Edinburgh who had studied under its author, or perhaps Haggerstone was simply the best mathematician available in the area. In any event, the choice was an important one.

---

26. See Arthur S. Langley, "Baptist Ministers in England about 1750 A.D.," *Transactions of the Baptist Historical Society* 6 (1918–19): 138–62.

27. Joseph Priestley, *Disquisitions relating to Matter and Spirit* (London: J. Johnson, 1777), dedication, vi.

28. See Caroline Robbins, *The Eighteenth-Century Commonwealthman* (Cambridge: Harvard University Press, 1959), 115–25. He would certainly read Gordon's *Tacitus* at Daventry Academy.

According to Timothy, Joseph once told Haggerstone that he was "the man who brought me out of the dark hole of Calvinism," at which Haggerstone was greatly concerned, asking himself, "Have I spoilt the finest genius in the world?"[29] But if Priestley's genius was "spoilt," it was not simply Haggerstone's Baxterianism that did it, though the *Memoirs* do credit the liberal turn of his conversation with the undermining of Calvinist prejudices. Nor was it the training in mathematics, though Maclaurin had been the "Infidel Mathematician" against whom Bishop Berkeley wrote and John Wesley was convinced that the study of mathematics would turn some students, at least, deist if not atheist. Even if one accepts the imputation, Priestley was in no danger from that cause for he never became a mathematician. Certainly he studied algebra and geometry—the latter from a copy of Isaac Barrow's *Euclid* of 1678 given him by Haggerstone—and later he studied trigonometry and fluxions, but not to advantage. He was to avoid the mathematical sciences and quantitative investigations throughout his scientific career.[30]

What Haggerstone did was to lead Priestley to the reading, "with but little assistance from him," of natural philosophy, logic, and metaphysics. During the years 1750–52, Priestley "made such a proficiency" in various branches of learning that, when he was later admitted to an academy, the studies of the first year and most of those of the second were excused. We do not know all the works he read during those years of essentially self-directed study, but Priestley remembered three of them, after a period of nearly thirty-five years: "Watts' Logic," "Gravesande's Elements of Natural Philosophy," and Locke's "Essay on the Human Understanding."

It is necessary to examine these works to find what it was that made them so memorable, for there are echoes of them in Priestley's subsequent writing: of Watts, in writings on rhetoric; of 'sGravesande, in the history of optics; and of Locke in so many instances that for Priestley, as for so many other Englishmen of the eighteenth century, it would seem that much of his thinking was a footnote to Locke. But without explicit citations, one

29. Timothy Priestley, *Funeral Sermon*, 40–41; Timothy had a remarkable "gift for remembering" exact details of a conversation for nearly half a century when they could give his narrative this kind of innuendo.

30. One of the few books to survive the vicissitudes of Priestley's library and end up in the United States was Barrow's *Euclid*, with an ownership inscription of Haggerstone and a note by Priestley that it had been given to him; see Thomas Stewardson, "The Destruction of Priestley's Library," *Notes and Queries*, 3, 4th ser. (1969): 64–65.

cannot credit any particular line of Priestley's thought to any particular passage, book, or author—not even Locke.

Between his reading of these three works, for example, and his later scholarship there intervened years of study, at school and on his own, and the reading of countless other books as well as his own experience and thought. One can, however, provide a general description of the books memorable to Priestley as representing his introduction to natural and to general philosophy, and then single out specific parts that may be supposed to have peculiar significance, as relating directly to Priestley's immediate problems and interests, or as introductory to subjects on which Priestley was later to devote substantial time and study.

Of the three, the most appropriate as an introductory text for someone in Priestley's circumstances was surely "Watts' Logic." Written by Isaac Watts as one of a series of texts on English grammar, astronomy, natural philosophy, logic, and methods of study, the *Logick; or, the Right Use of Reason in the Enquiry after Truth* was probably the most influential. First published in 1725, and used throughout the century at colleges of Oxford and Cambridge as well as in most Dissenting academies, it had reached an eighth edition by 1745. One of the books by which Watts, particularly through his friend and protégé Philip Doddridge, exercised a major influence on eighteenth-century British education, the *Logick* is a competent, simply written, compilation, derived from Le Clerc and the Port Royal logicians with a heavy admixture of Descartes and of Locke.[31]

As one might expect, it is permeated with that warm religiosity that characterizes the hymns for which Watts is today better known. Typical of his moral didacticism, but in an example that Priestley was sure to notice, is this illustration of a universal negative syllogism: "No Injustice can be pleasing to God; all Persecution for the Sake of Conscience is Injustice; therefore no Persecution for Conscience Sake can be pleasing to God" (284). But though Watts discussed the classic moods and figures of syllogisms, he believed emphasis on them had transformed the art of reasoning into a mechanical process. ". . . The Light of Nature, a good Judgment, and due

---

31. I have used Isaac Watts's *Logic . . . With a Variety of Rule to guard against Error, in the Affairs of Religion and Human Life, as well as in the Sciences* (London: T. Longman, T. Sewell, J. Brackstone, 1745); numbers in parentheses in the text are references to pages in this edition. See also Arthur P. Davis, *Isaac Watts: His Life and Works* (New York: Dryden Press, 1943) and J. W. Ashley Smith, *The Birth of Modern Education*, 144–47, for discussion of Watts and his influence in British education.

Consideration of Things tend more to true Reasonings than all the Trappings of Moods and Figures" (289).

Clearly this is not a book of formal logic, dealing only with the canons and criteria of validity of inference and demonstration. It is, in fact, a treatise on what would, formerly, have been the trivium: grammar, rhetoric, and logic; treating what Watts calls "the principle operations of the mind": (1) perception (2) judgment (3) argumentation (4) disposition; and the art of logic composed of rules made about these operations [4]. And these operations are discussed in a context that urged the reader, whatever his intentions, to acquire a rich variety of ideas, ancient and modern, foreign and domestic, relating to natural, civil, and religious concepts. "You will tell me perhaps, that you design the Study of the *Law* or *Divinity;* and what Good can *natural Philosophy* or *Mathematics* do you, or any other Science, not directly subordinate to your chief Design? But let it be considered, that all Sciences have a sort of mutual Connection; and Knowledge of all Kinds fits the Mind to reason and judge better concerning any particular Subject" (71–72).

Watts was a Calvinist, strongly opposed to deism and noted, in his day, for the moderate evangelicalism of his preaching. This is reflected in Watts's discussion of the "Causes of false judgment which are within ourselves," which include original sin. It also leads to moderate disagreement with Locke on the subject of innate ideas, when he insists that man has original notions of the difference between right and wrong. Watts is often inconsistent in his beliefs, reconciling contradiction in a good English insistence upon pragmatic considerations.

For Priestley, rethinking his religious commitments, Watts's treatment of sectarian arguments would surely be noted. "We set up *our own Opinions* in Religion and Philosophy as the Tests of *Orthodoxy* and *Truth;* and . . . judge every Practice of other Men either a Duty or a Crime, which we think . . . a Crime or a Duty in us, though their Circumstances are vastly different from our own" (204). ". . . every Religion has its Infant Votaries who are born, live and die in the same Faith without Examination of any Article" (215). "No Bishop or Presbyter, no Synod or Council, no Church or Assembly of Men (since the Days of Inspiration), hath Power derived . . . from God to make Creeds or Articles of Faith . . . and impose them upon our Understandings. We must all act . . . to the best of our own Light, and the Judgment of our Consciences" (227).

Nor can Watts's observations on language and its uses have been irrelevant to a young man already devoted to the study of language. Were these observations recalled in the years ahead when Priestley wrote and taught languages and rhetoric? To avoid error in pursuit of knowledge, it is necessary to take heed of the use of words and terms (51), yet (and here Watts again disagrees with Locke) language is not simply a means of communicating ideas. Metaphor and figures of speech represent ideas "with vivacity, Spirit, Affection, and Power . . . to move, and persuade" (66).

Because there are fashions in rhetoric, "many an excellent Discourse of our Forefathers has had abundance of Contempt cast upon it" (197). But there are also difficulties in communication because of historical or national differences. ". . . if we would gain a just and precise Idea of every . . . Expression, we must . . . consider the peculiar Idiom of the Language . . . the Time, the Place, the Occasion, the Circumstances of the Matter spoken of" (155), ". . . we must search the Sense of the Words . . . their Original and Derivation, what is their common Sense amongst Mankind, or in other Authors, especially such as wrote in the same Country, . . . same Age, . . . and upon the same Subjects" (94).

And as these last advices relate to historical as well as literary criticism, and Priestley was himself to engage in historical as well as historico-critical writing, Watts's treatment of that subject might also be remarked:

> If the Subject be *historical* or a *Matter of Fact,* we may then enquire whether the Action was *done at all;* whether it was done in *such a manner,* or by *such Persons* as is reported; *at what time* it was done; *in what Place;* by what Motive, and for what *Design;* what is the *Evidence* of the *Fact;* who are the *Witnesses;* what is their *Character* and *Credibility;* what *Signs* there are of such a fact; what *concurrent Circumstances* which may either support the Truth of it, or render it doubtful. (126)

Finally, there are frequent references to the nature of matter. This appears to be a major ontological question in eighteenth-century philosophy and science. It pervades Watts's book, as it dominated Locke's, and was a primary question in Priestley's subsequent inquiries. Substance is being, subsisting by and at the will of God, but variable in properties by inferior causes (11). Matter and mind (or spirit) are distinct substances: the one extended, solid or, impenetrable; the other capable of cognition. Some philosophers suppose that man can know only the properties of substances, there being an unknown (and unknowable) substratum supporting these properties and in which they inhere. But Watts believed "*Solid Extension*

. . . to be the very Substance of Matter . . . and a *Power of thinking* . . . to be the very Substance of all *Spirits*"; the other notion being a "mere Mistake which we are led into by the grammatical form and Use of Words; and perhaps our logical Way of thinking by *Substances* and *Modes*" (13n).

Modern philosophers oppose the followers of Aristotle, and chemists, making little use of the word element, suppose a single matter, "diversified by its various Shapes, Quantities, Motions and Situations" to make all the varieties found in the universe (14). Yet who can tell the shapes and positions, the figures of these corpuscles (43)? ". . . we are not . . . arrived at any sufficient Methods to discover the Shape of those little Particles of Matter which distinguish the various *Sapours, Odours,* and *Colours* of *Bodies;* nor to find what . . . Atoms compose *Liquid* or *Solids,* and distinguish *Wood, Minerals, Metals, Glass, Stone,* &c." (189).

John Locke's *Essay Concerning Human Understanding* is so widely known as a classic of philosophy that it might seem wasted effort to describe it. But the *Essay* most modern readers know has been abridged, and emphasizes the characteristics they are taught to expect from "The Father of British Empiricism." No doubt Priestley read the work because Haggerstone had recommended it, but he would not have approached it already knowing what was important in it. Assuming that Priestley would give equal attention to all its parts, and considering the time, place, occasion, and circumstances of the reader as well as the author, what might the young Priestley be expected to learn from Locke's *Essay?* Locke commenced writing the *Essay* early in the 1670s, a chance consequence of discussions in a group of friends. He continued it in exile in Holland, as "a diversion of idle and heavy hours," he tells us in his Epistle to the Reader, "written in incoherent parcels and after long intervals of neglect, resumed again." When it was published in 1690, it was prolix, repetitive, and might not have been so long, "but I am now too lazy, or too busy to make it shorter." Written in the baroque style of English still waiting for the improvements fostered by Addison and the Royal Society, the *Essay* is also largely a polemic; for its author, like those of the other great philosophic treatises of the seventeenth century, had been distressed by the Scholasticism of his formal education. One might well wonder how hard the eighteen-year-old Yorkshire reader had to struggle to make sense of the "peculiar Idiom of the Language" of more than a half-century earlier and what was to be made of arguments against opinions: of essences and substantial forms, universals, genera and species, for example, of which he had probably never heard.[32]

32. I have used John Locke, *An Essay concerning Human Understanding* (London: T. Longman, B. Law and Son, J. Johnson, et al., 1796), 20th ed., to which citations in

Some parts of Locke's treatise, Priestley must surely have noted, either because they were so central to the work they could not have been missed, or had been treated more simply by Isaac Watts, or because they were repeated and repeated in so many variations and contexts. One such section, of course, would be that for which the *Essay* is most famous, on the theory of perception and the empirical roots of human knowledge. ". . . ideas and notions are no more born with us than arts and sciences" (1-4-22); all ideas are derived by the mind from sensations, "produced in us . . . by different degrees and modes of motion in our animal spirits, variously agitated by external objects" (2-8-4); or by reflection of the mind, or by combinations of sensation and reflection.

This does not identify Locke as a sensationalist. Though he was an empiricist he was also a rationalist and sensations were acted upon by mind, in reflection and comparison, to produce complex ideas. Examples of ideas generated by the act of comparison are the relation of cause and effect, of time and space, identity and diversity, equality and excess, right and wrong. These are creatures of the understanding, not contained in existence of things, but superadded to data of experience.[33] Because men can form strong combinations of ideas, not allied in nature, they can develop "habits of thinking . . . trains of motion in the animal spirits, which once set a-going, continue in the same steps . . . which, by often treading, are worn into a smooth path, and motion in it becomes easy, and as it were natural" (2-33-6). And thus unnatural sympathies and antipathies are explained by an association of ideas in an introduction to Priestley of a concept of major later influence.

Another theme of Locke's *Essay*, as extensive as that relating to perception but more neglected in modern commentaries, is a concern with language. The sole purpose of language is to mark, or communicate man's thoughts to one another with all the dispatch that may be (2-22-5), but

---

parentheses in the text (e.g., [2-13-10]) refer to book, chapter, and paragraph, but I also compared this edition to earlier eighteenth-century versions to verify that all contain the same editorial variations from seventeenth-century editions. See also Maurice Cranston, *John Locke: A Biography* (London: Longmans, Green and Co., 1957).

33. See Thomas E. Webb, *The Intellectualism of Locke: An Essay* (Dublin: William McGee & Co., 1857) 94–95; Locke's *Essay:* 2-25-22; 2-26-3, 5; 2-27-1; 2-28-4. And, if this seems too idealist an attitude for Locke, the empiricist, note the comment by Richard I. Aaron, *John Locke* (Oxford: Oxford University Press, 1955), 27, that "much of the fourth book of the *Essay* might have been written by one of the Cambridge school [of Neoplatonists]." See also Georg Freiherrn v. Hertling, *John Locke und die Schule von Cambridge* (Freiburg im Breisgau: Herder, 1892).

languages constantly change, as change of custom and opinions bring new combinations of ideas requiring new names or a transfer of meaning of older ones (2-22-7). Even in the same country and period, different men will use words differently (as is evidenced by the Greeks), and when one considers the different notions, tempers, customs, ornaments and figures of speech, each influencing the signification given to words in different countries and ages, it is no wonder that confusion results more than improvement or information from the discourses of men (3-9-22).

Now ". . . common use, that is the rule of propriety, may be supposed . . . to afford some aid, to settle the signification of language," and so it does, for common conversation, "but nobody having an authority to establish the precise signification of words, nor determine to what ideas any one shall annex them, common use is not sufficient to adjust them to philosophical discourses" (3-9-8). And if one examines the nature of understanding and the development of language, it becomes clear that language cannot easily become a precise means of philosophical communication. Words relating to complex ideas ("mixed modes") are uncertain because these (such as honor, duty, incest) are of man's creation and have no real standards existing in nature to which the ideas refer, while names for simple ideas, of substances for example, have equally doubtful signification because the natural standards that they stand for "either cannot be known at all, or can be known but imperfectly and uncertainly" (3-9-11).

A consequence of this, almost inevitable, confusion of tongues is that controversies, which have laid waste the intellectual world, are generally not about realities, but are merely verbal, proceeding from the ill use of words (3-10-22). Volumes of interpreters and commentators on the Old and New Testaments attest to disagreements of this sort in religion. "Though every thing said in the text be infallibly true, yet the reader . . . cannot choose but be very fallible in the understanding of it." Surely Locke's solution to this problem would have interested the young Priestley: "Since . . . the precepts of natural religion are plain, and very intelligible to all mankind, and seldom come to be controverted; and . . . revealed truths . . . conveyed to us by books and languages are liable to the common and natural obscurities and difficulties incident to words; methinks it would become us to be more careful and diligent in observing the former, and less magisterial, positive and imperious, in imposing our own sense and interpretations of the latter" (3-9-23).

Another Locke example of verbal confusion is the "long agitated and . . . unintelligible question, viz. Whether man's will be free or no?" Priestley was debating this question with Peter Annet, about the time he was reading the *Essay;* it is hard to believe that he did not read with particular attention what Locke had to say on the subject. Liberty, says Locke, is the power to act or not as one wills. It is, therefore, a power of agents and cannot be an attribute of the will, which also is a power. Asking if a man's will be free is equivalent to asking if his sleep be swift or his virtue square (2-21-8, 10, 14).

It is true that man argues that he is "not free at all, if he be not as free to will, as he is to act what he wills." But man cannot avoid the act of volition, he cannot forbear to will (2-21-22, 23, 24). Yet no one counts it an abridgment of freedom that he is constrained to desire happiness, as God is under necessity of being happy. As man may mistake true felicity, he has power to suspend desire until the good or ill of action be determined. "He that has the power to act or not to act, according as such determination directs, is a free agent; such determination abridges not that power wherein liberty consists" (2-21-50).

One may well doubt that the young Priestley, in the throes of casting off Calvinist doctrines of predestination, would be satisfied by Locke's subtle equivocations and still believe that he would take note of the verbal gymnastics. Indeed he employed some of his own, in subsequent writings of doctrinal paradoxes in which man is morally responsible in a world of philosophical necessity.

Nowhere in Locke's *Essay,* however, does subtlety of argument lead more surely to seeming paradox and contradiction than in that sequence by which he concludes:

> . . . since our faculties are not fitted to penetrate into the internal fabrick and real essence of bodies; but yet plainly discover to us the being of a God, and the knowledge of ourselves . . . it will become us, as rational creatures, to employ those faculties we have about what they are most adapted to. . . . For it is rational to conclude that our proper employment lies in those inquiries . . . most suited to our natural capacities. . . . Hence, I think . . . that morality is the proper science and business of mankind in general. (4-12-11).

But the contradiction is more apparent than real and the paradox, if there, less obvious to an eighteenth-century reader than to one who reads Locke as "the father of British empirical philosophy." Throughout the *Essay,* Locke does use concepts of morality as examples of verbal confusions.

He does recommend natural religion as an escape from the futile animosities of theological controversy, but he does not say that misuse of moral words cannot be corrected. He does not necessarily connect morality with religion, and he bases natural religion on reason as well as observation.

His rejection of revealed religion is grounded on the "common and natural obscurities and difficulties incident to words" and upon his distrust of "enthusiasm" as a guarantee of inspiration. So long as "any proposition . . . we take for inspired, be conformable to principles of reason, or to . . . attested revelation, we may safely receive it for true, and be guided by it in our belief and actions" for "though perhaps it be not an immediate revelation from God, extraordinarily operating on our minds, yet we are sure it is warranted by that revelation which he has given us of truth" (4-19-15, 16).

There is no reason, then, why Priestley should regard, as perverse ingenuity, Locke's conclusion that, with right methods, morality could be made as clear and true as a proposition in mathematics, while ideas of the unknown properties of bodies, based on rational and regular experiments, are still but judgment and opinion, not knowledge and certainty. Thus natural philosophy is not capable of being made a science (4-12-8, 10). These views are not a result of verbal sophistries. Priestley could follow their development as a necessary consequence of the concept of matter, and of the ways man acquires ideas about it and about God and man, set forth early in the *Essay* and continuously pursued throughout it.

Sensations, Locke says, convince us that there are solid, extended (material) substances in the world, as reflection convinces us that there are thinking ones (2-23-29). Reasoning on sensations, we conclude that material substances possess primary qualities of bulk, figure, number, situation, and mobility; the power to act on our senses by means of primary qualities, to produce ideas of colors, sounds, smells, tastes, and so forth, which we call the sensible, or secondary qualities; and the power to act upon other bodies, by means of the constitution of primary qualities, to change their operations upon our senses (2-8-23). As the powers depend (in some unknown way) upon the primary qualities and only these are utterly inseparable from body (2-8-9), primary qualities are essential in the general definition of material substance. But are these qualities the abstract real essence of pure substance or is there a substratum, a "cause of the union of combinations of the simple ideas of qualities?" (2-23-2). On this subject, Locke is ambiguous, but if there is such a substratum, our faculties do not permit us to discover its nature.[34] Some of the powers of bodies can be explained on an assumption that sensible bodies are composed of insensible particles which themselves possess primary qualities whose variations of distribution and arrangement may be supposed to cause sensible qualities. This is the "corpuscularian hypothesis . . . which is thought to go farthest in an intelligible explication" of the active and passive powers of bodies (4-3-16). Heat and cold, for example, may be the increase or diminution of the motion of minute parts of our bodies, caused by the corpuscles of another body (2-8-21), or light as the sensation caused by a number of little corpuscles striking briskly on the bottom of the eye (3-4-10).

But such explanations are scientifically inadequate. Supposing the sensation of whiteness to be caused by a certain number of such globules with a certain degree of celerity in rotation about their centers, and a certain progressive swiftness; until we can demonstrate a quantitative relationship between number, celerity, and swiftness and the differences of sensation, we have "not so nice and accurate a distinction of their differences, as to perceive and find ways to measure their just equality or the least differences" (4-2-11, 12, 13). No doubt, could we discover the figure, size, texture, and motion of the minute constituent parts of any two bodies, "we should know without trial several of their operations upon one another as truly as we know the properties of a square or a triangle" (4-3-25). But our faculties are such that we cannot know these.

And even if we could know them, we would not have penetrated into the real essences of different substances, for we do not know the relationships between primary qualities of insensible particles and the sensible qualities which they cause (4-3-12). If the *Essay* can be said to have a litany, this is it. Over and over, Locke repeats, with every permutation and combination of contexts possible, that our faculties carry us only to a collection of sensible ideas; we do not, we cannot know the internal constitution of matter from which the qualities flow.

In vain we pretend to range things in sorts and classes by real essences that are far from our discovery or comprehension (3-6-9, 10). The properties of any sort of bodies are not easily collected and completely known by any way of inquiry of which we are capable (3-9-13). It is impossible for us to have any exact measures of the different degrees of simple ideas (4-2-

34. Locke's ambiguity on the "being of substance," a topic mentioned by Watts in the *Logick* also, is heightened by a controversy with Edward Stillingfleet, on that among other debatable issues in the *Essay*. Locke's answers to Stillingfleet, published in three "letters" in 1697 and 1699, are summarized in notes in most eighteenth-century editions of the *Essay*.

11). The active and passive powers of bodies, and their ways of operating consist in a texture and motion of parts that we cannot come to discover (4-13-16). The mechanical affections of bodies have no affinity at all with the ideas they produce in us (4-3-28).

Locke is far from wishing to dissuade men from the study of nature, as an occasion to admire, revere, and glorify its Creator or, if rightly directed, as of greater practical benefit to mankind than more ostentatious monuments of exemplary charity. He would not, however, encourage an expectation of knowledge where it is not to be had, the taking "doubtful systems for complete sciences," of unintelligible notions for scientifical demonstrations (4-12-12). In some of our ideas there are certain relations, habitudes, and connexions

> . . . visibly included in the nature of the ideas themselves. . . . And in these only are we capable of certain and universal knowledge. . . . But the coherence and continuity of the parts of matter; the production of sensation in us of colours and sounds, &c. by impulse and motion; nay, the original rules and communication of motion being such, wherein we can discover no real connexion with any ideas we have; we cannot but ascribe them to the arbitrary will and good pleasure of the wise architect. (4-3-29)

After Locke's litany of scientific skepticism, the stolid certainties of Willem Jacob 'sGravesande's *Mathematical Elements of Natural Philosophy, confirm'd by Experiments* may have been a relief, as they must have been a surprise to the young Priestley. Written in Latin in 1720 by the professor of natural philosophy at the University of Leyden, translated and published in first English edition in 1721, this was the first general text of Newtonian science to be published on the Continent and one of the earliest to be published in England. In the translation by J. T. Desaguliers, the English version reached a sixth edition in 1747 and, in one edition or another, was used in both universities and in Dissenting academies, to provide the "Introduction to Sir Isaac Newton's Philosophy" (the text's subtitle) for many a student.[35]

35. I have used W. James 'sGravesande, *Mathematical Elements of Natural Philosophy, confirm'd by Experiments; Or, an Introduction to Sir Isaac Newton's Philosophy* (London: W. Innys, T. Longman and T. Shewell, C. Hitch, and M. Senex, 1747), 2 vols., to which all page numbers in parentheses refer. See also Robert E. Schofield, *Mechanism and Materialism: British Natural Philosophy in an Age of Reason* (Princeton: Princeton University Press, 1970), esp. 140–46, for a discussion of 'sGravesande's work, in the context of his age.

It was to be Joseph Priestley's introduction to Newtonianism. Locke's *Essay* was written too early and published too soon to have been influenced by the *Principia* (1687).[36] Locke came to know Newton, but he died before publication of the *Opticks* (particularly the second, Latin, edition of 1706) made clear the possible significance of dynamic considerations to natural philosophy in general. His corpuscularianism was always that of Descartes and of Robert Boyle.

The situation of Isaac Watts is more complicated, but the result was much the same. Though Watts lived long enough to have taken note of Newtonianism, he had been educated by Thomas Rowe, of Newington Green, and Rowe was a Cartesian and an admirer of Henry More, the Cambridge Neoplatonist. Watts declared, in the preface to his *Philosophical Essays on Various Subjects,* that he had long seen reason to resign Cartesian opinions of heaven and earth, imbibed at the academy, at the foot of Sir Isaac Newton.[37] But that does not mean that he had accepted Newtonian doctrines of force and matter and, indeed, these do not appear in the *Logick,* where the corpuscular doctrine of matter remains Cartesian.

In any event, although Locke's *Essay* and Watts's *Logick* contain more natural philosophy than one might expect, neither are systematically concerned with that subject. 'sGravesande's *Elements* had also to introduce Priestley to the content and methods of physical science. It was, in many respects, a good introduction, for 'sGravesande was not only a gifted teacher, with a talent for creating experimental demonstrations, he was well aware of the essential difference between Cartesian science, for which mathematics was a model, and Newtonian science, for which mathematics was an essential structure. Mathematics, 'sGravesande agrees, relates only to ideas and not to things in being. But if things can be described quantitatively, they can be treated by "mixed mathematics" and the results adjudged by comparison with phenomena (xxix–xxx).

The laws of nature are the foundations of this kind of mathematical reasoning. Thus (by implication Locke is wrong) natural philosophy can become science (ii). This is particularly true respecting the "original rules and communication of motion." The greater part of the *Elements* is a discus-

---

36. In his third "letter" to the Bishop of Worcester (1699), Locke notes that "the judicious Mr. Newton's incomparable book" shows that bodies can act upon one another other than by impulse and declared that he will rectify that mistaken passage in "the next edition of my book," but the correction does not appear; see editor's note, book 4, chap. 3, para. 6.

37. Isaac Watts, *Philosophical Essays* (London: Richard Ford and Richard Hett, 1734), 2d ed., corr., v.

sion of mechanics: statics, simple machines, kinematics, simple dynamics, hydrostatics and hydrodynamics, and celestial motion (i.e., the macroscopic movements of sensible bodies).

'sGravesande does not, however, avoid the major ontological question of the period, the nature of matter, though here he agrees with Locke that we are probably ignorant of the substance in which properties inhere (iii–iv). As in the majority of eighteenth-century natural philosophy texts, the *Elements* begins with a section on Body in general. The primary properties of body are the usual extension (and a consequent divisibility), solidity, figure, and mobility. But to this last is added an "innate force" or inertia by which body resists change in motion in proportion to quantity of matter and velocity (4). Although not a customary quality in corpuscularian terms, inertia is scarcely a new concept and its inclusion here, in its quantitative form, is necessary for later treatment of dynamics under Newton's three laws of motion.

Perhaps, however, the peculiar designation of inertia as an innate quality serves to mask the subsequent introduction of further forces, which were very new indeed. For 'sGravesande continues his book with a treatment of such phenomena as cohesion, elasticity, capillarity, and ultimately of refraction and inflection of light and the motions of the planets by the introduction of other forces, of attraction and repulsion. These forces are not to be regarded as properties or qualities of bodies, "suppos'd to arise from the *specifick Forms* of Things"; they are rather principles, "*universal Laws* of Nature, by which the things themselves are form'd" (22).

The use of force as an explanatory concept says nothing of the cause of phenomena. Attraction means only a tendency of two bodies toward one another, repulsion a tendency to fly apart (16). This is good Newtonian doctrine, capped by a quotation from Query 31 of Newton's *Opticks:*

> To affirm that the several Species of Things have *occult specifick* Qualities, by which they act with a certain force, is just saying nothing. But from two or three Phenomena of Nature to deduce general Principles of Motion, and then explain in what Manner the Properties and Actions of all Things follow from those Principles, would be a great Progress made in Philosophy, tho' the Causes of those Principles should not yet be known. (22)

Along with the method by which experiment and mathematical deduction are necessarily linked, this attitude toward forces is what most clearly identifies the *Elements* as a Newtonian work, and in the quotation from Query 31 is outlined the challenge from Newton, to his contemporaries and followers, which was to set the pattern for much of British Newtonianism throughout the century.[38] Was the young Priestley perceptive enough to grasp the challenge? Perhaps he was; certainly there is an implication of possibility in 'sGravesande's treatment of problems of matter that contrasts sharply with Locke's reiterated pessimism. But one can also doubt that undirected reading of the *Elements* would have led Priestley to remember much of any of it, to say nothing of singling out any challenge for future investigation.

No doubt it was good for him to read of mechanics and astronomy as well as of heat, light, electricity, and sound; but the latter group, comprising most of the experimental natural philosophy of the century, is discussed in less than one-quarter of the whole and, of this small part, more than half treats the problems of geometrical optics. Possibly Priestley noted that in all those phenomena where the Newtonian forces were short-ranged (except that relating the elasticity of air to Boyle's Law), there was a disappearance of the close union between mathematical law and phenomena, for the mathematical form of the short-ranged force was unknown. More likely Priestley did not particularly notice the difference; one suspects that he skimmed the mathematics from choice as he scanted the experiments from necessity.

For 'sGravesande's mixture of mathematics and experimental demonstration had a major flaw to a solitary student in Priestley's circumstances. Without more discipline than Haggerstone was able (or perhaps prepared) to impose, the mathematical scholia could be passed over by a student without mathematical interests, while the experimental apparatus necessary for demonstration—a major feature of the *Elements*—was far too elaborate for Priestley to obtain. Timothy Priestley recalls his brother experimenting with levers and going into the fields with pen and paper to study astronomy; no doubt these are incident to Joseph's reading of 'sGravesande.[39] But the

---

38. See Robert E. Schofield, *Mechanism and Materialism,* passim, for a treatment of eighteenth-century British Newtonianism. Note that the quotation from Query 31 is not the usual English wording; it is likely that this is Desaguliers's rendering of 'sGravesande's printing of the Latin query.

39. Timothy Priestley, *Funeral Sermon,* 42–43. Timothy conflates these experiments with another (which he dates at Joseph's eleventh year!) on putting spiders into bottles to see how long they would live, and with one on melting steel with electricity. Joseph dates his first experiments on animal respiration much later and there is nothing in the *Elements* on vitiation of the atmosphere by respiration; while melting steel by electricity would require an electrical

air pumps, thermometers, optical benches, lenses and mirrors, even pulley systems, gears, and so forth, described and illustrated with such fidelity in the *Elements* were surely unobtainable to a young student in Yorkshire, even if he could afford them.

Possibly Priestley retained, from his reading of 'sGravesande, a conviction that mathematics could, by those interested, be employed to demonstrate phenomena discovered by experiment and that body was to be distinguished by the addition of force considerations to those qualities described by Locke and Watts. Perhaps he would remember 'sGravesande's assertion that "to follow the steps of Sir Isaac Newton" meant reasoning from phenomena and avoiding "feign'd Hypotheses," not simply accepting Newton's opinions (xi). And finally he may have remembered this first introduction to Newton's rule of reasoning: "That no more Causes of natural Things are to be admitted than such as are true and sufficient to explain their Phaenomena" (3); for this was a rule that guided his philosophical criticism most of his life. If he remembered these things, it would be enough; there were to be many years intervening, with ample opportunities to return to natural philosophy, before he was to experiment to any purpose himself.

Sometime during the period 1750–52, it was decided that Priestley should resume the design of becoming a minister. Who did the deciding is far from clear. Timothy's account of this episode is an evangelical melodrama, with pious family in victorious confrontation with a relation whose delight was luring bright young men from service to God. The presumptive villain of this drama cannot be identified. All that truly can be salvaged from Timothy's story is the impression, supported by Joseph's passive-voice description of the decision, that it was a family one. This is what would, in any case, be expected. He had been seriously ill; was he now sufficiently recovered for such an undertaking? He was committed to a commercial position in Lisbon; could that commitment be dissolved without difficulties? There would be financial obligations to be met; could these be dealt with?

Eighteenth-century English children customarily had their careers chosen for them, but surely Joseph was at least consulted in the course of this determination. It may seem odd that a person in a state of theological indecision should agree to become a minister, but perhaps Priestley's per-

---

machine and could have been performed no earlier than 1752 and more probably later than 1765.

sonal anxieties had, by then, been resolved. Early in Locke's *Essay* there is a fine rhetorical query:

> . . . who is there almost that dare shake the foundations of all his past thoughts and action, and endure to bring upon himself the shame of having been a long time wholly in mistake and error: Who is there hardy enough to contend with the reproach which is every where prepared for those who dare venture to dissent from the received opinions of their country or party: And where is the man to be found that can patiently prepare himself to bear the name of whimsical, sceptical, or atheist, which he is sure to meet with, who does in the least scruple any of the common opinions? (1-3-25)

Surely it would be too pat to assume that Joseph Priestley consciously nominated himself for the difficult role described by Locke. Yet, in fact, this was to be his role throughout his life. To be always in the minority, questioning established opinion regardless of obliquity. And consciously or not, this is the position for which he entered training as he left home for a ministerial education.

Appendix

To Mr. Annet on his New Shorthand

Annet again upon the field appears,
His plan matur'd and perfected by years:
No touch unnatural now disgusts the sight,
Where e'er you look, it tempts the hand to write.

Well may the World approve thy great Designs,
When with thy Zeal, an equal genius joins:
One vote for merit, still the muse will give,
And where thy Work is known thy Fame will live;
What, tho' unknown to thee, my vote I raise,
Merit, not Int'rest ever gains true praise;
Practice by which true Merit's ever known;

Practice has made thy useful Art my own;
Unerring Practice points thy Worth to view,
And worth acknowledg'd makes Applause but due;
Practice forsees thy lasting Use to Men,
And unborn Genius bless thy skilful Pen.

—J. Priestley

Joseph Priestley acknowledges these lines as his, in his *Reply to the Animadversions on the History of the Corruptions of Christianity* (Birmingham: for J. Johnson, 1783), 50–51, and declares that they were "written when I was a school-boy," which dates them during the years 1744 to about 1749, that is, between the ages of eleven and sixteen.

II

DAVENTRY ACADEMY, 1752–1755

When Joseph Priestley and his family determined that he was to become a minister, this meant, as a matter of course, that he should enter the Dissenting ministry. This, in turn, meant that he could not seek higher education at Cambridge or Oxford, even if he wished to do so. Graduation from the one, or even matriculation at the other, required formal assent to the Thirty-nine Articles of the Established Church. He could have gone to a Scottish university, or even to Holland or Switzerland, but the customary (and less expensive) alternative, by the mid-eighteenth century, was one of the academies founded in England for the education of Nonconformists, the so-called Dissenting academies.

Dissenting academies were, for practical purposes, an institution of the eighteenth century. The earliest of them had been established shortly after 1662 to provide for the training of those persons prohibited by the Act of Uniformity from attending the English universities. These were private schools, in operation only as the minister-tutor escaped persecution and closing with his death. In spite of efforts by church and state to close them, the academies flourished. With the Toleration Act of 1689 they came into the open and many acquired financial support from Dissenting organizations. Not till the middle of the eighteenth century, however, did they develop institutional existence independent of their founders. By then they were responsible for much of the secular education obtained by middle-class

Englishmen, as well as the theological training obtained by the majority of the ministers of Nonconformist congregations.

The English universities reached their lowest state during the eighteenth century. Hampered by formal curricular requirements established by the 1636 statutes of Archbishop Laud, isolated by deliberate policy from the economic and political life of the country, the colleges became irrelevant to the changing intellectual and cultural modes of Enlightenment England. Enrollments fell, university professors did not lecture, students did not study, examinations became a farce. No doubt it was possible to obtain an education at one of the colleges of Cambridge or Oxford and many people must have done so, but with considerable effort.

Meanwhile, most of the major scientific, philosophical, philological, historical, and literary works published in England during the last three-quarters of the century were written by men who had no connection with any of the English colleges or who wrote contemptuously of their college education. Many of these men received some part, at least, of their education at a Dissenting academy, whether or not they were Dissenters. The public character of the academies and the quality of their curricula made them increasingly popular for the education of the sons of country gentry and merchants. With the establishment in 1826 of London University, explicitly free of religious tests, a major raison d'être of the academies was gone. Except as theological seminaries, they vanished as independent institutions, but during the near century and a half of their existence there were founded more than seventy of these schools.

Many lasted only a short time; the early ones were handicapped by their need for secrecy and most were crippled by lack of institutional continuity. Some of them were, no doubt, bad; others, such as that attended by Timothy Priestley at Heckmondwike, never rose above pietist incubators for evangelical ministers. At many of them, however, a student could get an education equal to that of the colleges of Cambridge or Oxford. At some of them, the education was superior.[1] No one familiar with the comparative

---

1. Much has been written on the Dissenting academies and universities. See, for example, Irene Parker, *Dissenting Academies in England* (New York, Octagon Books, 1969, reprint of 1914 ed.); Herbert McLachlan, *English Education under the Test Acts* (Manchester: University of Manchester Press, 1931); and more recently J. W. Ashley Smith's immensely useful *Birth of Modern Education*. For the universities, see Christopher Wordsworth, *Scholae Academicae: Some Account of the Studies at English Universities in the Eighteenth Century* (Cambridge: Cambridge University Press, 1877); D. A. Winstanley, *Unreformed Cambridge* (Cambridge: Cambridge University Press, 1935); A. D. Godley, *Oxford in the Eighteenth Century* (New York: G.P. Putnam's Sons; London: Methuen & Co., 1908). Nicholas Hans, *New Trends in*

history of England's eighteenth-century universities and its Dissenting academies would conclude that Joseph Priestley was, educationally, the worse for his inability to attend one of the universities provided that he went to one of the better Dissenting academies. Almost by accident, the academy he attended was one of the best.

His relatives intended that Joseph be sent to the academy at Plaisterer's Hall, Stepney, run by the Reverend Zephaniah Marryat. Marryat died in 1754 and the academy was moved to Mile-End, where it continued under the direction of Dr. John Condor, divinity tutor.[2] As this academy survived several moves and eventually merged with other academies to form New College, London, it clearly served a useful purpose, but it may be doubted that this was the fostering of the intellect. John Kirkby's belief that his former student would have "a better chance of being made a scholar" (*W* 1/1:22) at a different academy finds support in extracts from the diary kept by Thomas Gibbons, appointed in 1754 to be tutor of logic, metaphysics, ethics, rhetoric, and pulpit style at Mile-End:

> 1/3/1755. My business as a Pastor is first to be taken care of. My business as a Tutor is only secondary. I design therefore . . . to prepare my Sermons as the first Work in the Week, and then to spare what other Time remains in preparing my Lectures.

> 26/5/1758. Lectured at Mile End . . . I finished the last lecture of the four years' course of Lectures at the Academy, & hereby I have acquired a Sett of Lectures for my Whole future life, or so long as I may continue in the Tutorship.[3]

If this appointment of 1754 reflected the quality of Marryat's academy in 1752, Priestley would not have found it intellectually stimulating.

The educational quality of the academy he was to attend was, however, not a primary consideration in the impasse that developed when Priestley

---

Education in the Eighteenth Century (London: Routledge and Kegan Paul, 1961) disputes the overriding claims made for Dissenting academies, but his interest is primarily in grammar and professional training schools. He admits "The leading men in the sciences, in philosophy, in social-economic studies, were seldom connected with the Universities" (15).

2. This is the origin of Priestley's anachronistic designation, in his *Memoirs,* of the academy to which he was to be sent in 1752, as the "academy at Mile-End, then under the care of Dr. Conder" (*W* 1/1:21).

3. Quoted in J. W. Ashley Smith, *Modern Education,* 196–97; used with permission. See also W. H. Summers, ed., "Dr. Thomas Gibbons' Diary," *Transactions of the Congregational Historical Society* 1 (1901–4): 380, 384.

refused to go to the school at Stepney. The point at issue was the severity of its Calvinist orthodoxy. The initial suggestion of Marryat's academy probably came from the Reverend Edward Hitchin, Deacon Joseph Priestley's London brother-in-law. Hitchin's doctrinal zeal was demonstrated four years later in organization of the Northern Education Society to support the academy at Heckmondwike (to which, not incidentally, Thomas Gibbons of Mile-End was appointed visitor and adviser.)[4] No doubt Hitchin knew that Marryat's academy, supported by the ultra-orthodox King's Head Society as well as the Congregational Fund Board, would adhere strictly to Calvinist doctrine. This was Joseph's objection: "If I went thither, besides giving an experience, I must subscribe my assent to ten printed articles of the strictest Calvinist faith, and repeat it every six months" (W 1/1:21).[5] But he could not give an "experience" and could not honestly have subscribed. He resolutely opposed going to Marryat's academy and might have had to give up hope of becoming a minister when his family was persuaded to approve another academy.

This account, while no doubt true, is incomplete, for the choice of Marryat's academy was not, finally, in Priestley's option. Attendance at an academy supported by the Congregational Fund Board required that the applicant's "Religious Disposition" be determined by investigation—which would have involved inquiries at his church.[6] Priestley could not have qualified, for his church had rejected him; his family was compelled to admit the proposal of another academy.

The proposal seems to have come from the Reverend John Kirkby, who "had no opinion of the mode of education among the very orthodox Dissenters, and being fond of me . . . was desirous of my having every advantage that could be procured for me" (W 1/1:22). Kirkby strongly recommended the academy of Dr. Philip Doddridge, at Northampton, founded in 1729 and one of the best known in England. The Priestleys

---

4. Summers, "Gibbons' Diary," 383; Gibbons, incidentally, attended Hitchin's ordination on 22 November 1759 (319).

5. The King's Head Society thought the Congregational Fund Board insufficiently orthodox and required additional doctrinal tests be imposed on pupils and schools it supported. The Society was, initially at least, entirely of laymen. Edward Hitchin could not have been a member, but one of its founders was a Mr. Hitchin, and the later connection with the Reverend Hitchin's Northern Educational was close; see John Waddington, *Congregational History, 1700–1800* (London: Longmans, Green & Co., 1876), 263–64; and R. W. Dale, *History of English Congregationalism* (New York: A. C. Armstrong and Son, 1907), 558–59.

6. See [T. G. Crippen, ed.,] "Congregational Fund Board," *Transactions of the Congregational Historical Society* 6 (1913–15): 209–13.

knew of it, for Joseph's stepmother was former housekeeper in Doddridge's family and "had always recommended his academy." And Northampton was less than three-quarters the distance to London and without the obvious temptations of the city.

Why had they not considered it in the first instance? The answer may lie in the influence of Edward Hitchin and in the criticism, beginning to be voiced in orthodox circles, that Doddridge's teaching methods encouraged the production of heterodox graduates. There was, however, no question of Doddridge's personal orthodoxy and his academy had long received support from the Coward Trustees, who required that students be "well instructed in the true Gospel doctrines, according as the same are explained in the [Westminster] Assembly's Catechism."[7]

The problem may only have been the uncertainty respecting the future of Doddridge's academy, for he had left Northampton and England in 1751, seeking relief in Lisbon for a consumptive illness and died there before the year was out. He had placed the academy in the charge of the assistant tutor, Samuel Clark, but no students had been admitted in 1751 and it might have been thought that the Northampton academy, like so many others, would close with the death of its founder. Happily that difficulty was resolved, and in a way that, peripherally at least, involved a Priestley family connection. The Coward Trust assumed full control of the academy and appointed the nominee of Doddridge's will, the Reverend Caleb Ashworth of Daventry, to be the new head. Not only was Ashworth properly orthodox, but his elder brother, Thomas Ashworth, was a Calvinistic Baptist minister and another of Deacon Joseph Priestley's brothers-in-law.[8] Whatever doubts there might once have been concerning Doddridge's Academy were thus, at least temporarily, satisfied and Joseph Priestley could depart on the next stage of his intellectual adventure.

When Priestley arrived at Daventry in September 1752, he entered a world in which "serious pursuit of truth" was the preoccupation of everyone, all topics were subject to continual discussion, and students were indulged in the greatest freedoms without doctrinal constraints or impositions

---

7. J. W. Ashley Smith, *Modern Education,* 149.

8. Arthur S. Langley, "Baptist Ministers," lists Thomas Ashworth as pastor at Cloughfold, 1751–55, when he moved to Gildersome, Yorkshire. S. Palmer, "Memoir of Dr. Caleb Ashworth," *Monthly Repository* 8 (1813): 693–96, recalls Daventry visits of Thomas Ashworth, "pastor of a Calvinistic congregation, at Heckmondwike"; and Frank Peel, *Nonconformity in Spen Valley* (127), refers to the Reverend Mr. Thomas Ashworth, Congregational minister of Gildersome, who was married to a sister of Deacon Priestley.

of tests and creeds (W 1/1:23). This environment of intellectual stimulus and freedom was new to Priestley and made his stay at the academy delightful, though it had few other advantages to recommend it. Possibly the landscape of Northamptonshire contrasted favorably with the bleaker dales of his native West Riding, but Priestley was essentially an urban person and seldom noticed the countryside. Daventry was a small market-town, center for a whipmaking industry, and sited at the intersection of major roads connecting Oxford, Leicester, Coventry, and London, making it an important coaching stop. On nearby Borough Hill, there were Roman remains and visiting gentry periodically staged horse races just out of town, but it was chiefly notable for the large number of inns serving the coaching trade. Its sixteenth-century grammar school was moribund, its parish church was dilapidated and had to be rebuilt in 1758, but its Dissenting congregation was active and must have welcomed the unexpected acquisition of an academy, with the excitement and trade incident to the presence of some fifty young students.

For the establishment of the academy at Daventry was indeed unexpected. In appointing Ashworth the new head, the Coward Trustees had not intended a move. Doddridge had nominated Ashworth as his successor to both the academy and the pastorate of the Dissenting congregation at Northampton. The congregation would have adopted his wishes, but Ashworth, well aware that his talents were neither popular nor scholarly, refused to become academy head without the continued assistance of the subtutor, Samuel Clark, and the Northampton congregation refused to accept, under Doddridge's successor, the continued services of Clark as assistant minister: ". . . he was not sufficiently popular and Calvanistical fully to satisfy the generality of them." The academy was, therefore, transferred to Daventry, where Clark "used to preach once in a month, with the consent of the people, who highly venerated his character, though his strain and manner were not quite to their taste."[9]

As the move to Daventry (see Fig. 2) had been unplanned, immediate facilities for the academy were inadequate. Ashworth erected a new building for the students, but its location was far from good, "being in a narrow street, close to a very public road, and opposite to a large inn."[10] The contrast with the grounds and buildings of any Oxford or Cambridge col-

9. P. R., "Brief Memoirs of the Rev. Mr. Samuel Clark," *Monthly Repository* 1 (1806): 617–22, esp. 618.
10. S. Palmer, "Ashworth," 695. The building is still standing and is now identified by a plaque noting Priestley's having studied there.

Fig. 2. Daventry Academy Building, as it appeared c. 1976. Photograph by the author.

lege is glaring, but so also is that in cost of attendance. It is likely that Ashworth continued, for the first few years at least, the fee schedule of Doddridge's academy. The cost of attending Daventry academy was then just over twenty-three pounds a year, including tuition, room, board, and fees for library purchases and maintenance of scientific apparatus; one could pay six times that amount in attending one of the colleges.[11] The Daventry fees were more than half the annual salary an aspiring young minister might hope initially to receive at his first church. On the grounds of expense alone, one can see why Priestley might prefer the less imposing academy to a university, even had he been able to attend one of these, but there were other compensations as well.

11. For Doddridge's Academy fees, see A. Victor Murray, "Doddridge and Education," in *Philip Doddridge 1702–1751: His Contribution to English Religion*, ed. Geoffrey F. Nuttall (London: Independent Press, 1951), 112; Irene Parker, *Dissenting Academies*, 82–84, notes that William Pitt expended roughly £132 in two terms at Trinity College, Oxford, in 1726–27. Poorer students would spend less, and there were jobs and endowed scholarships at the universities.

These compensations were, at first, only presumptive at Daventry, for the reputation of the academy that attracted Priestley had been earned under Doddridge. It was not obvious that Ashworth and Clark would continue in the same manner. Both, however, had trained under Doddridge and retained, unchanged for several years, his mode of teaching, texts, and curriculum. The most important of the practices continued at Daventry was that of doctrinal freedom.

This was so characteristic of Doddridge's academy that its passage to successor academies at which his former pupils taught is held to be the major positive influence of Doddridge on eighteenth-century English education. He was not personally ambivalent about theological doctrine. His published sermons and hymns leave no doubt of his essential Calvinism. His most popular work, *The Rise and Progress of Religion in the Soul*, first published in 1745, satisfied generations of evangelical Calvinists, ministers and laymen, with its fervently "practical" program for personal religious living. Yet Doddridge did not believe in the necessity of a rebirth experience, particularly for those persons who had had the benefits of a pious education, and he did not hold that any one sect had a monopoly on truth: "I pray God . . . to remove far from us those mutual jealousies and animosities, which hinder our acting with that unanimity which is necessary in order to the successful carrying on of our common warfare against the enemies of Christianity."[12]

No doubt these opinions had much to do with the lack of any doctrinal tests for admission to his academy. So also must his experience in 1733, shortly after the establishment of his academy at Northampton, of being cited by the Bishop of Peterborough for teaching without a license from the diocese. The case appears to have been prosecuted "purely to establish and vindicate the authority" of the bishop's spiritual court and, when carried to Westminster Hall, was dismissed by the judges there. But the effort illustrated the pernicious influence of religious authority. Doddridge heard of "near twenty such attempts . . . made within less than so many years, upon dissenting schoolmasters in this diocese," and his reaction was marked by his published sermon of 5 November 1735, "On the Absurdity and Iniquity of Persecution for Conscience' Sake in all its Kinds and Degrees."[13] Doddridge's academy accepted, as students, Anglicans, English and Scottish

12. P[hilip] Doddridge, *The Rise and Progress of Religion in the Soul* (London: W. Baynes, 1808), 17th ed., xviii–xix, xvi–xvii.

13. James Yates, "Ecclesiastical Proceedings against Dr. Doddridge," *Christian Reformer* 1, n.s. (1861): 552–57.

Presbyterians, Independents, and Baptists; and the policy was continued at Daventry under Ashworth.

Nor, to the scandal of the orthodox, did Doddridge consciously impose his opinions on students once they were admitted. The list of some two hundred persons who attended the academy between 1729 and 1751 indicates that the majority, as might be expected, became orthodox Dissenting ministers, but it also notes some as having conformed to the established church, at least one who became a Methodist and another who joined the Moravian Church, while several can be identified as having adopted some form of anti-Trinitarian belief.[14] The daughter academies at Daventry and Warrington had similar records.

Not all, probably not most, of these variant choices were made by students while still in their academies, though Priestley began his major heterodoxy there. Moreover, such decisions were made for many reasons (especially social and economic) other than doctrinal. What can be said of Doddridge's academy, and of its daughters, is that students there were provided with the background and information from which individual choices might be made. This is a common-enough claim for a twentieth-century university, but for an eighteenth-century academy to have reached some measure of its attainment is most unusual. Part of the credit for this achievement must go to Doddridge personally; another part goes to the two major traditions of Dissenting education that came together in the creation of a liberal curriculum and an eclectic, comparative teaching method developed at Doddridge's academy and passed along by tutors trained there.

Naturally the greatest immediate educational influence on Doddridge was that of John Jennings, at whose short-lived academy at Kibworth he had trained. Jennings had studied under Timothy Jollie, at Attercliffe, and Jollie under Richard Frankland, of Rathmell, who had graduated M.A. from Christ's College, Cambridge, and started one of the earliest of the Dissenting academies.[15] Frankland developed a course of studies that followed the lines of his Cambridge training, with a strong emphasis on philosophy, including natural philosophy, and an added concern for English literary style. The philosophy may have had a flavor of Neoplatonism in it, as Henry More was a fellow and tutor at Christ's College while Frankland

14. Thomas Stedman, "A List of the Pupils educated by P. Doddridge, D.D.," *Monthly Repository* 10 (1815): 686–88.

15. The details of academies and their curricula are primarily derived from J. W. Ashley Smith, *Birth of Modern Education*, passim.

was there, and Ralph Cudworth became a master of Christ's the year before Frankland obtained his M.A.

Jollie neglected philosophy and distrusted mathematics "as tending to scepticism & infidelity," but continued the emphasis on English style and elocution and introduced specific lectures on "practical religion": the sermon-making and delivery, scriptural exposition, and pastoral visitation necessary for theological students. Jennings returned to the fuller curriculum characteristic of Frankland, but retained Jollie's concern for practical religion. Doddridge expanded Jennings's curriculum to a five-year course for divinity students and this curricular framework was continued at Daventry. As the details of the framework are known, it is possible to determine, with some precision, what subjects Priestley studied and even to approach his own brief description of his studies at Daventry with some critical understanding.[16]

Because Priestley began his studies at Daventry in what was, essentially, the third year of a five-year divinity course, he missed most of the basic subjects: Latin, Greek, French, and Hebrew; geometry, algebra, trigonometry and conic sections; geography and history, logic and natural philosophy, of the regular formal program. He had already studied most of these, on his own, beyond the level of grammar school—which is why he had been excused "all the studies of the first year, and a great part of those of the second"—but he clearly felt that his preparation in them had been less than adequate. His *Memoirs* and the diary that he kept for 1754 and 1755 show him attempting to combine reading on the work of the "excused" years along with that of the last three.[17]

16. Evidence of curriculum continuity from Kibworth to Northampton to Daventry is circumstantial but strong. Jennings's course is described by Doddridge; see John Doddridge Humphreys, ed., *The Correspondence and Diary of Philip Doddridge, D.D.* (London: Henry Coburn and Richard Bentley, 1830), 2:462–75; and J. W. Ashley Smith, *Birth of Modern Education*, 112–19. Doddridge's course, paralleling Jennings, is discussed in Humphreys, passim, is described by a student for the period 1734–39, in *Memoirs of the Life and Writings of the Rev. Philip Doddridge*, by Job Orton (Salop: J. Cotton and J. Eddowes, 1766), 84–97, and treated by Ashley Smith. The course at Daventry is broadly described for 1752–55 by Priestley in *Memoirs* and indirectly in a diary; see the text and note 17. Finally, Thomas Belsham, tutor at Daventry from 1771 to 1789, outlines the course of study there, c. 1781, following Doddridge's, see John Williams, *Memoirs of the late Reverend Thomas Belsham* (London: for the Author, 1833), 224–26.

17. Most of the journal or diary kept by Priestley from 1752 to 1803 was destroyed, but the part for 1754 survived the Birmingham riot of 1791; see "Joseph Priestley's Journal while at Daventry Academy, 1754," *Enlightenment and Dissent* 13 (1994): 49–113, transcribed from shorthand by Tony Rail and Beryl Thomas, whose courtesy allowed me to examine it prior to publication; quoted extracts used here by permission. Extracts from that of 1755 were

For those three years, he and a classmate, John Alexander, rose early to read "every day ten folio pages in some Greek author, and generally a Greek play in the course of the week besides."[18] Logic he had already studied in the text by Watts used at the academy and felt no need to review. Mathematics also he had studied on his own. Mathematics was of secondary importance to Doddridge and therefore, to the successors whom he had trained. Nonetheless, Priestley read, at Daventry or, during vacation, at home: Saunderson's *Elements of Algebra*, Maclaurin's *Treatise on Algebra*, Benjamin Martin's *Biblioteca Technologica* on conic sections, and attempted via Martin, W. Owen's *Dictionary of Arts and Sciences*, and Ditton's *Institution of Fluxions* to learn fluxions. Journal entries include the following: for 3 September, "read several pieces about fluxions—learnt very little"; for 10 September, "Applied hard to fluxions, and in a great measure mastered them."

The description of the natural philosophy course, requiring seven lectures a week during the half-year, sounds as though the text was that of 'sGravesande, but the academy had the advantage of an extensive apparatus, purchased in 1732 with a grant from the Coward Trust.[19] How much of that course Priestley took is unclear, but he would have benefitted from the use of apparatus. He certainly heard Ashworth lecture on sound and experimented with a magic lantern and an equal-arm balance. He also read Newton's *Optics*, Barrow's *Universal Dictionary* on electricity, and Martin's *Philosophia Britannica* on Newtonianism.

The Daventry science curriculum included some natural history, which Priestley appears not to have studied, and considerable anatomy, which he did. He heard Clark lecture, participated in dissections of a rat, cat, dog, and the head of a drake. He also read James Drake's *Anthropologia Nova*, Cheselden's *Anatomy of the Human Body*, the anatomy selections in Cham-

---

published in *The Memoirs of Dr. Joseph Priestley to the Year 1795 . . . with a continuation by his son* (Northumberland [Pennsylvania]: John Binns, 1806), 1:178–84.

18. In his *Memoirs*, Priestley claims that "in the course of our academical studies . . . there was then no provision made for teaching the learned languages. We had even no compositions or orations in Latin" (W 1/1:26). But the 1754 journal records an examination oration on Ovid and Horace. Still, the 1755 extracts list independent reading, to remedy the supposed deficiency, in Horace, Ovid, Tacitus, Cicero, Homer, Demosthenes, and Herodotus, as well as from the Greek New Testament and a Latin Euclid.

19. On acquisition of the apparatus, see J. D. Humphreys, ed., *Correspondence . . . of Doddridge*, 3:109; for teaching using apparatus, 4:404, quoted by J. W. Ashley Smith, *Birth of Modern Education*, 134.

ber's *Universal Dictionary*, and, on medical chemistry, Boerhaave's *Chemical Lectures*.

Priestley criticizes the academy program as "defective in containing no lectures on the Scriptures, or on ecclesiastical history," but the issue again seems to be a matter of curricular timing. Reading and exposition of the Scriptures (sometimes from the Hebrew Old and Greek New Testaments) was assigned ten lectures a week through the first and second years of Jennings's course and appears to hold equivalent prominence in Doddridge's curriculum, but Priestley would have missed those lectures. Biblical commentary, Christian antiquities, ecclesiastical history (plus ancient philosophies), and history of religious controversies were subjects each assigned a lecture a week, through the third and fourth years; and Jewish antiquities, a favorite interest of Ashworth, was clearly contained in Daventry's curriculum in 1754. One suspects that Priestley, who was to devote considerable attention in later years to commentary and ecclesiastical history, might, in retrospect, regard a lecture a week for a half-year each, on these subjects as "defective."

Civil history and geography were also subjects taught in the first two years at Daventry and the diary again shows Priestley attempting to make up for some of his deficiencies. During 1755 Priestley read volumes 15 through 20 of the *Universal History*. One may assume that he earlier read the first fourteen volumes, but all twenty relate to ancient history. There is no indication that he acquired, at Daventry, even the slight knowledge of modern European history covered in lectures there.[20]

The only subject, other than religion, which was continued through every year of the academic program at Daventry was English, particularly as oratory (or pulpit style) and rhetoric. This was a subject that Priestley might not have studied on his own. Especially in its concern for English literature, it was to have a major influence on his personal development.

20. Priestley's diary for 1754 and 1755 serves to amend *W* 1/1:26, so far as Priestley's personal studies at Daventry are concerned. During those years Priestley read, in theology, nearly all of the New Testament in Greek, and one-quarter of the Old in Hebrew; at least two commentaries on the Scriptures, the Koran, Josephus, and several works on ecclesiastical history including King's *Constitution of the Primitive Church*, Neal's *History of the Puritans*, and a history of the Council of Trent. He read Butler's *Analogy*, Anthony Collins, Samuel Clarke on the Trinity, and Nathaniel Lardner. His History reading included Reland's *Antiquitates* and [George Sale, John Campbell, Archibald Bower, John Swinton, George Psalmanazer, et al.], *An Universal History, From the Earliest Account of Time: compiled from original authors* (London: T. Osborne, A. Miller, J. Osborn, 1747, 1748). Priestley does not identify the edition used, but the one cited concludes its "Ancient Part" with vol. 20, and the "Modern Part" was not available in 1755.

When he arrived at Daventry, he was learned in a number of languages, but of the literature of his own country he appears to have read only the didactic and homiletic. Indeed, by his own account, he must have been something of a rustic prig, having "a great aversion to plays and romances, so that I never read any works of this kind, except Robinson Crusoe, until I went to the academy. I well remember seeing my brother Timothy reading a book of knight-errantry, and with great indignation I snatched it out of his hands, and threw it away" (W 1/1:19).

Clearly he needed some cultural seasoning, and Daventry academy was the place to obtain it. Neither Ashworth nor Clark had the sensitivity to literary style that had marked both the writing and the teaching of Doddridge, but they retained the curricular emphasis on English that helped Doddridge's academy, and those that succeeded it, become so important in the development of English studies in Britain: the use of English as the general language of instruction, frequent assigned reading of English literature and criticism, formal instruction and practice in reading aloud and in declamation; formal composition of essays, criticism, homilies, and sermons; practice preaching and finally "student" preaching for neighboring churches.[21]

It was not possible for Daventry, with but three years in which to drill reading and elocution, to help Priestley much with his personal problems of speaking. He left the academy as he had entered it, with a Yorkshire accent and a stammer though both Ashworth and Clark tried to help with rehearsals, private reading, and lectures on delivering sermons. The stammer, according to journal entries, was a "cause of much distress": for 20 October, "Discouraged on account of my speaking with difficulty and low"; for 27 October, "Appointed Tayler to give me a hint when I spoke too fast . . . I spoke, as they tell me, pretty well." But Priestley came to believe that there were compensations, as it checked any tendency to be disputatious in company or "seduced by the love of popular applause as a preacher; whereas my conversation and my delivery in the pulpit having nothing in them that was generally striking, I hope I have been more attentive to qualifications of a superior kind" (W 1/1:28). No doubt Priestley meant by this, at least in part, that he concerned himself more with the content than the style of his delivery and Daventry certainly helped him in the acquisition of ideas.

21. Some subjects had previously been taught in English at Scottish universities and at other Dissenting academies, but Doddridge appears to be the first person to construct his entire curriculum in that language. For the influence of Northampton Academy and its successors on English studies, see J. W. Ashley Smith, *Birth of Modern Education*, passim.

It appears also to have helped him develop a mode of delivery and a literary style that compensated for imperfections of speech. Except for occasions of personal stress, when the stammer became a major handicap, he learned to control it in the pulpit by using a conversational manner and a language of almost colloquial directness which many hearers were, in time, to find appealing.[22] He also found compensation for the impediment of speaking in a vigorous use of his pen. Whether or not this alternate mode of self-expression was suggested to him by his tutors, they clearly aided him in developing it, teaching him, through literary example, the elements of style and drilling him in composition.

In the ten years between his entrance in Daventry and his teaching of belles lettres at Warrington Academy, Priestley was transformed from the boy whose literary reading had been limited to Robinson Crusoe into the man who could use examples drawn from English sources as various as Milton, Swift, the *Spectator* and *Guardian*, Young, and Thompson. That the transformation must have begun at Daventry is confirmed by Priestley's 1754 and 1755 "Entertaining" reading: fourteen plays of Shakespeare, Butler, Congreve, Dryden, Pope, Steele, Swift, Thomson, and Young are among the authors cited.

In the excitement of reading, it is a wonder there was time for writing. But under the heading "Business done from April 1st to June 23d . . . Composition," in the 1755 diary, is noted: "wrote an article on Edward's translation of the Psalms." This, Priestley's second publication (after the verses on Annet's shorthand) appeared in the *Monthly Review*.[23] The diaries also note the composition of many sermons and orations that were subjected to the "discriminating judgment" of Ashworth, who disliked fancy rhetoric and could "shew others what to avoid, not always exemplifying what was graceful."[24] And, at Daventry, Priestley began the writing of his *Institutes*

22. In 1781, Catherine Hutton described his pulpit manner: "mild, persuasive, and unaffected, . . . his sermons . . . full of sound reasoning and good sense . . . not an orator, he uses no action, no declamation; but his voice and manner are those of one friend speaking with another." Catherine Hutton Beale, ed, *Reminiscences of a Gentlewoman of the Last Century: Letters of Catherine Hutton* (Birmingham: Cornish Brothers, 1891), 28–29.

23. [Joseph Priestley], "A New English Translation of the Psalms from the original Hebrew . . . by Thomas Edwards," *Monthly Review; or, Literary Journal* 12 (1755): 485–89. Written with all the assurance of youth and inexperience, the review is chiefly a criticism of a theory of Hebrew meter invented by a Bishop Hare, defended and used by Edwards. Priestley cites an examination of Hare's hypothesis appearing in the *Universal History* (which he was reading at the time) and concludes that the nature of Hebrew renders Hare's reasoning uncertain and his hypothesis inadequate to the purpose of correcting Hebrew Scriptures.

24. T. Thomas, "Supplementary Hints to the Memoir of Dr. Ashworth," *Monthly Repository* 9 (1814): 10–12, 78–80; quote on 79.

of Natural and Revealed Religion, not to be published until 1772–74: "Mr. Clark, to whom I communicated my scheme, carefully perusing . . . it, and talking over the subject of it with me" (*W* 1/1:27).

By the date of his death in 1804, Joseph Priestley was to have published more than two hundred books, pamphlets, and articles. The quality of writing throughout such an extraordinary number of publications—"Many more, Sir," Priestley once told an impertinent inquirer, "than I should like to read"[25]—could not but be uneven, though from first to last he always wrote in a simple and direct prose. "My object," he declared, "was not to acquire the character of a fine writer, but of an useful one." Sometimes his writing achieved a measure of literary distinction, but it was uniformly prolix, frequently repetitive, and increasingly involved mannerisms, such as too-frequent use of the phrase: "I flatter myself," which can be exasperating if one reads too much of his work at a time. Nonetheless, his style is never distasteful and suits the purpose of admonition and instruction for which it was designed.

As a writer, Priestley won the approval of two not-inconsiderable critics of English prose, Charles Lamb and William Hazlitt. Lamb, no doubt, was attracted as much by the content of Priestley's writings as by their mode and, in writing to S. T. Coleridge in the mid-1790s, when Coleridge was also in sympathy with their content, his praise was perhaps excessive. Yet, with his acute sensitivity to mood and to benign simplicity of expression, it is hard to believe that Lamb would write of a prose style he disliked: "You would be charmed by his *sermons,*" or "You have no doubt read that clear, strong, humorous, most entertaining piece of reasoning," respecting Priestley's *Examination* of Scottish commonsense philosophers, or "I rejoice, and feel my privilege with gratitude, when I have been reading some wise book such as I have been reading [Priestley on philosophical necessity] in the thought that I enjoy a kind of communion, a kind of friendship even, with the great and good."[26]

William Hazlitt made the nature of Priestley's appeal more explicit:

> . . . he took in a vast range of subjects of very opposite characters, treated them all with the same acuteness, spirit, facility, and perspi-

25. [Samuel Rogers], A. Dyce, ed., *Recollections of the Table-Talk of Samuel Rogers* (New York: D. Appleton and Co., 1856), 122.

26. [Charles Lamb], E. V. Lucas, ed., *The Letters of Charles Lamb, to which are added those of his sister, Mary Lamb* (London: J. M. Dent and Sons, and Methuen and Co., 1935), 1:11, 78, 87.

cuity, and notwithstanding the intricacy and novelty of many of his speculations, it may be safely asserted that there is not an obscure sentence in all he wrote. . . . He was one of the very few who could make abstruse questions popular . . . on a par with Paley with twenty times his discursiveness and subtlety. . . . Dr. Priestley was certainly the best controversialist of his day, and one of the best in the language . . . [his] *Controversy with Dr. Price* is a masterpiece not only of ingenuity, vigour, and logical clearness, but of verbal dexterity and artful evasion of difficulties . . . in boldness of inquiry, quickness and elasticity of mind, and ease in making himself understood, he had no superior.[27]

The clarity, boldness, and vigor of Priestley's writing made it an admirable vehicle for his ideas, but it was on the acuteness, novelty, and subtlety of his speculations that contemporary attention was most focused, not on their mode of expression. These speculations were begun at Daventry, particularly in the student debating society, amid the heady ferment of discussion that Priestley loved: "In my time, the academy was in a state peculiarly favourable to the serious pursuit of truth, as the students were about equally divided upon every question of much importance, such as liberty and necessity, the sleep of the soul, and all the articles of theological orthodoxy and heresy; in consequence of which, all these topics were the subject of continual discussion" (W 1/1:23).

He came to believe in controversy as a utilitarian method of discovering truth, but for all his own repeated changes of opinion, he was never to adopt a nominalist or relativist position as to the ultimate nature of that truth. For at Daventry he began also to construct an idealist ontology to combine with utilitarianism and the rationalist empiricism he had learned by reading Locke.

The metaphysics taught at Daventry was explicitly linked to a study of "Christian Evidences" in a lecture course of which Doddridge had been particularly proud. Consisting of some two hundred and thirty lectures, the course was introduced during the third year, dominated the fourth, and monopolized the fifth of the five-year curriculum. The subjects taught included the powers and faculties of the human mind, the existence and attributes of God, the nature of virtue and civil government, the character

27. William Hazlitt, "The Late Dr. Priestley," *The Atlas*, 14 June 1829; reprinted in [William Hazlitt], P. P. Howe, ed., *Complete Works of William Hazlitt* (London: J. M. Dent and Sons, 1934), 20:237–38.

of the soul, evidences for revelation and vindication of Scriptures, and the doctrines of scriptural religion. Structurally, the course derived from a similar one taught by John Jennings, having the same curiously geometrical format of axioms, definitions, propositions, corollaries, scholia, and lemmas.

This geometrical mode would seem a clumsy way of discussing theological questions, but Priestley wrote of it: "if these definitions and axioms be laid down with due accuracy and circumspection, they not only introduce the easiest, the most natural, and cogent method of demonstrating any position, but lead to an easy method of examining the strength or weakness of the ensuing arguments."[28]

For Priestley, such a course, in content and structure, provided just that opportunity to argue and contest each point, which, he became convinced, led to discovery of truth. It was, in fact, designed to bring forward conflicting opinions on every issue thought to be important, with an array of authors being cited to support each opinion, so that truth could be recognized and defended against error. Students were expected to copy the notes from which the lectures were read, add the materials given in lecture, and look up, in the academy library, all the references so that they were thoroughly familiar with the issues.[29]

Although this practice also owed something to Jennings, who had cited varying opinions in his classes, Doddridge's explicit emphasis on "comparative theology" suggests the influence of his warm friend Isaac Watts, with whom he frequently consulted on educational matters. Watts recommended the method of stating both sides of every argument with references to the principal authors of each side, and praised it in his influential study-guide *Improvement of the Mind; or, A Supplement to the Art of Logic* (1741–51). Although the method originated with Theophilus Gale who passed it to Thomas Rowe, the teacher of Watts, it came to be regarded as a particular characteristic of Doddridge's teaching. It was continued at Daventry and adopted at the several academies at which his pupils were to teach.

28. Joseph Priestley, *A Course of Lectures on Oratory and Criticism* (London: J. Johnson, 1777), 46.

29. In 1763, Samuel Clark, Priestley's former tutor at Daventry and then a Dissenting minister in Birmingham, edited those notes—essentially "Heads of Lectures"—as *A Course of Lectures on the Principal Subjects in Pneumatology, Ethics, and Divinity: with References to the most considerable Authors on each Subject*, by the late Reverend Philip Doddridge (London: J. Buckland, J. Rivington, R. Baldwin, L. Hawes, W. Clarke and R. Collins, W. Johnston, J. Richardson, S. Crowder and Co., T. Longman, Longman, B. Law, T. Field, and H. Payne and W. Cropley, 1763). These lectures were several times reedited, with changes

It did not go without criticism. More orthodox ministers thought it generated a controversial spirit, without providing guidance in true religious principles, producing indifference to truth and willing reception of corrupt and depraved doctrines.[30] Criticism came from the liberal side as well. Timothy Kenrick, who studied at Daventry in 1775, declared that "Doddridge's lectures . . . [were] singularly unfavourable to the impartial discussion of controverted opinions" as their tendency was to inculcate popular and reputedly orthodox tenets, with opposing statements merely tacked on as scholia or lemmas.[31]

The bulk of contemporary evidence, including that of Kenrick's own career (he left Daventry an Arminian and an Arian), does not support the liberal criticism, while Priestley certainly found the lectures stimulating: "The general plan of our studies, which may be seen in Dr. Doddridge's published lectures, was exceedingly favourable to free enquiry, as we were referred to authors on both sides of every question, and were even required to give an account of them. It was also expected that we should abridge the most important of them for our future use" (W 1/1:23–24). Perhaps Ashworth became more strictly orthodox in his teaching between 1755 and 1775. More likely the tide of liberal theology had changed so much (particularly under the influence of Priestley's writings) by the time Kenrick wrote that what had once seemed daring and controversial no longer did so.

Certainly most of the controversies addressed in the *Lectures on Pneumatology* were those that had exercised English Presbyterians and latitudinarian Anglicans during the first third of the century: the challenge of deism, the so-called atheism of Hobbes and Spinoza, the rise of Arianism, and the general problem of determinism and free will. This is hardly surprising. Though Doddridge's teachings had not yet begun when the major excitement of these controversies had subsided, these were the questions of his own student days. What is unexpected, given the encomia of the broadmindedness of his teaching, is the narrow range of opinions, theological, philosophical, and political, represented in the authorities most often cited in his *Lectures*.

in the list of references, but this 1763 edition must be very like the unpublished version that Priestley read under Clark at Daventry.

30. Paraphrased from the criticism of R. Hall and of Bogue and Bennett quoted by J. W. Ashley Smith, *Birth of Modern Education*, 141.

31. Timothy Kenrick, *An Exposition of the Historical Writings of the New Testament, with Reflections subjoined to each Section* (Boston: Monroe and Francis, 1828, from 2d London ed. of 1824), 11.

The philosophical-theological opinions considered are primarily those of such men as René Descartes, Isaac Watts, John Locke, Jean Le Clerc, Henry More, Richard Baxter, and the Boyle lecturers, Richard Bentley and Samuel Clarke. And in those sections related to civil affairs and government, the selections are not of wider range: Algernon Sidney, Samuel Puffendorf, Hugo Grotius, Locke, Benjamin Hoadley, Walter Moyle, the "Old Whig" tracts. More even than the practice of citing "opposing" authorities, these selections identify the Gale-Rowe-Watts tradition of Dissenting education, for they represent, or develop out of, the Arminian and Cambridge Platonist thinking which Gale had introduced at Newington Green.[32]

The Cambridge Platonists, particularly including Benjamin Wichcote, John Smith, Henry More, and Ralph Cudworth, were nominally adherents of the established church, but they did not believe in a religion of dogma, hierarchy, or ritual. Their greatest affinities, philosophically and theologically, were with the hermetical Florentine Academy of late fifteenth-century Neoplatonists and with their own contemporaries, the Dutch Arminians of the second half of the seventeenth century.[33] Like the Arminians, the Platonists rejected the doctrines of election, predestination, and irresistible grace. Wary of "enthusiasm" and obscurantism, they believed that the exercise of man's reason—"the candle of the Lord"—could be a guide to religious truth. But in sharp and explicit opposition to Spinoza, whose mode of idealism they did not comprehend and whose pantheism they abhorred, they held that rational religion was confirmed, and even guided, by the Sacred Scriptures, whose inspiration was evidenced in the fulfillment of prophecy and the working of miracle.

Committed to a belief in the individual's use of "God's right reason" in interpretation of both natural and scriptural foundations of religion, they

32. Theophilus Gale devoted much of his life to a scholarly project of Christian Cabbalism, *The Court of Gentiles*, intended to prove that the works of Plato must be dependent upon a tradition of Hebrew learning. This belief is consistent, at least, with the Neoplatonism shown by Gale's student and successor as head of Newington Green, Thomas Rowe, who was also an admirer of Descartes, but moderated his admiration as Henry More had done. For a detailed study of Neoplatonic elements of Doddridge's teaching, see Robert E. Schofield, "Joseph Priestley, eighteenth-century British Neo-Platonism, and S. T. Coleridge," in *Transformation and Tradition in the Sciences: Essays in Honor of I. Bernard Cohen*, ed. Everett Mendelsohn (Cambridge: Cambridge University Press, 1984), 237–54.

33. Of the literature on the Cambridge Platonists, the best short account is probably Ernst Cassirer's *The Platonic Renaissance in England* (Austin: University of Texas Press, 1953). J. W. Ashley Smith's *Birth of Modern Education* first called my attention to the Platonic element in Dissenting education, while Rosalie Colie's *Light and Enlightenment: A Study of the Cambridge Platonists and the Dutch Arminians* (Cambridge: Cambridge University Press, 1957) is a wealth of information about English-Dutch relationships.

favored social and political toleration of all Protestant Christians. They could not approve a theocratic state nor a civil government claiming authority over religion. The test of a Christian was not beliefs, but morality of life, and the basis of morality was an innate moral sense, the manifestation of divine presence in every man.

As rationalists and philosophical idealists, they had welcomed Cartesian philosophy, especially as a weapon against the materialistic atomism of Hobbes. Henry More was largely responsible for introducing Cartesianism into Cambridge as Thomas Rowe was responsible for its introduction into Dissenting education. Much of this Cartesianism, especially its mechanical natural philosophy, physiology, and psychology, survived well into the eighteenth century; it is to be found in Locke and Watts, and Descartes is frequently cited in Doddridge's *Lectures*. But Descartes's dualism implied a physical universe functioning independently of God and this was unacceptable to the Cambridge Platonists. Instead of a matter rigorously separated from the action of spirit, they posited a material universe under control of an immanent God, through some intermediate principle: say, More's *spiritus naturae* or Cudworth's *natura plastica*.

The Cambridge Platonists, Henry More and Ralph Cudworth, exchanged books and admiring correspondence with the Dutch Arminians, especially Philip van Limborch and Jean Le Clerc; the latter published a series of extracts, in French, of Cudworth's *True Intellectual System of the Universe* beginning in 1703. Limborch and Le Clerc befriended John Locke during his Holland exile and continued their stimulating exchange of ideas by correspondence after Locke's return to England, particularly after he acquired the patronage of Lady Damaris Masham, Cudworth's daughter. Le Clerc arranged for Locke's first publication, the *New Method for Making Common-Place Books*, in 1686; translated his *Reasonableness of Christianity*, and dedicated the second edition of the *Logica* to him.

Together, the Cambridge Platonists, Dutch Arminians, and John Locke inspired the Anglican latitudinarians. They also inspired the increasingly liberal wing of English Dissent, the English Presbyterians, first led by Richard Baxter. And when, at the end of the seventeenth century and early in the eighteenth century, Newtonian physico-theologians like Richard Bentley and Samuel Clarke attempted, as Boyle Lecturers, to confute the coffee-house devotees of Hobbes and Spinoza, they drew on the inspiration of More and Cudworth in their interpretation of Newtonian forces of attraction and repulsion as physical evidences of an immanent God.

The most frequently cited authorities in Doddridge's *Lectures* includes all these: Cambridge Platonists, Dutch Arminians, Anglican latitudinarians, English Presbyterians, and Newtonian physico-theologians. Any group that ranged in eclecticism from the Hebrew testament and Plato to Descartes and Locke would necessarily comprehend a wide variation of opinions, if only in specific interpretations of its general agreement. Although the passages cited by Doddridge from the various books do emphasize agreement, a curious and omnivorous reader might find much in the remainder of those books to challenge orthodoxy.

It was in these circumstances, Priestley tells us, that he "saw reason to embrace what is generally called the heterodox side of almost every question" (*W* 1/1:34–35). Sometimes that heterodoxy seems comparatively minor, serving only to show that Priestley was capable of a consistency not shown in his text. Accepting as he did, for example, Doddridge's Propositions 5 and 6, that there are neither innate ideas nor propositions in the mind (references, of course, to Locke), Priestley could not accept the Demonstration to Proposition 48, supposing "a moral sense, implanted in our natures, or an instinct, like that of self-preservation" which is the foundation of our knowledge of moral law (referenced to Francis Hutcheson).

This denial did not mean, to Priestley, that there was no absolute moral law, nor that the law was an arbitrary fiat of God, knowable only by revelation or special illumination. His explanation of the source of man's knowledge of morality illustrates his ability to combine an idealist ontology with an empirical, utilitarian, epistemology. God created man to live happily in a world ordered by moral law. True morality is therefore, the only kind of life that is compatible with the whole nature of man. Man can, therefore, deduce moral law from his experiences (personal and historical) living in it.[34]

It is difficult to know how much Priestley's idiosyncratic position in natural philosophy owed to the idealist ontology he was taught at Daventry. He was not to do any scientific work for ten years after leaving the academy and not to write philosophically about it, in any detail, for more than twenty. At that time, however, his spiritually materialist opinion was in marked agreement with some ontological propositions presented in Doddridge's *Lectures*. Corollary 2 of Definition 3 declares, for example, "We can have no conception of any substance distinct from all the properties of the

---

34. See part 2, sect. 1, "Of the rule of right and wrong," in Priestley's *Institutes of Natural Religion* (London: J. Johnson, 1772), probably written while he was still a student at Daventry.

being in which they inhere; for this would imply that being itself inheres, and so on to infinity." Priestley could also have learned there that matter is not self-existent or necessary, but was created by God, for a world of beauty, harmony, and order; and that its continued existence and the connections between its solidity and gravity, and between motion and thought, are owing to the continual exertion of the will of God: the perpetual omnipotent agency of a self-existent being.[35]

The natural philosophy (Definition 6: "that branch of learning which relates to body, giving an account of its various phaenomena, and the principles on which the solution of them depends") presented in the *Lectures* is chiefly the mechanical, corpuscularian philosophy of Descartes, as modified by the Boyle lecturers, Bentley and Clarke, and described in John Ray, William Derham, Bernardus Nieuwentyt, or the Abbé Pluche. Each of these is more interested in evidences of design and final cause than in physical process. There is no direct reference to Newtonian forces and scarcely any to Newton himself. The arguments against a Cartesian plenum and in support of vacuum (Demonstration of Proposition 23) are referenced, for example, to Bentley, Clarke, and Baxter, to Samuel Colliber, author of pantheistic tracts, and to John Howe, Presbyterian minister and Platonist friend of Henry More and Richard Baxter.

Priestley says his "attention [at Daventry] was always more drawn to mathematical and philosophical studies" than was that of his best friend there (W 1/1:26). He had entered with information on natural philosophy derived from Locke, Watts, and 'sGravesande. This had introduced him to the corpuscular philosophy and led him from the Lockean position that man could not determine, in scientific exactitude, how corpuscles might act to produce phenomena to the suggestion, in 'sGravesande, that they probably did so through the agency of forces of attraction and repulsion, though the causes of those forces were yet unknown. Surely the proposition of the *Lectures on Pneumatology* implying that the forces were manifestations of the will of God would be of more than casual interest to Priestley. And before he left Daventry he was to find a recapitulation of the entire argument, from material corpuscles through the exertions of an immanent God, in Rowning's *Compendious System of Natural Philosophy*.[36]

---

35. Propositions 21, 23, 27, and 32, and the demonstrations accompanying them.

36. John Rowning, *A Compendious System of Natural Philosophy* (London: Sam. Harding, 1737–43); published in parts, separately paginated. The set I have consists of preface, appendix, and index, 1743; part 1, 3d ed., 1738; part 2, 3d ed., 1737; part 2, cont., 1736; part 3 and part 3, cont., 2d ed., 1743; part 4, 1743; part 4, cont., 1743. Rowning's book is discussed in

John Rowning's was one of the last, and best, of the general eighteenth-century texts on Newtonian corpuscular natural philosophy. Subsequent texts tended to specialize in subject (optics, for example, or electricity) or to adopt materialist, imponderable fluid, explanations of phenomena, or both. Priestley would have found, in Rowning, very little he had not already seen in 'sGravesande (though there is an added and portentous section on fermentation) but what he found was presented in a simpler form, without the confusion (within the text) of mathematical demonstration, and without the interruption of multiple experimental demonstrations.

More important, Rowning's work is explicitly presented in a corpuscular form, using Newtonian forces, while 'sGravesande's is not. Where, for example, 'sGravesande declares, simply, that the speed of sound varies as the elasticity of the air, Rowning attempts to demonstrate that this follows from the fact that the "Time . . . in which each Vibration of the Air is performed, depends on the Degree of Repulsion in its Particles" (2:49–50). All matter (respecting its substance) is, for Rowning, ultimately homogeneous, and its particles are characterized as solid, extended, divisible, mobile, and inert. As substance is essentially inactive, for the explanation of phenomena there must be added three principles: the attraction of gravitation, the attraction of cohesion, and repulsion.

The principles are not, indeed cannot be, the result of any mechanical cause: ". . . they are the very Reverse; and consequently can be no other than the continual acting of God upon Matter, either mediately or immediately" (xxxix). It is a mark of the "wonderful Wisdom and Contrivance of the Supreme Being" that He should choose "so short and easy a Method of producing so great a Variety of Effects" (xxxviii). So much is this the case that when Rowning is unable to find an explanation for phenomena that employs homogeneous particles and his three force principles (as in the case of electricity and magnetism), he chooses rather to give no explanation, confident that when all the circumstances of the phenomena are completely known, these principles will be found to be sufficient. He even rejects Newton's own explanation of the colors produced by films, involving as it does the concept of the aether (soon to become a prototype for the imponderable fluids) and of "easy fits of reflection and transmission," in the confidence that the "Time will come, when the Principles of *Attraction* and

my *Mechanism and Materialism: British Natural Philosophy in An Age of Reason* (Princeton: Princeton University Press, 1970), 34–39.

Repulsion will be found alone sufficient to account for this perplexing *Phae-nomenon*" (3:167).

Priestley's reading of Rowning during 1755 should have reinforced the ontology of Doddridge's *Lectures* and provided him with hints for future religious disputation and scientific research. Perhaps it did so, but there is no sign that it had any such immediate effect. In the *Institutes of Natural Religion*, he writes:

> As the matter of which the world consists can only be moved and acted upon, and is altogether incapable of moving itself, or of acting, so all the *powers of nature* . . . can only be the effect of the divine energy, perpetually acting upon them. . . . an energy without which the power of gravitation [for instance] would cease, and the whole frame of the earth be dissolved.[37]

This sounds like a derivation out of his Daventry reading, but the *Institutes*, while first written at the academy, were not to be published for seventeen years and by 1772 Priestley was beginning the exposition of a metaphysics of his own. During 1772 he was also to publish his *History and Present State of Discoveries relating to Vision, Light and Colours* (the *History of Optics*), in which he describes, for the first time a theory of matter proposed by the Abbé Roger Joseph Boscovich, in which mathematical-point particles are surrounded by concentric spheres of repulsion and attraction. Now Rowning writes, in his *Compendious System*, of the possibility that each particle of a body might be surrounded by such spheres, so as to attract and repel each other alternately at different distances (2:5 n–6 n]. It is hard to believe that Priestley would have totally forgotten this suggestion, on hearing about Boscovich, had he been seriously reading Rowning's natural philosophy at Daventry. Yet there is no reference to it in the *History of Optics*, though he cites Rowning in the preface for some of the plates used.

It is entirely possible that Priestley had forgotten the details of a work he had read so many years earlier, for at the time of reading he was concerned with heresies of greater importance than that of being a scientific mechanist when his contemporaries were using materialist explanations. By the year he left Daventry, he had adopted the major heterodoxies of Arian-

---

37. Joseph Priestley, *Institutes of Natural Religion* (London: Joseph Johnson, 1782), 2d ed., 35.

ism and determinism, or (as he was to write it) "philosophical necessity."
Of his becoming an Arian, Priestley's *Memoirs* have next to nothing to say.
Now this belief: that Christ the Son was a created being and therefore
subordinate to God the Father, was but a way station to Priestley's ultimate
humanitarian Unitarianism. Nonetheless, the Arian position was sufficiently anti-Trinitarian formally to exclude him from the Act of Toleration;
it caused him trouble with his family and first congregation, and was the
position he held for at least a dozen years. One should think its adoption
sufficiently important to justify more than the remark, in the *Memoirs*, that
the "extreme of heresy among us [at the Academy] was Arianism" (W 1/
1:25), and the subsequent acknowledgment that his congregation at Needham Market "soon found that I was an Arian" (W 1/1:30).

Doddridge's academy was frequently accused of nurturing Arians and
Socinians, but, in the context of eighteenth-century religious polemic, this
might mean no more than that its graduates differed in some way from the
accusers. Actually, records of graduates of Northampton, or Daventry,
show that neither produced more Arians or Socinians than might be expected of a liberal theological program at a time of doctrinal examination.
Samuel Clark, however, appears to have been one of the Arians graduated
from Northampton and, the more engaging of the tutors at Daventry and
taking always the side of heresy in every question (W 1/1:23), he may have
influenced young Priestley to do the same.

While Doddridge's *Lectures,* at least as edited by Clark, are quite explicit
on the impossibility of more than one self-existent being (Proposition 39,
with seven Scholia), the section devoted to the divinity of Christ and of the
Spirit (most of part 7, of some twenty-five pages) is scarcely definitive
against an Arian interpretation. While the tendency may slightly favor
Trinitarianism, and is quite clearly against complete Unitarianism, the conclusion may fairly be summarized by two of its statements:

> If it be asked *how* these divine persons are *three,* and how *one:* it
> must be acknowledged an inexplicable mystery: nor should we won-
> der that we are much confounded when enquiring into the curiosities
> of such questions, if we consider how little we know of our own
> nature and manner of existence. (Scholium 1 of Proposition 130)
>
> Considering the excellent character of many of the persons . . .
> whose opinions were most widely different, we may assure our-
> selves, that many things asserted on one side and on the other relat-

ing to the trinity, are not fundamental in religion (Corollary 1 of Proposition 132)

Under the circumstances, we may best conclude that Priestley, having arrived at the academy in self-conscious revolt against the religious opinions of his childhood, found no reason there to halt that revolt short of Arianism.

This conclusion, however, makes his adoption of determinism the more singular, for that same revolt had led him, by the age of sixteen, to precisely the opposite opinion; that is, to belief in free will. Determinism, in the religious circles of Priestley's boyhood, would not have been thought heterodox. The liberal, philosophically minded Protestants were, however, placed in a quandary. With their denial of election and predestination, they had to reconcile free will and omnipotence, benevolence and eternal damnation.

John Locke had evaded the issue in a semantic quibble. Locke's disciple, Anthony Collins, maintained a determinist position in his *Philosophical Inquiry concerning Human Liberty* (1717), referenced in Doddridge's *Lectures*, which Priestley was to edit for a reprinting in 1790. But it was not Collins who reconverted Priestley to a doctrine of necessity. Nor was it the *Lectures on Pneumatology*, which argued in favor of free will, primarily on the grounds that God could not otherwise reward or punish diversity of action (Proposition 16). Priestley's renewed belief in determinism was a consequence of his reading David Hartley's *Observations on Man*, to which he was referred by Doddridge's *Lectures*—though not in connection with the discussion there of free will.

The *Observations* is a theological treatise in three stages.[38] The first treats the mechanistic physiological processes by which the brain, as material substance, receives and retains sensations of the physical universe. The second is a discussion, in analogy with the first, of how the mind, as spiritual substance, perceives and constructs complex ideas by association. Together these constitute part 1, on the "Frame" of man, and form the base for part 2, on man's duties and expectations.

Priestley regarded discovery of Harley's work as an event that "produced the greatest, and in my opinion, the most favourable effect on my general turn of thinking through life" (W 1/1:24). The mechanistic physiology was

---

38. David Hartley, *Observations on Man, his Frame, his Duty, and his Expectations* (London, 1749). I have used the reprint (Hildesheim: Georg Olms, 1967). Hartley trained for the clergy at Jesus College, Cambridge, but declined subscription to the Thirty-nine Articles and

joined with Rowning-Boscovich corpuscles in his metaphysical work, *Disquisitions on Matter and Spirit* of 1777. Associationist psychology was a prominent element in the aesthetics of his *Lectures on Oratory and Criticism*, delivered at Warrington Academy in the 1760s; his objections to Scottish Common Sense Philosophy were argued from a basis of associationism, and he even edited a selection of that part of the *Observations*, published with introductory essays, in 1775.

It was, however, part 2 of the *Observations*, described as "a kind of unitarian Christian 'Platonism' with a pseudo-Calvinistic flavor," that had the greatest and most immediate influence on Priestley.[39] "It established me in the belief in the doctrine of Necessity, which I first learned from Collins," Priestley wrote, "it greatly improved that disposition to piety which I brought to the academy, and freed it from the rigour with which it had been tinctured. . . . I do not know whether the consideration of Dr. Hartley's Theory contributes more to enlighten the mind, or improve the heart; it affects both in so supereminent a degree" (W 1/1:24).

Priestley's previous concern over free will and determinism had led him, at the age of sixteen or seventeen, into a correspondence with Peter Annet; now he commenced a correspondence with David Hartley.[40] The correspondence was a brief one, as Hartley died in 1757, and the letters have not survived. Had they done so, they might easily reveal, as a reading of the *Observations on Man* does only doubtfully and with difficulty, why Priestley became so ardent a disciple of Hartley. Turgid and prolix, almost baroque in style, with little sense of proportion in argument, eclectic and so derivative that scarcely anything in it can have been entirely new to Priestley, the *Observations* was, nonetheless, second only to the Bible in Priestley's estimation; compared to its author, thought Priestley, both Locke and Hume were inferior philosophers.[41]

---

practiced medicine instead. See my *Mechanism and Materialism*, 198–200, for a discussion of Hartley's physiology.

39. The description is that of Robert Marsh, "The Second Part of Hartley's System," *Journal of the History of Ideas* 20 (1959): 272. For a discussion of the "Christian 'Platonism'" of Hartley, arguing that the determinism that justified description of this part of the *Observations* as "pseudo-Calvinistic," is consistent with the Cambridge school, see Schofield, "S. T. Coleridge, Joseph Priestley, and 18th-Century British neo-Platonism."

40. The correspondence is referred to in Priestley's *Letters to a Philosophical Unbeliever*, 2d ed. (Birmingham: J. Johnson, 1787), 1:92, and the first letter from Hartley mentioned in the journal for 1754.

41. He described Hartley's work as "still greater" than Locke's in his *Proper Objects of Education in the Present State of the World* (London: J. Johnson, 1791), 6; and declares Hume "not even a child" when compared to Hartley, in *Letters to a Philosophical Unbeliever* 1:126.

Perhaps the eclecticism itself was some of the appeal to a reader who was to become one of the notable polymaths of eighteenth-century England. If Priestley was, as he says, even then interested in natural philosophy as well as theology, it surely delighted him to find, in the chapter deriving assent to propositions from associations, the statement: "And thus it seems, that Optics and Chemistry will, at last, become a Master-Key for unlocking the Mysteries in the Constitution of natural Bodies, according to the Method recommended by Sir *Is. Newton*" (1:352). Or to discover, in a chapter on the truth of the Christian religion, an explanation of the deluge based upon changes in the spheres of attraction and repulsion surrounding particles of water (2:106).

However, there is more in the appeal of Hartley's eclecticism than the occasional juxtaposition, in argument, of unexpectedly congruent elements. Hartley created a coherent and satisfying theological system out of just those views: mechanistic material process, Lockean empirical psychology, and Christian final cause; which Priestley found (or was, under the influence of the *Observations*, to find) both essential and undeniable in a description of the universe. Hartley had further had the courage to accept the theological implications of these views and the ingenuity to argue an escape, consistent with his principles, from the Calvinistic conclusions to which those principles seemed to lead.

From the natural attributes of God (e.g., his omnipotence and omniscience), it is clear that man cannot possess philosophical free will: the power to do either one thing, or its contrary, when the previous circumstances are identical (2:66). This is equally clear from the nature of man, whose actions result "from previous Circumstances of Body and Mind, in the same manner, and . . . Certainty, as other Effects do from their mechanical Causes" (1:500). But God is infinitely benevolent, tho' finite man has difficulty perceiving this: ". . . the Appearance of Things to the eye of an infinite Being must be called their real Appearance in all Propriety" and man must suppose himself in the center of God's system and "reduce all apparent Retrogradations to real Progressions" (2:28–29).

From this vantage, it is clear God has created the world and man so that man must come naturally to perceive the will of God; that is, the development of moral sense is a consequence of the doctrines of mechanism and

---

The valuation next to the Bible is in *Examination of Dr. Reid's Inquiry . . . Dr. Beattie's Essay . . . and Dr. Oswald's Appeal* [Examination of Common Sense Philosophy], 2d ed. (London: J. Johnson, 1775), xix.

association (1:492–99). Moreover, as man seeks pleasure and avoids pain and since doing good is pleasurable, while doing ill is painful or comes to be, man will naturally come to associate God, the source of all good, with all his pleasures: ". . . the Idea of God, and of the Ways by which his Goodness and Happiness are made manifest, must, at last, take place of, absorb all other Ideas, and He, himself become, according to the Language of the Scriptures, *All in All*" (1:114).

The doctrines of mechanism and association, therefore, also provide man with the power of generating the causes and proper associations (e.g., meditation, religious conversation, reading religious books, prayer) that enable him to comply with the will of God (2:53). They are, therefore, not merely consistent with, but the causes of, "Practical free-will," the powers of calling up Ideas, of deliberating, suspending and choosing. The exertions of this "practical free-will" are, however, the results of associations from the circumstances of life. The Scriptures show that man cannot obtain ultimate happiness without experience of pain: ". . . such is the present Frame of our Natures and Constitution of the external World, which affects our Organs, that we cannot be delivered from . . . Sensuality and Selfishness . . . and advanced to Spirituality and Disinterestedness . . . but by the perpetual Correction and Reformation of our Judgement and Desires" (2:18–19).

Some men, in their finite lives, may not sufficiently experience the circumstances that enable them to achieve pious and virtuous lives conducive to salvation. In Hartley's opinion, reason and Scriptures combine to make it probable that the "Evils of a future state will have the same Tendency . . . as those of this Life, *viz* to . . . perfect our Natures, and to prepare them for ultimate unlimited Happiness in the Love of God, and of his Works" (2:419). Though the "future punishment of the Wicked will be exceedingly great both in Degree and Duration, i.e. infinite and eternal, in the real practical Sense to which alone our Conceptions extend" (1:viii), its purpose is to continue preparation of those unprepared in this life for their ultimate salvation.

Even as a student, Priestley did not accept everything Hartley wrote. Within a year of his graduation, for example, he had written a pamphlet earnestly denying the plenary inspiration of the New Testament, though Hartley had argued the inspiration of the whole. Nonetheless, Hartley's *Observations* was to provide Priestley, throughout life, with a general system of thought and belief in which the widely varying activities of an exceedingly diverse career were to find coherence. And though Priestley "gave up my freedom with great reluctance," he was to find in the doctrine of philosophical necessity, what Hartley had promised: the basis of serenity, a "great support in grievous Trials and Sufferings."[42]

Indirectly, Daventry Academy provided Priestley with a philosophy of life, as it directly provided masculine companionship that shared his interests, gave him a fund of culture and ideas, and schooled him in ways of using his mind. No college or school could have done better; perhaps only Doddridge's Northampton Academy would have done as well. Doddridge was apparently a brilliant teacher, prompting students to think for themselves and spurring them to accomplishment. A list of graduates from Northampton Academy includes a number of men who later distinguished themselves. The list of graduates of Daventry under Ashworth includes, perhaps, as many worthy ministers and tutors, but only Priestley and possibly William Enfield were to achieve the eminence of some three times their number from Northampton.[43]

Nevertheless it appears that some part of Priestley's productivity as a scholar, and therefore of his quality and eminence, was uniquely owing to the character of Daventry Academy under Ashworth. Where Doddridge had been popular, and sometimes lax in discipline, Ashworth was severe and "often drew upon himself the ill-will of many of the students" (W 1/1:25). The difference was not in daily routine, for there the schools were apparently much the same: rise at 6:00, assemble by 6:10, private study till "family" prayers at 8:00, when a student would read and expound a scriptural passage; breakfast, followed by lectures from 10:00 till 2:00, each lecture commencing with an examination of the previous day's lecture and of the readings referred to in it; dine at 2:00, then private study or recreation till "family" prayers again at 7:00; tutorials and then supper till 9:00, gates locked at 10:00, students in rooms by 10:30.[44]

But Ashworth was a slave to "the love of order." He "knew the value of time, seldom . . . wished to relax." "His hours, his moments were regularly

42. Hartley, *Observations*, 1:505. For Priestley's reluctant acceptance of determinism see his *Doctrine of Philosophical Necessity Illustrated*, 2d ed. (Birmingham: for J. Johnson, 1782), xxvii.

43. Graduates of Northampton are listed in Stedman, "Pupils educated by Doddridge"; graduates from Daventry under Ashworth are named in Thomas Belsham, "A List of Students educated at the Academy at Daventry," *Monthly Repository* 17 (1822): 163–64, 195–98.

44. Irene Parker, *Dissenting Academies*, 86, 147–52. Job Orton exhorts Ashworth, in a letter of 13 November 1752, to maintain order and discipline as Doddridge sometimes had not, see [Job Orton], "Letters to Dissenting Ministers," in *The Practical Works of Job Orton* (London: Thomas Tegg, 1842), 2:499–500.

appropriated. A train was laid in the morning [with] . . . no deviation in the course of the day. Every thing had its own place. He would rather erase than retain a just idea if it did not belong to the series he was then pursuing. With him every thing was reduced to a system."[45]

Some pupils may have been made "diffident, fearful and backward" in this atmosphere. Priestley learned in it to channel his ardency into systematic scholarly discipline. To this description of Ashworth at Daventry, compare that of Priestley, just after he had left the academy:

> Mr. N[orman] often heard Mr. T[ailor] predict the future eminence of his friend, not merely from his great application, but from the most undeviating adherence to plan in every thing he did. So exact was he in the division of his time, that he accustomed himself to study with a watch on the table, and however interesting the subject engaged in might be, he never suffered one branch of literature to encroach upon the period allotted for another. (W 1/1:31 n)

Given Ashworth's discipline at Daventry, it is not surprising that Priestley managed to learn and accomplish so much during his years at the academy. What is surprising is that this did not turn him into a drudge, grinding out his days at Daventry in work. But the same journal that details his progress in learning reports also the youthful exuberance of a young man (he was then only twenty-one) living free of the constraints of home and in the company of young men like himself. This is the only intimately personal view of Priestley available to us and it reveals a playful spirit to combine with the intensity and dedication long known to be his. It is easier now to appreciate the charm that won and held friends as various as Richard Price, Benjamin Franklin, Theophilus Lindsey, and Thomas Jefferson.

Clearly academic discipline did not prevent the students from having a good time; perhaps, even, that discipline was relaxed for students in the final year of their program. The journal reports evening parties at a private home, playing at crambo, cross-questions, or blindman's bluff with the daughters of the house and their guests (some of the students kissed them "all the while") or celebrations at the Wheatsheaf Inn, across the way. There were also "feasts" in student rooms ("several enjoyed too much liquor") and even horseplay in anatomy class, pelting one another with parts of a

---

45. This description of Ashworth is a summary of part of T. Thomas, "Supplementary Hints to the . . . Memoir of Dr. Ashworth."

dissected cat. There was skating, bowling, fencing, walks to and around Borough Hill. There was music, singing and instrumental (1 May: "played the lyre in my room") and amateur theatricals, even a visit to the theater in Birmingham.

And, of course (for these are young, unmarried males) there were girls: girls gossiped about, girls visited from the academy and at home, flirted with. There is a hint, on vacation home, of a mild courtship with a Miss Carrott, on whom he called at least six times in a week: 6 June, "they left me and Miss Carrott together about two hours; 10 June, "all forenoon with Miss Carrott." The episode ends without explanation after a call on 6 August. Back at the academy, a new interest—"the cuddliest creature I ever beheld"—appears, and then disappears after a church service. Student life could be sweet, even in the eighteenth century.

Priestley's academy program ended in 1755 and the carefree student had to disappear into the Dissenting minister. Students at Dissenting academies could not apply for particular positions; they had to await "calls" from congregations, usually prompted, in the first instance, by the divinity tutor. And thus it was that Caleb Ashworth inadvertently provided also the last element needed in the formation of the formidable scholar-disputant he had helped, unknowingly, to train. For Ashworth passed to Priestley the application for a minister sent by the people of Needham Market, Suffolk, and there, "remote from my friends in Yorkshire, and a very inconsiderable place" (W 1/1:28), he was to be tempered by adversity, misunderstanding, and bigotry.

III

NEEDHAM MARKET
AND NANTWICH,
1755–1761

The call of Joseph Priestley to the Dissenting chapel of Needham Market, Suffolk, was a mistake, by Ashworth, Priestley, and the congregation alike. In retrospect, it seems clear that Priestley "should" have gone as junior minister to some large, liberal, urban congregation in the North or Midlands. There his background would have suited; his passion for scholarship, his developing concern for the structure of Dissenting worship, and his interest in the religious instruction of young people would have found support and complement in the seasoned social wisdom and more popular homilectics of a vigorous senior minister. Such a position may not have been open, however, and had it been, Priestley would not likely have been called to fill it. Congregations of that description could select from several candidates and did not usually invite to their prestigious pulpits young and untried ministers lacking the personal influence of friends or relatives.

Moreover, the characteristics of Priestley's later ministry are not easily discerned in the young candidate. Caleb Ashworth, even if he saw them, was too inexperienced in the placing of academy graduates to be able to match candidate and post. More likely, Ashworth was worried about finding any position at all for his ugly duckling and was grateful when Needham Market invited him to become assistant minister.

Priestley preached as a candidate at Needham and was interviewed by the congregation before receiving their unanimous invitation to assist their aged minister, the Reverend John Meadows, "with a view to succeed him when he died" (W 1/1:29). Nonetheless, it is hard to imagine their selection of a person less suited to the location or to the church. In the eighteenth century, before improved transportation and communications had moderated some of the distinctions, widely separated regions of England were almost foreign countries to one another. And such were the differences between Suffolk and the West Riding of Yorkshire that Priestley's accent, even unthickened by a stammer, could not have been easy for natives of Needham Market to understand.

The town itself, on the Gissing River roughly halfway between Stow Market and Ipswich, had once been a prosperous woolmarket, but the plague had nearly destroyed it in 1685. By mid-eighteenth century it had become the butt of a Suffolk folk-saying: to be on the road to Needham was to be a pauper. Not until the river was made navigable in the nineteenth century was prosperity to return to Needham Market; during the eighteenth century its economic base and its population steadily eroded.

The Dissenting congregation in Needham Market could trace its descent continuously from the Act of Uniformity and for more than half of the intervening ninety-three years it had the same minister. Now seventy-nine, infirm and indomitable, the Reverend John Meadows had assisted the first minister at Needham, succeeded him in 1701, married and settled as a strong influence in the small community. He was probably the wealthiest man in the congregation, with an estate and additional property in ten communities scattered across Suffolk. The chapel, a handsome red brick building of two stories resembling houses on the long main street of Needham Market (see Fig. 3) had been built under Meadows's leadership in 1716.

At that time the congregation had numbered approximately three hundred; by 1755 it had dwindled to just under one hundred. Much of the decrease was, no doubt, owing to the declining population of Needham Market and to the drastic decline of the Dissenting interest during the first half of the eighteenth century during which Dissenting congregations across England decreased by nearly 50 percent.[1] Yet some of it, surely, was the

---

1. Alan D. Gilbert, *Religion and Society in Industrial England: Church, Chapel and Social Change, 1740–1914* (London: Longman, 1976), 16. This change preceded the subsequent rise of evangelical dissent associated with the Methodist revival movement.

Fig. 3. Needham Market Congregational Church, as Priestley would have known it, from a painting now in the United Reform Church, Needham Market. Photograph by permission of the East Anglian Daily Times, Ipswich.

falling away of people for whom the ministrations of the aging, increasingly conservative, Meadows was no longer appealing.

They could find, in John Taylor of Stow Market or Thomas Scott of Ipswich, younger, more vigorous, and more liberal Dissenting ministers within four miles up- or downriver. Their leaving, no doubt, increased the loyalties of those that remained in Meadows's congregation. After fifty-four years, however, Meadows needed assistance. The congregation had been unable to attract a junior minister the previous year, for they could offer but forty pounds a year and that only with the assistance of both the Presbyterian and the Independent funds. They too must have been grateful when the young candidate minister accepted their invitation.[2]

2. For information respecting Needham Market and the chapel there, and for his advice and encouragement, I am deeply grateful to the Reverend W. Vine Russell, minister at the United Reform Church, Needham Market. See also John Browne, *History of Congregationalism and Memorials of the Churches in Norfolk and Suffolk* (London: Harrold and Sons, 1877), 493–502, 348, 521, 530; and *Looking back over three centuries: 1662–1962* (Needham Market:

Priestley should have known better. At Heckmondwike he had witnessed the early stages of an attempt to replace a superannuated minister. Even with Kirkby's cooperation (until his death in 1754), the struggle within that congregation went on for eight years and was not resolved without formal expulsion of some of the members in 1768. Priestley may not have been aware of the continued struggle at Heckmondwike, however, and he "flattered myself that I should be useful and happy" in Needham Market when he accepted the invitation.

But, in his impatience to commence at last the profession toward which he had been headed since he was eight years old, he was less than candid in his discussions with the congregation. They must have had some notion of his unorthodoxy when he asked if they could make up his salary without assistance from the Independent Fund, as he "did not choose to have any thing to do with the Independents" (W 1/1:29). He seems not to have told them, though, that the Independent Fund Board would not, if asked, contribute to his support. The "rules and orders" of that Board declared: "(3) That nothing be allow'd to any Minister, though he be Congregational in his Sentiments, till there has been Satisfaction given to this Board of his Abilitys for the Ministry and of his unblameable Conversation, and of his Approbation by the Church to which he is or was related before he came into the Ministry."[3]

Nor did he tell the congregation that his heterodoxy had gone beyond Arminianism (sufficient, in itself, to bar him from that "church to which he . . . was related before he came into the Ministry") into Arianism. There is no need to credit Timothy Priestley's "memory" of his brother's admission, "I did all that I could; I so far hid my cloven foot, that I taught the Assembly's Catechism; but they found me out." That "memory" is belied by Joseph's character, by the memories of his associates while at Needham Market, and by the events of his ministry there.[4]

He did, however, follow the standard practice of liberal ministers in attempting to keep theological controversy out of the pulpit, preaching rather on "practical religion," the moral life enjoined by Scriptures; and he used Isaac Watts's Calvinistic *Catechism* in the instruction of the children of the congregation. There was a "general dissatisfaction" though, with "the freedom with which he delivered his opinions" and the congregation soon

---

Needham Market Congregational Church, 1962), a tercentenary booklet sent me by the Reverend Mr. Russell.

3. T. G. Crippen, "Congregational Fund Board," 210.

4. Timothy Priestley, *Funeral Sermon*, 37; and J. T. Rutt's refutation, W 1/1:30 n–31 n.

found that he was an Arian, especially when he began a series of lectures on the theory of religion, based upon the *Institutes,* which he had composed at Daventry, in which he "came to treat of the Unity of God, merely as an article of religion" (*W* 1/1:30). The old minister "took a decided part" against him and, although the principal families remained, his congregation fell off rapidly. The most he ever received of his promised salary "was in the proportion of about thirty pounds per annum, when the expense of my board exceeded twenty pounds," and generally the amount was far short of that.[5]

This should not have mattered, practically, for his aunt had promised that if he became a minister, "she would leave me independent of the profession" (*W* 1/1:33). No doubt this promise had contributed to his willingness to accept a position with so small a salary. But now his "orthodox relations," in the person of brother Timothy, took a hand. Timothy tells the unsavory tale with a complacency remarkable in its revelation of evangelical self-righteousness. "His aunt," Timothy writes, "though one of the fairest characters in the world, was not aware how much her nephew differed from her, till I informed her." Apparently she could not believe him so a trap was set for Joseph on his visit home before going on to Needham Market. The aunt required that he read aloud to her from the devotional literature of James Hervey.

The writings of Hervey were extremely popular among people influenced by the early stages of the Methodist movement, but Priestley's reluctance to read them may have derived simply from the sophistication of his literary tastes acquired at Daventry. Hervey's works have been described as illustrating "most effectively the fantastic and affected style which the most sincere writers . . . seemed to assume. . . . ideas impoverished and the expression at once affected and commonplace."[6] Nonetheless, Joseph's evident distaste for the task convinced Mrs. Keighley that Timothy was right. She ceased her remittances and were it not for occasional extraordinary grants from London charities, Priestley might not have subsisted.

Nor was this the end of sordid family intrigue. When Mrs. Keighley died in 1764, she had cut her ward off with "only a silver tankard as a token

5. Note that an ordinary London laborer might, by 1789, expect wages of this amount or more; see, for example, George Rude, *Paris and London in the Eighteenth Century* (New York: Viking Press, 1971), 52.

6. W. H. Hutton, "Divines," chap. 15 in *The Age of Johnson,* vol. 10 of *The Cambridge History of English Literature,* ed. A. W. Ward and A. R. Waller (Cambridge: Cambridge University Press, 1933), 413–14. Timothy Priestley, *Funeral Sermon,* 41.

of her remembrance." Priestley understood she had exhausted her estate by liberality to others, especially a deformed niece who lived as her companion and was dependent on what the aunt had left to bequeath. By then Priestley was beyond the need for assistance and his aunt, having "spared no expense in my education," had done "more for me than giving me an estate" (W 1/1:33). Priestley "freely consented to her leaving all she had to my cousin."

He may not have known, however, that part of that "liberality to others" that reduced her estate, had been to the chapel at Heckmondwike, at the instance of his orthodox relations. In 1759, it was determined that a new chapel be built and, over the strenuous objections of some of the congregation, the new structure was placed on land given by Mrs. Keighley at the urging of the Priestley family.[7] Thus, part of Joseph's sequestered inheritance was given to that church whose rejection of him at nineteen had substantially contributed to setting him on the path leading to estrangement of his family. No wonder he was later to write that he had been "incommoded" by the congregation; a less charitable person might have been considerably more outspoken.

Living in the "low despised situation" at Needham Market, Priestley found his stammer "increased so much, as to make preaching very painful." His first visit to London was made in hopes of finding relief from this problem through a method to cure all speech defects advertised by a Mr. Angier. The "cure" cost twenty guineas, which he had to beg from his aunt, took a month, and gave only temporary benefit—after which he spoke worse than ever. This difficulty in speaking not only made his position at Needham more difficult; it blocked his chances of moving to a better place and even interfered with his exchanging occasionally with neighboring ministers. One of these acknowledged that the "more genteel part of his hearers" always absented themselves when they heard that Priestley was to speak, though they had no objections to his doctrines. Years later, when Priestley had achieved reputation, he was invited to speak from that pulpit and had the wry satisfaction of eliciting commendations for a discourse those hearers had formerly despised, though his elocution had not much improved.

With pulpits apparently closed to him, and his income reduced below subsistence, Priestley determined to open a school, though he had often said he "would have recourse to any thing else for a maintenance in preference to

7. Frank Peel, *Nonconformity in Spen Valley*, 147–49.

it" (*W* 1/1:41). The circumstances should have been favorable, for if the free endowed grammar school at Needham Market still existed, it was not very active; but his printed and distributed proposals to "teach the classics, mathematics, &c. for half-a-guinea per quarter, and to board . . . pupils for twelve guineas per annum" met no response. Priestley was insufficiently orthodox to teach pupils at Needham Market, even when they lacked other schools.[8] He next proposed giving public lectures on various branches of the sciences and here had some small success. One series of twelve lectures on the "Use of the Globes"—popular astronomy, geography, and the methods of erecting and correcting sundials—had sufficient hearers (ten, at half-a-guinea each) to pay for the globes. The work contributed nothing to his livelihood, but it was of use to the community and might have helped Priestley build a stock of apparatus; he would have continued the lectures had he remained there.

Although his material circumstances were bad, he had resources to help him survive the generally hostile community. First, and most important, was his Hartleyan determinism, which sustained him with the firm conviction "that a wise Providence was disposing every thing for the best" (*W* 1/1:34). Second, there were his professional activities. His personal ministerial services were rejected. Nonetheless he wrote a sermon a week and he continued his studies, which were chiefly theological. Typically, when his heterodoxies were challenged, Priestley commenced an investigation of them and ended more heterodox than before; that is what happened at Needham Market.

He had left Daventry "with a qualified belief of the doctrine of *atonement*" (*W* 1/1:35). The belief that the death of Christ was a sacrifice in atonement for man's sins was supposedly based on Scriptures, but was incongruent with the system of natural theology Priestley had adopted at the academy. As revealed religion cannot contradict natural religion, when each is correctly understood, he set himself the task of examining the scriptural bases of the doctrine of atonement. Collecting all the relevant texts from the Old and New Testament, he arranged them under a variety of heads, and concluded that the doctrine "had no countenance either from scripture or reason" (*W* 1/1:36).

These conclusions were set down in a treatise, which he sent to a friend (probably Samuel Clark), who sent it, in turn, to Caleb Fleming and

---

8. Nicholas Carlisle, *Concise Description of Endowed Grammar Schools*, 2:530, reports that no response was made to his inquiry, early in the nineteenth century, respecting that

Fig. 4. Joseph Priestley's
globe. Courtesy
Harris Manchester
College, Oxford.

Nathaniel Lardner in London. With Priestley's permission, Fleming and Lardner published about half of what he had written as an anonymous pamphlet: *The Scripture Doctrine of Remission. Which sheweth that the Death of Christ is no proper Sacrifice nor Satisfaction for Sin: but that pardon is dispensed solely on account of Repentance, or a personal reformation of the Sinner.*[9] With the appearance in 1761 of this, his third, publication,

---

grammar school. No copies have been found of Priestley's printed proposals for his school, or for the public lectures on science mentioned below.

9. [Joseph Priestley], *Scripture Doctrine of Remission* (London: C. Henderson, R. Griffiths, T. Beckett, and P. A. deHondt, 1761), 95 pages, 8 vo. This is one of the rarest of Priestley's published writings. The copy I have used is that in the Bodleian Library, Oxford. There was a subsequent republication, with frequent modifications and transpositions of material, in Priestley's "An Essay on the One Great End of the Life and Death of Christ, Intended more especially to refute the commonly received Doctrine of Atonement," which first appeared over the signature "Clemens," in Priestley's journal of religious controversy, the *Theological Repository* in 1769. The subject remains one of religious controversy, as can be seen, for example, in Martin Hengel, *The Atonement: The Origins of the Doctrine in the New Testament* (London: SCM Press, 1981).

Priestley commenced his long career as a regularly publishing author. From that date till his death, there was never more than a year between publications of some of his writings; usually there were several each year.

Most of the *Doctrine of Remission* is occupied by numerous quotations from the Old and the New Testaments (roughly equal in number) to prove, first, that ". . . in all the books of scripture, neither in the Old or the New Testament, neither the Divine Being himself to the patriarchs, neither Moses, nor the prophets, by his direction to the Jews, nor Christ or his apostles, to the Christians, ever assert, or explain the principles on which the doctrine of atonement is founded" (37). And, second, to support the Arminian position that the general tenor of the Scriptures shows pardon for sin and ultimate salvation to depend upon sincere repentance, moral character, and the good works of the sinner.

Reason and common sense also show the doctrine of atonement is utterly incompatible with the true principles of moral government by a naturally benevolent and placable God. Man, with his imperfections, is unable to discern the hearts of men. As the end of all government is the happiness of the community (75), magistrates must exercise severity even on apparently reformed offenders, lest they harbor designs further to injure society. But an all-seeing God, not laboring under apprehensions, must freely pardon all the truly penitent as the most effectual motive to induce men to regulate their lives well.

True, some passages of Scripture have been interpreted as supporting the doctrine of atonement. One, a passage from St. Paul, depends upon a faulty translation from the Greek. Moreover, St. Paul's discourses must be understood as arising out of their immediate context, in combating Jewish prejudice. To suppose his writings relate to general moral principles, rather than to discussions of particular questions, is to make him "advance notions, the most shocking and inconsistent in the world."[10]

The most unusual part of Priestley's criticism of the doctrine of atonement relates, however, to the metaphorical language of the Scriptures, whose allusions, analogies, and figures of speech have given rise to such a

---

10. *Doctrine of Remission*, 38 n–41 n. This suggestion that St. Paul's reasoning was determined more by circumstance than revelation was but one of many in the original text of the *Doctrine of Remission*, but Lardner did not approve of such criticisms and most were omitted in publication. Priestley remained "satisfied . . . that his [Paul's] reasoning was in many places far from being conclusive" and commenced a treatise on the "imperfections of the sacred writers," which he sent to the press from Needham, but withdrew on the advice of Andrew Kippis (*W* 1/1:38). It was later rewritten and published in the *Theological Repository*.

variety of inconsistent interpretations as to justify rejecting anything so vague as intended to demonstrate a major article of religion. Throughout Priestley's text there are frequent references to the problem of correctly interpreting the language of people so different from Europeans in time and culture. He remarks, for example, that the apostles had used phrases of the Jewish religion, most familiar to them and to their hearers, as fittest for argument with their countrymen, but it was high time to lay aside such figurative expressions, suited only to their own times, and strip "our pure religion" of "some part of the disguise they have forced upon it" (94).

This is a continuation of the discussion of philosophy of language to which the first nine pages of the introduction are devoted. It is a poverty of all languages that when they go from sensible things to compound ideas, they adopt figurative and analogous expressions. Now these enrich, color, and enliven our speech and may even facilitate the discovery of truth, but they may also lead to obscurity and even falsity. For by association, com-pounded and figurative expressions come to pass for simple ones and the nature of the thing described is forgotten in adopting allusion for reality. This is particularly true when Europeans read the writings of "Asiatics," who are extreme in their use of metaphor and figures of speech. Distance of time and differences of manners confuse our understanding of the Scrip-tures. We must reduce the artificial or compounded words and figurative passages to their simple components to achieve understanding.

Although written early in his career, the *Doctrine of Remission* is typical of Priestley's later work in calling upon a variety of different, and some-times unexpected, elements of contemporary thought in asserting the con-clusions of a theological argument. It detracts little from the quality of his work that those elements are scarcely original. The derivation of the philosophy of language from his reading of Watts's *Logick* and Locke's *Essay concerning Human Understanding* is obvious. To these he added the doctrine of association from David Hartley and citations to Plato, Samuel Clarke's Boyle Lectures, the *Universal History,* and even a paraphrase of Shakespeare's *Tempest,* all of which he had but recently read as a student at Daventry Academy. Not surprisingly, Priestley's first treatise was a direct fruit of his educational experience.

Another treatise, drawn up but not published by Priestley while at Need-ham Market, was "on the doctrine of *divine influence.*" When its substance appeared in 1779, as an ordination sermon, there were surely changes in emphasis as well as in form. Still, it is tempting to suppose some parts of it relate to Priestley's time at Needham. Discussing the notion of "peculiar providence," or (infrequent) divine intervention in the material world, he gives an example drawn from the Newtonian belief that God must, occasionally, reestablish the harmony of celestial motion:

> We shall learn to respect the laws of nature the more, if we consider the extraordinary provision that the Author of Nature has made to preserve their uniformity, and to supersede the necessity of the frequent violation of them which he has done by means of occasional and seasonable miraculous interventions. In fact the proper use of miracles has been to make more miracles unnecessary.[11]

Possibly this has reference to the arguments of the Leibniz-Clarke correspondence to which Priestley had been directed at Daventry by Doddridge's *Lectures*.[12] It could well also be a reflection of his lectures on popular astronomy, in the uses of the globes, which he was delivering at the time the first version was written.

Though none of his writing was published while he was at Needham Market, Priestley's circumstances there would not have bettered by their appearance. That was not the community nor the congregation to encourage scholarship or the reach toward original thinking of which Priestley was capable. To realize his potential, it was necessary that he find a less constraining position. For that purpose he drew upon his third, and most immediately practical, resource, his genius for friendship.

All his life Priestley was to disarm even his critics with sincerity and charm and the friends he made he kept. The Reverend William Graham, for example, whom he met at his aunt's table in 1750–51, befriended the youth and remained a friend of the man. They corresponded, often in Latin; Priestley dedicated the *Disquisitions on Matter and Spirit* (1777) to him and when Graham died in 1782, he left Priestley his manuscripts, polyglot bible, and £200. At Daventry, Priestley was friendly with both tutors, carrying on "more or less of a correspondence" with Caleb Ashworth until his death in 1775 and a much more frequent correspondence with Samuel Clark "till the very week" of his death in Birmingham in 1769.

11. Joseph Priestley, *The Doctrine of Divine Influence on the Human Mind, Considered in a Sermon* (on Matt. 13:3–20) (Bath: R. Cruttwell, for J. Johnson, 1779), 9.

12. The Leibniz-Clarke correspondence was first published in 1717; there is a modern edition, edited by H. G. Alexander, *The Leibniz-Clarke Correspondence* (Manchester: Manchester University Press, 1956).

Of the students at Daventry, Priestley names seven who were his particular friends; four of these remained his friends later in life. He corresponded with John Alexander, often in Greek, till Alexander's death at Birmingham in 1765. Caleb Rotherham and Radcliffe Scholefield were correspondents and associates in many of Priestley's activities; the latter became a colleague when Priestley moved to Birmingham (1780). Isaac Smithson, who went from Daventry to Harleston, Norfolk, when Joseph went to Needham Market, accompanied him on his first trip to London and there introduced him to Andrew Kippis and to George Benson, John Alexander's uncle; these two procured the occasional £5 from various charitable funds at their discretion, which kept Priestley going when his congregation failed to pay his salary. Some of these friends, old and new, combined in an attempt to relocate Priestley to a more suitable position.

The deaths, in 1751 and 1754, of Caleb Rotheram (Sr.), Philip Doddridge, and Ebenezer Latham threatened Dissenting education in the North and Midlands with the dissolution of academies at Kendal, Northampton, and Findern. That at Northampton moved, as we have seen, to Daventry, where it continued under Ashworth, but the situation in 1757 was so uncertain that a new academy was proposed for the vicinity of Manchester. At the first meeting of subscribers, June 1757, it was agreed that Warrington, Lancashire, should be the location. A group of tutors was selected, including Samuel Dyer of London for languages and belles lettres. Dyer, however, declined the invitation and the "Committee for the Conduct of the Academy" requested nominations for the empty post.

One of these nominations, sent in time for the third meeting of the committee, 5 January 1758, by George Benson, with supporting letters from Samuel Clark, Andrew Kippis, and John Alexander, recommended the "Rev. Mr. Priestly [sic] of Needham in Suffolk": "A Young Gentleman about 25 of most unexceptional Character; of Steady Attachment to the Principles of Civil & Religious Liberty, and remarkable for a Degree of Critical & Classical Learning not common in one so young."

By the next meeting, of 23 January 1758, the committee had found Priestley's "amiable character & considerable learning" confirmed by several persons who knew him, but, as they wrote Benson, they were "not without some apprehensions" that subscribers to the academy might think him too young to be a tutor, insufficiently known in the world, and with too little experience in life and manners. The committee "are informed, too, that he has some Hesitation & Interruption in his manner of Speaking, whether it be so considerable as to be worth any regard; or how far it might be likely to have an unfavourable Effect in Forming the Voice & Manner of the Students, of wch they think with you of the Highest Importance, they are not able to Judge."[13] Not surprisingly, the committee selected another nominee, John Aikin, a former pupil and assistant of Doddridge, and a person already experienced as a tutor at an academy at Kibworth. Aikin accepted the position, but Priestley had been brought to the committee's attention for the future.

Meanwhile, other efforts had been made to move Priestley and with greater success. A distant relation of Priestley's mother procured an invitation for him to preach as candidate for a position at Upper Chapel, Sheffield. Thomas Haynes, senior minister there and an old friend of Priestley, favored his candidacy, but the congregation rejected him because, they said, he was "too gay and airy," but more probably because they did not want to be served by two Arians.[14] Haynes, however, had some influence with his former congregation at Nantwich, Cheshire, who were looking for a minister. He wrote a friend there: "I think I could move a young Gentleman in Essex [Suffolk] viz. Mr. Priestley, who is a man both of genius and learning, but it is too far to come on trial, nor can you afford to pay him for his journey, nor will, at all adventures, give him an invitation on my opinion of him."[15] As Haynes had married into a prominent family of the Nantwich congregation, his recommendation was successful. Priestley was invited to preach at Nantwich "for a year certain," and went there in September 1758, going from Ipswich to London by sea because he hadn't money to go by land.

Within a generation of his leaving Needham, memory of Priestley's residence there was all but gone.[16] There remains no record of his service to its Dissenting chapel but that in his *Memoirs*. John Meadows had died in January 1757 and Priestley's departure left the congregation without a minister. The Reverend John Farmer was invited to succeed and accepted, but

13. MS Warrington; Minutes of the Proceedings of the Trustees of Warrington Academy; Warrington Municipal Library, 54–55, 59; by permission, Cheshire County Council, Warrington Library.

14. See Alexander Gordon, *Cheshire Classis Minutes*, 178. Haynes had corresponded with Priestley as early as 1750 (see Chapter I), had moved to Sheffield in 1745 and become an Arian sometime after 1750. He lived just long enough to come to Priestley's aid, dying in December 1758.

15. W 1/1:42 n; letter dated 12 December 1757.

16. The Reverend Elias Fordham was unable to learn anything about Priestley's Needham Market period when he went to Stowmarket in 1783; by 1831, and J. T. Rutt's inquiries, "more than seventy years has obliterated all traces of the youthful preacher's habits and intercourse" (W 1/1:31 n, 29).

did not stay the year. When Farmer left, Needham Market Chapel closed its doors, as had so many other churches in the area.[17]

For a time the building was even used as a theater for traveling companies—the ultimate indignity for a Calvinistic chapel. Priestley's friends in the neighborhood were chiefly Dissenting ministers and, except for Scott of Ipswich, they all left the region the year that he did. Of his other friends: the rector of Stowmarket—"whilst he lived we were never long without seeing one another"—killed himself in a fit of depression; Mr. S. Alexander, the Quaker who gave Priestley freest access to his library, cannot further be identified; nor can the family with whom he boarded and "from which . . . [he] received much satisfaction."

There is a Priestley cottage, barn, and wood in nearby Barking, but no evidence links their name with that of Joseph Priestley. Faint legend holds that Joseph lived in Pillar House, Needham, but again there is no support for the association. Joseph Priestley lived for three years in Needham Market and made not the slightest mark on its consciousness. Needham Market, however, made its mark on Priestley. It had tempered his convictions by its bigotry and disciplined his eagerness to serve by its indifference. The time had clearly come for him to move on to a community where he might be appreciated.[18]

It was not at once clear, however, that Nantwich was to be anything but a temporary retreat in which Priestley might lick his wounds and prepare for endeavors elsewhere. Nantwich was a small town; not as small as Need-

---

17. The chapel at Harleston, Norfolk, for example, closed in 1758 when its minister, Isaac Smithson, left for a church in Nottingham. John Farmer, older brother of Hugh Farmer, well-known, independent minister of London, was a strict Calvinist but could not get along with people and died "deranged."

18. In 1795 the chapel was reopened as a Congregational Church; in 1838 the building was replaced. Not until the twentieth century was there any desire to claim any pride in Priestley's association with town or congregation, and by then it was too late. The church now has a Priestley Hall, but that is a recent creation. The portrait of Priestley in the church (apparently a copy of that by Fuseli) was placed there by the parish council, which acquired it by gift from a vicar of Needham Market, the Reverend W. G. Hargrave Thomas. John Browne's *History of Congregationalism . . . in Norfolk and Suffolk* is bitterly anti-Priestley as late as 1877 and blames the failure of the earlier church on his heterodoxy (500–501). My thanks are due to the Reverend W. Vine Russell for personal communications and for the tercentennary booklet of Needham Market Church, from which my information is chiefly derived. In 1770, Priestley directed a parcel with his *Answer to Mr. Venn* to a Mr. Goode at Needham Market; this may be a clue to his host family there; see Priestley's letter to John Canton, 28 February 1770 in *Scientific Autobiography of Joseph Priestley*, ed. Robert E. Schofield (Cambridge: MIT Press, 1966), no. 26; source quoted or cited by permission of the publisher.

ham Market, perhaps, but the census of 1801 records only 824 houses and a population of 3,463 for the township. There is no reason to believe there had been a marked decline over the previous half-century. Like Needham, Nantwich's economic base had eroded. From very early it had been a center for production of salt from local brine pits. By the time of Henry VIII there were as many as three hundred saltworks operating there. The salt had never been of high quality, however, and the discovery of rock-salt beds and failure to extend the Weaver River navigation beyond Winsford (approved by Parliament in 1733–34 but never carried out) diverted the major part of the salt trade elsewhere.

By 1774 only two saltworks were in operation at Nantwich and the situation cannot have been much better when Priestley was there, for the trade was then confined to salt carried in horse-packs to neighboring counties. Nantwich was still a market-town. It was on a major highway between Chester and the pottery district, Stone, Lichfield, and London; and there were small shoe and lace-making industries.[19] The town was not dead, that is, but neither was it flourishing.

Nor was the Nonconformist congregation that had called Priestley to its service. In 1715 the congregation is said to have numbered three hundred and when the new chapel was dedicated in 1726 it would accommodate 230 hearers. By the time Priestley arrived, however, there were but sixty regular members and a great proportion of these were "travelling Scotchmen," though not at all "Calvinistical" (W 1/1:43, 45).

Nevertheless, Priestley was happy in Nantwich and found there an opportunity for service. Cheshire is one of the northern Midland counties and is included in the Midland dialect region that also encompasses the West Riding of Yorkshire. The people of Nantwich might be expected, therefore, to be more sympathetic to a minister from neighboring Yorkshire. And this sympathy was important, for Priestley's stammer had so increased that he "once informed the people that I must give up the business of preaching." By the end of his second year there, he had, however, succeeded in getting the better of the defect, by taking great pains, "making a practice of reading very loud and very slow every day."

The people of his congregation also were not obsessed with the Trinitarian controversies "which had been the topics of almost every conversation in Suffolk." This was probably not, as Priestley seems to have thought,

19. James Hall, *A History of the Town and Parish of Nantwich, or Wich-Malbank, in the County Palatine of Chester* (Nantwich: for the author, 1883), 3, 265, 387–98, and passim.

because people in Cheshire were more liberal theologically, but because the Trinitarian battle had been fought out before Priestley arrived. In 1745 the regional organization of Dissenting ministers (the Cheshire Classis) had split on the issue of subscription to the doctrine of the Trinity and the divinity of Christ. Richard Meanley, Priestley's immediate predecessor at the chapel in Nantwich, had sided with representatives of the old (English) Presbyterian principles of nonsubscription to creeds and he and his supporters in the congregation retained control of the Presbyterian Chapel on Hospital Street.

It is thought that Meanley may have been a "Socinian," but whatever his beliefs, he paved the way for Priestley's heterodoxies.[20] By the time Priestley arrived, the remaining members of the congregation were committed to theological liberalism. There was, however, in a congregation that small and with few children, very little scope for activities as a minister—he even neglected the composing of new sermons—and less incentive for theological scholarship. He occupied his time in Nantwich, as he had latterly tried to occupy it in Needham, in teaching school.

When Priestley established his school, soon after he arrived, there were already two schools in Nantwich. The endowed grammar school, founded as early as 1572, had eight places free to boys of the parish and taught others for a fee. Its curriculum appears to have the standard: Latin, arithmetic, and geography, with additional charges of five shillings for instruction in reading, writing, and merchant accounts. There was also a "Blue-Coat" charity school, begun about 1710, under the auspices of the Society for the Propagation of Christian Knowledge (S.P.C.K.) "to instruct the Children in the Principles of the Church of England." There were places for forty boys in the Blue-Coat school, who were issued uniform clothing and fed in addition to being taught reading, writing, the grounds of arithmetic, and elements of trade (e.g., shoemaking).[21] Neither of these schools were, it appears, open to the children of Dissenting families and both were limited in their curricula.

20. Hall, *History of Nantwich*, 387; Gordon, *Cheshire Classis*, 192; and William Urwick, "Deanery of Nantwich," in *Historical Sketches of Nonconformity in the County Palatine of Chester*, ed. Urwick (London: Kent & Co.; Manchester: Septemius Fletcher, 1864), 124–25. The term "Socinian" is supposed to mean a follower of the Unitarian doctrines of Fausto Sozzini, but in eighteenth-century practice it generally meant any person who denied the pre-creation divinity of Christ. It was, that is, a more extreme form of Unitarianism than Priestley's Arianism.

21. Derek Robson, *Some Aspects of Education in Cheshire in the Eighteenth Century* (Manchester: for the Chetham Society, 1966), 112–19, 134, 167, 187; Hall, *History of Nantwich*, 378.

Priestley opened his school in a typical Nantwich "old-town" half-timber building, fronting on Hospital Street, with upper room extending over the gateway into the Presbyterian chapelyard. There for three years, six days a week, from seven in the morning till four in the afternoon (with a one-hour interval for dinner) and no holidays save "red-letter days," he instructed about thirty boys and, in a separate room, about half-a-dozen young ladies. He taught Latin—and perhaps some Greek—combining these lessons with geography (ancient and modern), with careful reference to maps, "for by this means . . . pupils will indirectly acquire much knowledge." He collected a small school library of suitable books of natural and civil history and accounts of travels, which students were permitted to read (with maps before them) as a favor. Presumably he taught some mathematics (though this is not mentioned) and he purchased some "philosophical instruments: a small air-pump, electrical machine, &c.," which he taught students in the upper classes to use and keep in order: ". . . by entertaining their parents and friends with experiments, in which the scholars were . . . the operators and . . . the lecturers, too, I considerably extended the reputation of my school."[22]

He also taught English grammar, writing a text for the use of his students. A critical knowledge of English was "absolutely necessary to all persons of a liberal education," though its study was neglected in most grammar schools.[23] The printed version of his text was not available for Priestley's use at Nantwich, but its substance surely reflects the nature of his grammatical teaching there. It also reveals vigorous common sense, a freedom from the irrelevant apparatus of Latin grammar, and an insistence upon a utilitarian criterion of usage as the only viable standard of the English language. There was a consequent ability to observe with accuracy and describe without preconceptions the structure of the language used in his day. These were the qualities that have earned him the description of "one of the great grammarians of his time."[24]

22. W 1/1:43 and MS letters to Caleb Rotherham, 18 May, 7 January 1767, Priestley Papers, Dr. Williams's Library, London (quotations and citations by permission; hereafter Williams Colln.), printed in W 1/1:64–65, 67 (part only).

23. Joseph Priestley, *The Rudiments of English Grammar; adapted to the use of Schools. With Observations on Style* (London: R. Griffiths, 1761), viii. The reviewer of the *Rudiments*, in Griffiths's *Monthly Review* 26 (1762): 27–31, agreed (27) that "the study of our own Tongue has hitherto been most shamefully neglected in our public schools."

24. Students of linguistics and of the history of English grammar are in near-unanimous agreement in naming Priestley a major figure in the study of English; the quoted description is that of Ivan Poldauf, *On the History of Some Problems of English Grammar before 1800* (Prague: Philosophy Faculty, Karlovy University; LV. Prispevky k Dejinan Reci a Literatury Anglike [Prague Studies in English], 1948), 231.

Priestley did not deny that he made use of other English grammars, which existed in sufficient numbers to make suspect the claim that instruction in English was being neglected in schools.[25] None of these had achieved national distribution or reputation, however, and he hoped that his was sufficiently original in materials and their disposition to excuse him from charges of plagiarism (iv). Examination suggests that much of the work's originality stems from Priestley's characteristic virtues in scholarship. He was an innovator in the teaching of grammar primarily because, in that as in theology, he states more distinctly, with more illustrations, and urges with less caution and reserve, principles that others had previously proposed and were even proposing at the time.

His insistence on the independence of English grammar from that of Latin—implicitly in the text and in a reference in the preface to the "remarkable simplicity of its [English] structure, when compared with . . . most other languages, ancient or modern"—is, for example, a continuation of a campaign dating as early as John Wallis's *Grammatica Linguae Anglicanae* of 1653. The fight was continued through the eighteenth century in the work of many (though by no means all) other English grammarians, including Lindley Murray who cites with approval Priestley's distaste for Latinization in his influential *English Grammar* of 1795. Few of the other grammarians were prepared, however, to go so far in their efforts as he.

Priestley was willing to use technical terms, derived from Latin grammar, because "they admit of easy definition, the language of Grammar is observed to be aukward without them," and young persons are helped in their studies by their use (iii–iv); but he would not extend that use to English where the terms did not explicitly apply. As he believed the technical terms referred to inflectional form and not to function, when English words failed to change inflection, he denied the application of inflectional terms. Thus, for example, he insisted that there exist only two cases: nominative and genitive, in English and only two tenses: present and preter (or past). To describe other functional variations for verbs, where there was no change in termination, Priestley developed an elaborate taxonomy of compound verbs with three orders, depending on use of the radical, present participle,

---

25. Charles C. Fries, "The Rule of Common School Grammars," *PMLA* 42 (1927): 221–37, gives a *sample* listing of more than twenty English grammars published between 1586 and 1761, including Benjamin Martin's of 1754, which Fries cites (222 n) as complaining, like Priestley, about neglect of school English. Poldauf, *Problems of English Grammar*, 109, lists three grammars, of 1734, 1735, and 1738 (none of those mentioned by Fries), as the first "true school grammars" for English readers.

or past participle of the principal verb; and three classes, depending on the use of one, two, or three auxiliary verbs. Thus the progressive future perfect: I shall have been hearing, is described as a triple compound verb of the second order; and the future perfect: I shall have been heard, is a triple compound of the third order. Priestley's contemporaries ignored this proposal and only recently has there been any appreciation of what he was attempting.[26]

Among students of language development, Joseph Priestley is perhaps best known for his resolute urging of usage as the standard for English. No doubt his distaste for the idea of "a publick Academy, invested with authority to ascertain the use of words" (vii) is related to the liberal Dissenter's characteristic suspicion of authoritative bodies (he calls such an academy "a Synod" and describes it as "unsuitable to the genius of a free nation") as well as to his conviction that it was "ill calculated to reform and fix a language." Priestley may have derived his belief in "all-governing custom" from his reading of Locke and Watts, each having suggested that language is based on convention and changed with time, but Priestley's position is not the historical relativism of twentieth-century linguistics. Consistent with his fundamental philosophical view, he believed that there is a correct form of the language, which, however, can only be achieved by experience, over time: "We need make no doubt but that the best forms of speech will, in time, establish themselves by their own superior excellence. . . . A *manufacture* for which there is great demand, and a *language* that many persons have leisure to read and write, are both sure to be brought, in time, to all the perfection of which they are capable" (vii).

The emphasis on usage is one with which many of Priestley's contemporaries agreed, at least in principle, for eighteenth-century English grammarians were less explicitly and self-consciously prescriptive than they are some-

---

26. Argument about the number of cases in English continues; see F. G. Cassidy, "Case in Modern English," *Language* 13 (1937): 240–45, which cites Otto Jesperson declaring that English possesses only the genitive and nominative cases. Respecting tense, even disagreeing with George L. Kittridge's grammatical conservatism, one sympathizes with his *Some Landmarks in the History of English Grammars* (Boston: Athanaeum Press, 1906), 10–11, where Priestley's scheme is described as "going too far" and "perplexing to the last degree." Sterling Leonard, *The Doctrine of Correctness in English Usage* (Madison: University of Wisconsin Press, 1929), 52, however, cites with approval Priestley's recognition that there is no future tense inflection in English. Poldhauf, *Problems of English Grammar*, 294–95, states that James Pickbourne's *Dissertation on the English Verb* (1789), "a truly great work," was the first to pay attention to progressive forms of the verb, but Ian Michael, *English Grammatical Categories and the Tradition to 1800* (Cambridge: Cambridge University Press, 1970), 417, says that Priestley did so in 1761 (and was earlier done in 1685).

times described. Even Samuel Johnson, in the grammatical preface to the *Dictionary* (1755), which Priestley acknowledges using, disavows the notion of a language academy and shows more effort to describe than to fix the language. Nonetheless, Priestley is practically the only grammarian consistently faithful to utilitarian principles. Having declared for usage, he does not, as most other grammarians (e.g., George Campbell, in his *Philosophy of Rhetoric*, 1776) did, proceed to provide authoritative canons for correctness. It is this consistency, rather than any linguistic relativism, which justifies the encomium "the doctrine of usage is so fundamental to all sound discussion of linguistic matters that it is important to recognize the man [Priestley] in whom it first found real expression."[27]

Because Priestley believed that the best forms of language would be, and could only be, established in the marketplace of usage, it was essential that the forms in use be accurately observed. One finds, therefore, that his description of English grammar is remarkably free of the errors of prescription and preconception that mar the grammars of his contemporaries. Indeed, in a number of instances, he appears to have noticed for the first time forms that were later to be recognized as accepted standard. The "conventional" classification of eight parts of speech, for example, "seems to have begun with Joseph Priestley," as do the present rules for formation of the plural possessive.[28]

Included in the first edition of the *Rudiments*, and therefore, probably a part of Priestley's English language instruction at Nantwich, are some "Observations on Style" and examples of English composition in prose and poetry selected from the English Bible, Addison, Bolingbroke, Samuel

---

27. Albert C. Baugh, *A History of the English Language* (New York: Appleton-Century-Crofts, 1935), 343. The chief authority on standards of usage is still Sterling Leonard, *Doctrine of Correctness*. Scott Elledge, "The Naked Science of Language," in *Studies in Criticism and Aesthetics, 1660–1800: Essays in Honor of Samuel Holt Monk*, ed. Howard Anderson and John S. Shea (Minneapolis: University of Minnesota Press, 1967), 266–95, criticizes Leonard's description of eighteenth-century English grammarians as prescriptive, but his counterexamples focus on good intentions of prefaces, ignoring textual bad practices. Elledge agrees on Priestley's consistency and most authorities stress the singularity of his consistent practice.

28. The observation respecting Priestley's eightfold classification, for which see *Rudiments* (1761), 2 n, is that of Charles Carpenter Fries, *The Structure of English: An Introduction to the Construction of English Sentences* (New York: Harcourt, Brace and Co., 1952), 66; but Michael, *English Grammatical Categories*, 231–32, regards Priestley's classification as inadequately radical, given his recognition that the numbering of parts of speech is an arbitrary operation. The system of classification that prevailed through the eighteenth century was not Priestley's, but that plus a ninth part: the article. Leonard, *Doctrine of Correctness*, 264, credits Priestley with the plural possessive form.

Johnson, Pope, Shakespeare, Swift, and Edward Young. The examples were selected quite as much for their subject and sentiment as for their literary quality, being calculated to lead students "into a just and manly taste in composition, and also to empress their minds with the sense of what is rational, useful and ornamental in their temper and conduct in life" (65). The "Observations on Style" is a preview that Priestley would expand in his *Course of Lectures on Oratory and Criticism* (1777), with particular emphasis on content and clarity as more important than ornament in writing: ". . . to spend the best part of our time in literary criticism, and in pouring over authors that have nothing to recommend them but the beauties of modern style, when the sublime studies of *Mathematics* and *Philosophy* lie open before us, is most preposterous" (62).

Here also is Priestley's first acknowledged use of the doctrine of association, but in a low key more reminiscent of Locke and Watts than of Hartley. He insists that perceptions (or ideas existing in the mind) are associated, in their very rise and first impression, with the words that denote them and is convinced that "by enlarging our acquaintance with men and books, we increase our stock of words . . . [and] at the same time make a proportionable augmentation of our ideas" (46–47).

As Priestley devoted more of his time and efforts to the teaching of languages and belles lettres, his treatment of English grammar and style became more explicit and detailed, but the material published in the 1761 edition of the *Rudiments* was an admirable preview of his capabilities. One may reasonably assume that it contributed to the invitation that came to him, in June 1761, from the trustees of Warrington Academy to fill the post of tutor in languages and polite learning, at a salary of £100 per annum, with a house and £15 a year each for boarders. The *Rudiments* was printed, for publication late in 1761, at the press of William Eyres of Warrington and the manuscript for it had earlier been seen and accepted by the publisher, Ralph Griffiths, a personal friend of Thomas Bentley, Liverpool merchant and trustee of Warrington Academy.

In 1758, when John Aikin had been selected for that post, Priestley had been thought too young and inexperienced. In 1761, with Aikin moving to the tutorship in divinity left vacant by the death of John Taylor, the situation had changed. Possibly the existence of the manuscript text in grammar provided some of the evidence that Priestley was "an able master of the learned Languages; and a considerable proficient in many of the modern." In addition to his knowledge of language, however, it could now also be said that Priestley "has . . . attended to the duties of a private school with indefatigable labour; and has discovered a singular genius for the management of youth."[29]

Priestley claims "that in no school was more business done, or with more satisfaction, either to the master or the scholars, than in this of mine." Certainly the reputation of teacher and school was high. James Tomkinson of Dorfold Hall, a Nantwich attorney, employed Priestley to tutor his children after school hours, though the family belonged to the established church.[30] One pupil, William Wilkinson, came to the school from his father's home near Wrexham, in North Wales. There was ample justification for the invitation and the resolution to invite Priestley to the post at Warrington was unanimous.

After due consideration, Priestley accepted. It cannot have been an easy decision. He was happy and successful at Nantwich, had found great satisfaction in his teaching (contrary to his earlier expectations) and discovered it to be financially rewarding as well—more so, perhaps, than the Warrington position would be. His school probably brought him an income between £60 and £75 per year, with a prospect of improvement. To this he could add the fees from tutorial work and the salary from the church, which surely was equal, at least, to that received at Needham Market.[31] There was, it is true, little satisfaction in his ministerial work and only two of the families of the congregation were congenial. One of these was that of John Eddowes, a prosperous tobacconist-grocer, at whose house and shop on High Street, Priestley lodged and boarded. This must have been a happy arrangement, for it elicited the only example of lighthearted behavior ever reported of the post-Daventry Priestley. He sometimes, it is said, jumped over the counter of Eddowes's shop and it was there that he was induced

29. Both the quotations, on his being an able master of languages and a manager of youth, are from *A Report of the State of Warrington Academy, By the Trustees at their Annual Meeting June 25th. MDCCLXI*, 2; copy in the Warrington Library, Warrington, Lancsashire. My thanks are due Mr. C. A. Carter, Borough Librarian on the occasion of my visit to Warrington.

30. Of the six Tomkinson children: James, nineteen in 1758 and probably not a student of Priestley, became a clergyman; Henry, then seventeen; Edward, fifteen; Catherine Maria, eleven, who married the Reverend George Cotton, dean of Chester in 1771; Elizabeth and Margaret; see Hall, *History of Nantwich*, 389.

31. Estimate of school income based on fees of half-a-guinea per quarter per pupil proposed for Priestley's school at Needham Market. There is contradiction respecting Priestley's salary offer from Warrington. His *Memoirs* are explicit on the proposal, but Herbert McLachlan, "Warrington Academy: Its History and Influence," *Chetham Society: Remains* 107, n.s. (1943; citations and quotations by permission): 38, declares that, in 1758 the trustees had raised the salary to £120. Perhaps the difference is to be resolved by suggesting that £100 is what was offered the inexperienced candidate.

to learn to play the flute—not well, but sufficient to contribute to his amusement for many years. "I would recommend," Priestley writes, "the knowledge and practise of music to all studious persons; and it will be better for them, if like myself, they should have no very fine ear, or exquisite taste, as by this means they will be more easily pleased, and be less apt to be offended, when the performances they hear are but indifferent."[32]

He had made other friends in the neighborhood as well. The masters of the other schools in Nantwich, the Reverend I. Rathbone of Nantwich Grammar School and Mr. Thomas Davies of the Blue-Coat School, do not appear to have been friendly, nor was the rector of Nantwich. But the Reverend Mr. Brereton, vicar of Acton, shared Priestley's philosophical interests. Priestley occasionally spent a night at Brereton's house. When he left Nantwich, Brereton gave him a telescope as a token of friendship and several years later Priestley was arranging the purchase of some "glasses" (probably optical parts) for Brereton's use.[33]

He had another philosophical friend in the Reverend William Willets, Dissenting minister at Newcastle-under-Lyme, with whom he frequently spent some leisure time in discussions of theology and of optics and magnetism. Willets's chief contribution to Priestley's well-being was, however, an introduction to his brother-in-law, Josiah Wedgwood, the potter and soon-to-be partner of Thomas Bentley. There were other friendly Dissenting ministers in the area, including a Dr. Harwood of Congleton, who also had a school. He used to exchange pulpits with Priestley "for the sake of spending a Sunday evening together every six weeks in the summertime." Nevertheless, Priestley decided that employment at Warrington would be "more liberal and less painful . . . [and] also a means of extending my connexions." He left Nantwich for Warrington in September 1761, having been there for just three years.

The move was very different from his previous one. He did not, this time, leave behind him a legacy of bitterness or indifference nor carry with him a burden of failure. The church continued to struggle along and, after

32. W 1/1:44; Priestley's jumping exploits are reported in Alexander Gordon, *Cheshire Classis Minutes*, 178.

33. See MS letter, Priestley to John Seddon, 19 May 1762, Wedgwood Papers, reproduced by courtesy of the director and university librarian, the John Rylands University Library of Manchester (printed, in part, in *Christian Reformer* 10, n.s. [1854]: 629); F. W. Gibbs, *Joseph Priestley: Adventurer in Science and Champion of Truth* (London: Thomas Nelson and Sons, 1965), 15.

many vicissitudes, repaired its more-than-a-century-old building in 1849, with a celebration of Priestley's services there.[34]

As late as 1829, one of the pupils of Priestley's school was still alive to reminisce happily about his teacher and school, but the community had the visible evidences of that school, still in existence, better to remind them of Priestley's contributions to the town. Priestley was succeeded at the Presbyterian Chapel by John Houghton, a student of Doddridge, who continued the school and even wrote a text of English grammar for its use.[35] Houghton was succeeded, in 1771, by Richard Hodgson, who had graduated from Daventry Academy two years later than Priestley. Hodgson conducted the school until his departure in 1799 or 1800. Reference to the school is then lost until 1804, when David William Jones served the Nantwich Unitarian Chapel and is recorded as having "kept a school there" to 1815. Again there is an hiatus in the records, but the Reverend James Hawkes found a school in existence, in the building in which Priestley had started it, when he arrived in Nantwich in 1823. Hawkes continued the school there until the building was pulled down. He then moved it to the vestry of the chapel, where it continued till Hawkes's death in 1846.

When Priestley left Nantwich for Warrington, he carried with him pleasant memories, useful experiences, and the manuscript of his book on grammar to launch him on the national scene as an educator. He left behind him a congregation that prized his memory and a school that would continue to serve the community for another eighty-five years.

34. The chapel was little changed in exterior appearance; its interior was remodeled, but the changes carefully preserved and adapted the oak pulpit from which Priestley had officiated; see "An Interesting Revival of Unitarianism at Nantwich in Cheshire," *Christian Reformer* 6, n.s. (1850): 55–60. It remained the only chapel in which Priestley served that was still standing as late as 1953; it has since been destroyed.

35. Hall, *History of Nantwich*, 91–192; John Houghton, *A New Introduction to English Grammar: In the simplest and easiest Method Possible* (Salop: for the author, by J. Cotton and J. Eddowes, 1766). Houghton's text is a conservative one, praising Robert Lowth and James Harris's *Hermes* and making no reference to Priestley or to the *Rudiments*.

IV

WARRINGTON ACADEMY, 1761–1767

Language, Rhetoric

Priestley's hopes for an employment in Warrington "more liberal and less painful" than at Nantwich were well founded. The town itself was the largest and most prosperous of any in which he had lived. Located on the Mersey River, halfway between Manchester and Liverpool, it had been a settled community since before the Romans; its population in 1781 numbered 8,791, more than twice that of Nantwich twenty years later. One of the earliest manufacturing towns of Lancashire County, with glass- and potteryworks, metal-smelting, and linen- and sail-making looms, it was also a transport and trading center. The bridge over the Mersey at Warrington had been there as early as 1495; over it passed the major coaching road to Chester and London. One could, by 1757, go from Warrington to London, on the twice-weekly coach, in just three days and two nights. The river, at tide, was navigable to Warrington by small coastal vessels and from Warrington to join the Irwell and thence to Manchester by barges and lighters. In 1761, when Priestley arrived there, the Bridgewater Canal, passing within a mile and a half of Warrington, was opened and canal packets carrying passengers as well as coal and other freight were soon providing another source of wealth to the community.

Although the town had remained royalist during the Civil War, Dissenting interests were early established there. The rector of Warrington had been one of the ministers ejected at the Act of Uniformity in 1662; in 1673 he was licensed to preach for Dissenting worship. By 1703 a second chapel was opened, on Cairo Street, at which the Reverend John Seddon, a major figure in the establishment of Warrington Academy, was the second minister. Warrington possessed a Blue-Coat school and Boteler Grammar School, reviving under the direction of the Reverend Edward Owen, soon to be rector of Warrington. There was also, in the town, one of the notable provincial presses, run by William Eyres, who printed (and sometime published) school texts by Priestley and other academy tutors, as well as such important works as Priestley's *History of Electricity* (1767), Thomas Pennant's *British Zoology* (4 vols., 1768–70) and John Howard's *State of Prisons in England* (1777). And, under the guidance of John Seddon, one of Britain's earliest subscription (circulating) libraries began in 1760. Even without Warrington Academy, the town might have had some small claim to its cognomen, the "Athens" of north England.

But, of course, it was the academy, its tutors and their families, and its students, that helped to sustain the activities that gave Warrington that claim. And it was at the academy where Priestley was to achieve his first successes. At Warrington, he was, for the first time, to find sufficient scope for a demonstration of his formidable talents as a scholar-teacher. This was partially because he possessed the intelligence, the courage, and the imagination to exploit the opportunities made available to him, but it was also because Warrington Academy was structured to afford those opportunities.

The academy had been planned to avoid the circumstances that had caused the recent closing of so many other, liberal, Dissenting academies in the north of England. Rather than depend upon the leadership of a single minister (and fail when he died) or upon the sponsorship of any of the funding bodies of denominational organizations (which might withdraw support for failure to adhere to specific doctrine), Warrington Academy was to be funded by private subscription, directed by trustees and an elected "Committee of Management," and staffed with full-time tutors. It was the first of the academies financed that way and inevitably there were problems that ultimately led to its failing. In its early years, however, these problems seemed surmountable and success seemed assured.

By four years' persistent negotiation, Seddon and his friends had won Manchester and Liverpool to the compromise of Warrington as the academy's site. After consideration of nominations by trustees and subscribers, such as Thomas Bentley, wealthy merchant of Liverpool and soon-to-be partner to Josiah Wedgwood, and John Roebuck, manufacturer and entrepreneur of Birmingham and Prestonpans, a selection of tutors had been made, each of whom would attract financial, theological, and intellectual support for the academy: Dr. John Taylor (1694–1761), popular minister of the Octagon Chapel, Norwich, and distinguished author of such learned works as his *Hebrew Concordance* (2 vols., fol., 1754), to be tutor in divinity; the Reverend John Aikin (1713–88), former student of Doddridge and head of his own school at Kibworth, to be tutor in languages and belles lettres; and to fulfill an early declaration that the academy was not only for the education of Dissenting ministers, but also intended to "give some knowledge to those who are to be engaged in commercial life," the Reverend John Holt (d. 1772), once a Presbyterian minister in London but long head of a large mathematical and commercial school at Kirkdale, near Liverpool, to be tutor of mathematics and natural philosophy.[1]

The academy had commenced operations in October 1757, with gifts and annual pledges of £469 and the donation of several private collections to the library. It began in rented houses for the tutors, at which students were to board for £15 per year, and a large brick house at the northwest end of the bridge, with garden and a terrace walk on the banks of the Mersey, for a library and a common hall. Five students enrolled the first year and each succeeding year more were added than left, so approximately sixty students had been enrolled by the time Priestley arrived, with thirty attending during his first year there. Clearly, by the early spring of 1761, the academy was a going concern. Then John Taylor died.

Although Taylor had been the most distinguished of Warrington's tutors and his presence was responsible for much of the initial support it received, his death was not, to many of its trustees, an unmitigated misfortune. He was sixty-three when he reluctantly accepted the call to Warrington and had long been deferred to as a minister and scholar. His solemn charge to the divinity students, that they accept no doctrine he taught unless they found it justified by revelation or the "reason of things," was in the tradition of liberal dissent. But, in practice, Taylor found it difficult to entertain objections from immature pupils and seems never to have referred them to any authority but his own.[2]

---

1. Information about Warrington Academy, and to a lesser degree about Warrington itself, is derived chiefly from William Turner's "Historical Account of Warrington Academy," originally published in the *Monthly Repository* 8–10 (1813–15), and republished as *The Warrington Academy*, with an introduction by G. A. Carter (Warrington: Library and Museum Committee, 1957; quotations and citations by permission of G. A. Carter). Also Herbert McLaughlan, "Warrington Academy," and Arthur Bennett, "Glimpses of Bygone Warrington," *Proceedings of the Warrington Literary and Philosophical Society* (1898–99).
    2. Turner, *Warrington*, 10–11.

Nor were his relations much easier with the Committee of Managers, and especially with its secretary and resident agent, John Seddon, more than thirty years younger than he. At first merely impatient with constraint, he then decided that a conspiracy had been organized against him. During the last years of his life, he kept academy personal relationships disturbed and turned many original subscribers against the school. His death did not entirely solve the problems he had created, but it made possible a renewed tranquillity within the academy and opened a position for a tutor destined to surpass him in long-term reputation. The trustees moved John Aikin to the divinity post and selected Joseph Priestley to fill the vacancy left by Aikin's move.

Priestley arrived in Warrington in September 1761 and at once fit happily into an increasingly harmonious academic community. More of an age (twenty-eight to Seddon's thirty-five) than Taylor, or even Aikin or Holt, Priestley and John Seddon quickly became friends. Seddon found it necessary, early in 1762, to travel to London, Bristol, Liverpool, and Manchester to try to repair the damage Taylor and his friends had done to the academy's reputation. While he was away, arranging particularly for financial support from the Presbyterian Fund, Priestley seems to have acted as Seddon's deputy.

It was Priestley who calmed anxieties among students and arranged for petitions supporting the academy. It was Priestley who successfully negotiated renewed employment of the academy's "provider of the commons" (a member of the Taylor clique) and arranged for a visiting lecturer in chemistry. And it was Priestley who acted as Seddon's "curate" in the Cairo Street chapel, preaching sermons and even presiding over three burials; he also did minor personal chores for Seddon's household.[3] By 4 June 1762, he had also become a member of the governors of the subscription library. During the years he remained in Warrington, he attended nearly every monthly meeting of that board, helping establish rules of library operation, selection of books, assessment of fines, etc.[4]

Early in May 1762, Priestley moved into a new house, one of two the Trustees had built especially for the tutors, along with a common hall and

3. See Priestley's letters to John Seddon, 9 April, 10 April, 1 May, 6 May, 19 May 1762, in [Robert B. Aspland], "Brief Memoir of Rev. John Seddon of Warrington, with Selections from his Letters and Papers," *Christian Reformer* 10, n.s. (1854): 224–40, 358–68, 618–29; 11, n.s. (1855): 365–74 (Priestley letters, 625–29, of vol. 10).

4. MS Minutes of the Circulating Library, Warrington, 1760–67; Warrington Municipal Library; quotations and citations by permission of the Cheshire County Council, Warrington

library, when their landlord unreasonably raised rents and refused long-term leases for their original buildings (Figs. 5a and 5b). Fronting onto the courtyard formed by the three buildings, in a complex still called "Academy Place," between Mersey and Buttermarket streets, the three-storied brick house was "neatly fitted up, handsomely sashed to the front, with a flight of five steps to the entrance . . . four rooms on a floor, cellared under, with convenient kitchens, yards and out-offices."[5] On 18 May, despite some early objections by Seddon, Priestley was ordained at the Warrington Provincial Meeting of the Ministers of the County of Lancaster, because, as he wrote to Seddon: "I am going to have a dearer, more important stake in this world than I have ever yet had in it. I can sincerely say, I never knew what it was to be anxious on my own account; but I cannot help confessing I begin to feel a good deal on the account of another person. The hazard of bringing a person into difficulties which she cannot possibly have any idea or prospect of, affects me, at times, very sensibly."[6] And, on 23 June 1762, in the parish of Wrexham, Wales, Joseph Priestley married Mary Wilkinson, daughter of Isaac Wilkinson, ironmaster, and sister of William Wilkinson, whom he had brought with him to Warrington as a student from Nantwich.[7]

Nothing in Priestley's correspondence or his published writing, gives any hint of his courtship and his *Memoirs* records the event, retrospectively, with a personal reticence typical of the eighteenth century: " . . . I married a daughter of Mr. Isaac Wilkinson. . . . This proved a very suitable and happy connexion, my wife being a woman of an excellent understanding, much improved by reading, of great fortitude and strength of mind, and of a temper in the highest degree affectionate and generous; feeling strongly for others, and little for herself. Also greatly excelling in every thing relating to household affairs, she entirely relieved me of all concern of that kind,

---

Library. I owe my awareness of these Minutes to the courtesy of Mr. G. A. Carter, librarian of the Warrington Municipal Library when I visited there.

5. T. E . Thorpe, *Joseph Priestley* (London: J. M. Dent, 1906), 50.

6. Priestley letter to John Seddon, 1 May 1762, *Christian Reformer* 10, n.s. (1854): 627–28. The advantages of ordination were that it opened more prospects should Priestley need, or want, to return to the Dissenting ministry. A disadvantage was that canon law, which many "Old Presbyterians" still respected, held that an ordained minister could have no other occupation save farming and teaching; see Andrew M. Hill, "The Death of Ordination in the Unitarian Tradition," *Transactions of the Unitarian Historical Society* 14 (1967–70): 198–208.

7. Alfred Neobard Palmer, *A History of the Town and Parish of Wrexham* (Wrexham: Woodall, Minshall and Thomas, 1893), part iv, 279.

Fig. 5(a). Warrington Academy Buildings and (b) Joseph Priestley's House at Warrington, c. 1762. Drawings from Arthur Bennett, *The Dream of a Warringtonian* (1900).

which allowed me to give all my time to the prosecution of my studies, and the other duties of my station."

Six years after the marriage, Priestley was to write a general description of family life which leaves no doubt that his was a happy one: " . . . our wives and children are, in general, inseparably connected with us and attached to us. With them all our joyes are doubled, and in their affections and assiduity we find consolation under all the troubles and disquietudes of life."[8] And, from other sources and impressions, we learn that Mary Priestley, at the age of twenty, must have been as charming and capable as she was capable and engaging as a matron. She took charge of Priestley's new house, making it a home for him and for academy pupils boarding there. Her relations with the pupils seems to have been firm and older-sisterly; for two of them, at least, Benjamin and William Vaughan, affection for Mary and Joseph Priestley developed into lifelong friendship. Moreover, she brought cheerfulness to the life of serious, sober-sided Priestley and softened the social impression of his intellectual intensity. She taught him cribbage, which they played together nightly. She entertained his friends— and boxed the ears of brother Timothy when, on a singular visit to Warrington, he criticized Joseph's heterodoxy. She visited with him when he called on academy trustees.

The Priestley family took part in the evening entertainments, described by the nineteen-year-old Anna Laetitia Aikin, when the daughters of academy faculty and staff, and a few of the older students, took part in plays, charades, card games, dancing, and even writing scraps of verses to be hidden in Mary Priestley's sewing basket, read and identified as to author. Great was the surprise when one of these was traced to Joseph Priestley, who had no "pretension to the character of a poet": "Mrs. Barbauld (neé Aikin) has told me that it was the perusal of some verses of mine that first induced her to write any thing in verse. . . . Several of her first poems were written when she was in my house, on occasions that occurred while she was there" (W 1/1:54). One of her earliest poems, "On Mrs. P's leaving Warrington," was written on the occasion of the Priestleys' move to Leeds and others (e.g., "A Mouse's Petition," and "an Inventory of the Furniture in Dr. Priestley's Study") were written when she visited them in Leeds, for

8. W 1/1:49; Joseph Priestly, *An Essay on the First Principles of Government; and on the Nature of Political, Civil and Religious Liberty* (Dublin: James Williams, 1768), 90. For a sympathetic description of Mary Priestley, see Alexander Gordon, "Joseph Priestley," in his *Heads of English Unitarian History, with appended Lectures on Baxter and Priestley* (Bath: Cedric Chivers, 1970), 123–24.

the Aikin family and the Priestleys remained friends long after their meeting in Warrington.[9]

It was a happy group and to it was shortly added a Priestley daughter, Sarah, born 17 April 1763 and named, no doubt, for Joseph's aunt and former guardian, Sarah Keighly, in a last and unsuccessful gesture of reconciliation: Mrs. Keighly died the following year, still unforgiving of Priestley's apostacy. Doctrinal differences did not disturb relationships within the academy, however, for Seddon, Priestley, Aikin, and Holt were each "necessarian" and Arian. Priestley and Aikin had studied theology from the same text, Doddridge's "Lectures on Pneumatology," which Aikin was to use at Warrington. Aikin had even proposed Samuel Clark, Priestley's favorite tutor at Daventry and the one who had encouraged him to take "the side of heresy in every question," for the Warrington divinity post to which Aikin was himself elected. The three tutors, and Seddon, representing the Committee of Managers, met weekly to drink tea and talk over issues of mutual interest and concern.

These were stirring and difficult times for England. The Seven Years' War had begun the year before the academy was founded and George III succeeded to the throne the year before Priestley went to Warrington. There was an economic slump and a drought in 1762, followed by inflation. The war's end in 1763 brought additional financial problems for industrial areas such as Warrington. Restructuring of political loyalties was taking place.[10] There was a bad harvest in 1764 and another drought in 1766. None of these problems or events seem to make any appearance in the writings or extant correspondence of the four attending the weekly academy conferences. Yet the academy was in some financial difficulty, with subscribers defaulting on their pledges, and two, at least, of the four were politically involved and aware. Aikin was timid and Holt preoccupied by mathematics

9. The manuscript of "On Mrs. P's leaving Warrington" was thrown into the chaise as Mary Priestley was departing; it was never published, but a short extract "Address to Mrs. Priestley" appeared in Barbauld's posthumous memoirs; see Grace A. Ellis and Co., *A Memoir of Mrs. Anna Laetitia Barbauld, with Many of her Letters* (Boston: James R. Osgood, 1874), 1:28; a manuscript of the whole was found in Liverpool City Library by William P. McCarthy, and published in his edition of Barbauld's poetry. The "petition" and "Inventory" are to be found in most collections of Barbauld's poems: for example, *The Words of Anna Laetitia Barbauld* (New York: Carvill, Bliss and White, et. al., 1826), 11:67–69, 79–81. As late as 23 October 1831, John Aikin's grandchildren, Lucy and Arthur Aikin, vigorously defended Priestley against the ignorant criticism of W. E. Channing; see Anna Laetitia Le Breton, ed., *Correspondence of William Ellery Channing, D.D. and Lucy Aikin, From 1826 to 1842* (Boston: Roberts Bros., 1879), 93–95.

10. See, for example, John Brewer, *Party Ideology and Party Politics at the Accession of George III* (New York: Cambridge University Press, 1976), passim, esp. 110.

and natural philosophy, but Seddon was something of a radical, if his possession of a reprinted collection of *Cato's Letters; or, Essays on Liberty, Civil and Religious* (4 vols., 1734) means anything, and Priestley would soon be involved in politico-theological debate.[11]

The academy four spent their conferences talking of higher, or more immediate, issues: "the doctrine of atonement, concerning which Dr. Aikin held some obscure notions," wonderment at the Socinian beliefs of a Mr. Seddon of Manchester. One of their topics, surely, was the Widows' Fund Association, suggested as early as 1762 by Philip Holland to John Seddon, for relief of widows and children of Dissenting ministers and still in independent existence as late as 1885. This was not exactly an academy affair, but Aikin, Holt, and Priestley were members of the first Fund Committee and Priestley, who qualified for the fund by his ordination, preached the sermon at the formal establishment of the association, in Manchester on 16 May 1764. His was the only sermon ever published by the Fund Association and it was Priestley's first published sermon.

The sermon, itself, *No Man Liveth to Himself*, was not unusual, being the typical rational Dissenter's emphasis on practical morality: "Nothing can be more evident than that the dictates of conscience strongly enforce the practice of benevolence . . . and the pleasures of benevolence certainly constitute the greatest part of those pleasures which we refer to the moral sense" (48). It was, however, written with more attention to literary quality than most of Priestley's later writings and the dedication, "To Wives of Dissenting Ministers," shows an uncommon concern for women's education and suggests that he was thinking of his own wife and daughter: "It is part of your wisdom to teach them [your children] or get them instructed (especially your daughters) in those arts by which they may be enabled to procure themselves a decent provision by the labours of their own hands, at the same time that they must be educated in such a manner as to be proper companions for those who live by the labour of others."[12] And two of its sections pass to his hearers Priestley's own formula for happy living:

With our affections and our faculties . . . engrossed by a worthy object, we scarce need fear being ever dull, pensive, or melancholy,

11. See Caroline Robbins, *Eighteenth-Century Commonwealthman*, 120–25, 392–93. H. McLachlan, "Warrington Academy," 12, identifies Cato's Letters as having been in Seddon's library.

12. Quoted by Herbert McLachlan, "Warrington Academy," 108, from *No Man Liveth to Himself, a Sermon Preached before the Assembly of Protestant Dissenting-Ministers, of the Counties of Lancaster and Chester, met at Manchester May 16, 1764, to carry into Execution a Scheme for the Relief of their Widows and Children and Published at their request* (War-

or know what it is to have our time hang heavy upon our hands. (74)

Then the human mind, roused to the most intense exertion of all its faculties, burdened with no consciousness of guilt, referring itself absolutely to the disposal of its God and father, . . . acquires a fervour of spirit, a courage, fortitude, and magnaminity, tempered with the most perfect serenity, and the greatest presence of mind, that is sufficient . . . to bear a man through every difficulty. (75)

Surely, however, the major topics of those weekly consultations must have been academy business, planning, and review of day-to-day operations: which books to request for the academy library (Seddon was its librarian) and what scientific apparatus was needed for Holt's lectures; which students were disciplinary problems (usually sons of West Indian planters) and how to handle them; and, most important, what the subjects of the two curricula—for nonprofessional and for divinity students— should be, who was to teach them, and what books should be used. So far as Priestley was concerned, the early stages of curricular assignments should have been simple enough. He had been selected to teach languages and belles lettres and to these subjects he gave his immediate attention. The demands of his position are recorded in the Warrington Trustees' Report, at the end of Priestley's first academic year:

IN THE LATIN CLASS the *Classics* are read; *Latin Compositions* made; and a Course of *Roman Antiquities,* and *Mythology* gone through.
IN THE GREEK CLASS are read the *Greek Authors,* and a Course of *Greek Antiquities.*
THOSE young Gentlemen who learn *French* are taught to read, and write that Language; and go through a Course of Exercises calculated to prepare them to converse in it.

---

rington: W. Eyres, 1764). When the sermon was republished as one of *Two Discourses; I. On Habitual Devotion, II. On the Duty of not living to Ourselves; Both Preached to Assemblies of Protestant Dissenting Ministers, and published at their Request* (Birmingham: Piercy and Jones, for J. Johnson, 1782), the dedication had been dropped as had the appended "Rules of an Association." Priestley ended his subscription in the Fund in 1775, shortly after becoming companion-librarian to Lord Shelburne. In his preface to *Two Discourses,* Priestley acknowledges the occasion of this sermon's delivery and its earlier publication with so small a printing that he had been urged to republish. He also claims that Mrs. Barbauld's poem, "An Address to the Deity," was inspired by her hearing (as Anna Laetitia Aikin) this sermon. References here from the version of 1782.

THE ENGLISH GRAMMAR is taught to the younger Students, and they are trained up in a regular course of *English Compositions.*

BY the Tutor in this Department are read Lectures on Logick; the *Theory of Language, and Universal Grammar, Oratory and Criticism;* the *Study of History* and *Anatomy.*

HE also directs the public *Academical Exercises,* consisting of *Translations* from *Greek, Latin,* and *French* Authors; and *Orations,* or *Dissertations,* which are delivered alternatively in *English,* and *Latin,* or *French;* wherein a particular Attention is paid to the Manner of Reading and Speaking.[13]

This was a heavy teaching load, but one for which Priestley, so far as languages were concerned, was scarcely unprepared. He had, after all, been making a particular study of languages since he was a boy, when he learned Latin, a little Greek, Hebrew, French, Italian, "High Dutch," and taught himself some Chaldee, Syriac, and a little Arabic. It had, as Priestley acknowledges, been the "attention I had given to the learned languages when I was at Daventry," which prompted Samuel Clark to recommend him to Warrington and, at Nantwich, he had taught languages and written his text on English grammar.[14]

Yet there is some ambiguity in his later references to this task. In his *Memoirs,* he declares "at the time of my removal to Warrington, I had no particular fondness for my profession." The ambiguity is compounded by a letter of May 1766 to the Reverend Caleb Rotheram, responding to an inquiry about his teaching. Priestley tells Rotheram what texts he has used: "Holmes' Latin Grammar, not because I altogether liked it, but because I thought it easy for beginners . . . the London Vocabulary, a few of Clarke's Translations . . . a few of Sterling's editions. . . . Several of the collections for the use of Eton school are excellent, as are their four books of *Exercises,* beginning with *Exempla minora* and ending with historical examples." But he distinguishes between teaching at Nantwich and at Warrington: "All my experience in teaching school was small, for I was schoolmaster only three

13. *A Report of the State of the Warrington Academy, By the Trustees at their Annual Meeting July 1st–MDCCLXII,* 2–3. Aikin soon took over the logic course and Priestley taught anatomy one year only, but later added the teaching of elementary Italian, according to a letter, 14 February 1766, to Caleb Rotheram, Priestley Corres., Williams Colln.

14. Chapter I and W 1/1:47. See Chapter II for reference to his daily reading of Greek at Daventry with John Alexander, later, according to the *Dictionary of National Biography* to become known as "one of the best Greek scholars of his time."

years"; and he writes: "My *English Grammar* was not ready time enough for me to make trial of it. It has been out of print two or three years, and I shall not consent to its being reprinted. *Lowth's* is much better, but I question whether it will signify much to teach any English Grammar."[15]

The letter suggests that Priestley's *Rudiments of English Grammar; Adapted to the Use of Schools* was not suitable for use at the academy. Yet he complained, in that work, that many grammar schools failed to teach English grammar. The Trustees Report states that younger students at Warrington were, in fact, getting such instruction. That he used the *Rudiments* at Warrington is clear from the "Notes and Observations" appended to the new edition when Priestley did, after all, consent to its reprinting in 1768. This appendix was derived "from the notes which I collected at Warrington; where, being tutor in the languages and *Belles Lettres,* I gave particular attention to the English Language, and intended to have composed a large treatise on the structure and present state of it. But dropping the scheme in another situation, I lately gave such parts of my collections as I had made no use of to Mr. Herbert Croft, of Oxford, on his communicating to me his design of compiling a dictionary and grammar of our language."[16]

It is true that Priestley quickly composed and had printed for his Warrington classes, *A Course of Lectures on the Theory of Language and Universal Grammar* and the notes "collected at Warrington" may have developed from this work.[17] It is difficult to believe, however, that he did not explicitly teach English grammar to the more advanced classes, as well as the younger ones, for he insists in those Lectures that the "rules of Grammar, when persons are capable of using them, do very much facilitate

15. W 1/1:50; Priestley to Rotheram, 18 May 1766, Priestley Corr., Williams Colln.

16. W 1/1:45–46; Croft announced his intentions of publishing a "New Dictionary of the English Language" in 1787, to correct Johnson's *Dictionary* which, he declared, was "defective beyond all belief, and defective far beyond what I shall have time in this letter to assert, much less to prove." The plan reached to *An Unfinished Letter to the Right Honourable William Pitt,* ed. R. D. Alston, Facsimile Reprints in English Linguistics 1500–1800, no. 71 (Menston, Eng.: [1788], Scholar Press, 1968), from which, 9, the quotation is taken. In 1792, Croft sent out *Proposals for publishing, in May next, Croft's Johnson's Dictionary,* but publication did not take place. In the list of persons appended to the *Unfinished Letter,* to whom Croft felt himself obliged for assistance is "Dr. Priestley."

17. Joseph Priestley, *A Course of Lectures on the Theory of Language and Universal Grammar* (Warrington: W. Eyres, 1762). The *Lectures* were printed but not published; yet other Dissenting academy tutors used them. Eyres did not have Greek or Hebrew fonts, leaving blanks in the text to be filled by student users. My copy is bound with shorthand notes of introductory lectures of the Reverend Mr. Kippis, at Coward's Academy in 1763. According to Priestley's letter to Seddon, 1 May 1762, *Christian Reformer* 10, n.s. (1854): 625–29, a quarter of the *Lectures* had then been printed.

the acquiring of this art [of language], and are of great use in order to make a person more exactly and extensively acquainted with a language that was learned at first without their assistance" (4). And, teaching English grammar at least to "the younger students," why should he not have used his own text while it remained in print? For he did not really think Bishop Robert Lowth's *Short Introduction to English Grammar,* published but one month after the *Rudiments,* was "much better" than his own work. When the second edition of the *Rudiments* appeared, Priestley noted his obligations, and also "very considerable" differences in "plans, definitions of terms, and opinions," to Dr. Lowth (xxiii), as later editions of Lowth's *Short Introduction,* in turn, take over substantial parts of Priestley's descriptions and evaluations of usage, while retaining their stricter logical sense of the language.

Lowth's was but one example of later borrowings from Priestley, for with its publication in 1761 the *Rudiments* had commenced a long influential role in the history of English grammar. There were nine English editions during Priestley's lifetime (1761, 1768, 1769, 1771, 1772, 1784, 1786, 1789, 1798), including a pirated Dublin edition, and a translation into French (1799). In 1798, it was also the base for a privately printed text, *Elementary Principles of English Grammar collected . . . chiefly from Dr. Priestley* and, more than half a century after its first appearance, its reputation was still sufficient to justify exploitation in yet another version, *Dr. Priestley's English Grammar Improved* (London, 1827).

Priestley did not take part in the acrimonious debates between avowed grammarians (e.g., James Harris, Horne Tooke, James Pickbourne) during the latter part of the eighteenth century, perhaps because he had gone on to very different problems. Nor was his grammar one of the most popular of the day: during the same period in which the *Rudiments* went through nine editions, Lowth's *Short Introduction* had twenty-two; while by end of the century, Lindley Murray's *English Grammar, comprehending the Principles and Rules of the Language* (York, 1795) swept nearly all others from English institutions where grammars were used. But practically all eighteenth-century English grammars and discussions of English language after 1768 were influenced by Priestley's work and acknowledged that influence. Murray, for example, avows his debt to Harris, Johnson, Lowth, and Priestley and repeatedly reminds students of his agreement with Priestley about false Latinization of English grammar.[18] Moreover, in America,

---

18. See Scott Elledge, "Naked Science of Language," 284, which also says: "It was Priestley more than anyone else who showed him [Murray] how to adapt the naked science to the purposes of life."

the most influential nineteenth-century grammar was Noah Webster's *A Grammatical Institute of the English Language* (1782–85, and many subsequent editions) and Webster makes his sympathy with and debt to Priestley's grammatical ideas quite clear.

It was the 1768 edition of the *Rudiments*, enriched by Priestley's years of study and teaching at Warrington (subsequent editions being substantially the same) that most influenced these grammarians.[19] The substance of the "Observations on Style," appended to the 1761 *Rudiments,* were included in his lectures on oratory and criticism. They were therefore replaced, in later editions, by the "Notes and Observations, For the Use of Those who have made some Proficiency in the Language." These "Notes and Observations" constitute nearly 60 percent of the whole and represent Priestley's views of English as it was then used and a collection of more than five hundred quotations from various authors, as examples of different usages. Priestley says that he collected his examples from "modern writings, rather than from those of Swift, Addison, and others, who wrote about a half century ago" so that "we may see what is the real character and turn of the language at present" and, by comparing with earlier writers, better perceive the way the language was tending and which tendencies should perhaps be avoided.[20] Nor does he believe one should look only to the "best" writers, grammarians, and critics, to determine the "real present state of any language," but take examples from "books which may be supposed to be written in a hasty manner, when the writers would not pay much attention to arbitrary rules." They would then "indulge that natural propensity, which is the effect of the general custom and genius of the language, as it is commonly spoken" (xii). Finally, though he does not "scruple" to say which exemplified grammatical form he prefers, he insists that "this is to be understood as nothing more than a conjecture, which time must confirm or refute" (xvii).

Modern students of linguistics and of the history of English grammar have an enthusiasm for Priestley, the scientist who was also a student of languages. Theirs is an attitude that is, perhaps, not entirely based upon

19. The edition I used is: Priestley, *The Rudiments of English Grammar, Adapted to the Use of Schools; with Notes and Observations, For the Use of those Who have made some Proficiency in the Language* (London: for J. and F. Rivington, T. Lowndes, S. Crowder, T. Becket and Co., and J. Johnson, 1771), which appears substantially the same as the 1768 edition, and so far as I could establish, with those that followed.

20. Preface, xi; note, however, that he does cite, among his examples, Shakespeare, Milton, the Bible, Addison, Swift, Pope, Dryden, and Locke.

Priestley's positive contributions to their field.[21] In 1857, Goold Brown declared: "The treatises of the learned doctors Harris, Lowth, Johnson, Ash, Priestley . . . owe their celebrity not . . . to their . . . fitness for school instruction as to the literary reputation of the writers."[22] Brown was a romantic and a petulant critic, but that need not vitiate his criticism. One may reasonably doubt that Priestley's grammatical writings would have achieved so much currency with modern grammarians had he not gone from grammar to science (though he had yet to do any of note when he wrote on languages). Nonetheless, when so much of linguistics has involved a genetic study of language forms and structures, it is also reasonable that scholars should turn to Priestley who believed that English was in a process of development, and specially to his 1768 "Notes and Observations," which contain so many illustrations of current uses.

Certainly many examples of linguistically significant early uses were first cited or discussed in that work. Ivan Poldhauf observes that Priestley was one of the chief eighteenth-century figures in consideration of the agreement between naturally composite nouns (e.g., clothes or lungs) and their verbs (185–94); that he was the first to describe the tendency to incorporate foreign elements into a language by giving them regular terminations or varying terminations with different meanings (e.g., indexes of books, indices of numbers) (58); that he was the first to point to the use of the definite article for ultimate emphasis (47); and that, in the use of superlatives and comparatives, Priestley first made the distinction clear between old-elder-eldest (76).[23] W. F. Bryan cites Priestley as the first to recognize in English the passive force of the active verb in certain expressions (111–12) and the pronounial, pro-adjectival, pro-adverbial use of the word "so": "The word *so* has sometimes the same meaning with *also, likewise, the same.* . . . They are happy, we are not so, i.e., not happy" (137).[24] And Stirling Leonard found worthy of note Priestley's near-singular recognition that, with class dialects, correctness means nothing more than the social impressions—formality, ease, rusticity, illiteracy, and so forth—desired;

---

21. Examples of this pervasive attitude are Ivan Poldhauf's "the scientific character of his [Priestley's] attitude towards language," *Problems of English Grammar*, 128; and Scott Elledge's "The mind and method of a scientist mark these lectures," "Naked Science of Language," 206.

22. Goold Brown, *The Grammar of English Grammars, with an Introduction Historical and Critical* (New York: Samuel S. and William Wood; Samson Low, Son and Co., 1857), 140.

23. Poldhauf, *Problems of English Grammar*, 239, 237, 195, 251.

24. W. F. Bryan, "A Late Eighteenth-Century Purist," *Studies in Philology* 24 (1927): 368–69.

the examination of the real effect, in negative constructions, of the use of or, nor (201); and the attempt to define a principle of distinction about the placing of apostrophes before the "s" in substantives (57, 67–68).[25]

That Priestley should be the source of so many early examples of usages validates the high opinion of modern grammarians for his skill as an observer, but the "scientific objectivity" of his observations could not cause his reputation as one of the most important philosophical grammarians of his age.[26] For the attitudes that prompt the description "philosophical," one must turn to other parts of the *Rudiments* than the "Notes and Observations" and to Priestley's other book on languages, *A Course of Lectures on the Theory of Language, and Universal Grammar.*

The text of the 1768 *Rudiments* does not differ substantially from that of 1761; there is a new, more general, beginning, with a definition of language as the means of conveying ideas and a new, and better, definition of grammar as "a collection of observations on the structure of . . . [a language], and a system of rules for the proper use of it." A few minor questions and answers have been added to the dialogue form, some few of the 1761 footnotes have been moved up into the body of the text and the greater number have been displaced to become the basis for the appended "Notes and Observations." The significant change is in the preface, now twice as long as in the first edition, and containing a shortened version, with specific application to English, of some of the discussion of the *Lectures on . . . Language and Universal Grammar.* Although the 1768 preface refers to the *Lectures* as a "more extensive view of language in general" in a work "printed some years ago for private use . . . which I propose to correct, and make public," they were never published.[27] There is some suggestion that subsequent English writers on language during the eighteenth century knew the work, but the only clear line of transmission of many of Priestley's

25. Stirling Leonard, *Doctrine of Correctness*, 180, 92–93, 198.

26. The phrase is that of Poldhauf, *Problems of English Grammar*, 167, but the idea is found in other linguistic studies, for example, Murray Cohen, *Sensible Words: Linguistic Practice in England, 1640–1785* (Baltimore: Johns Hopkins University Press, 1977), passim, but esp. 101, 122, 128.

27. Scott Elledge, "Naked Science of Language," 286, believes the *Lectures* were used by later grammarians. The *Lectures* were published in 1824 as part of volume 23 of John Towill Rutt's edition of *The Theological and Miscellaneous Works of Joseph Priestley* and the volume was separately issued in 1826 and 1833. Volumes of the Rutt edition are always suspect, however, for the editorial changes made in them and this volume is no exception; the 1824 version does not agree with that of 1762.

ideas on the nature of language in general is that in the 1768 preface to the *Rudiments.*

Lectures 13 and 14 (187–218), "Of the Complex Structure of the Greek and Latin Languages," for example, make clear (what he has implied in earlier lectures) Priestley's conviction that "Northern" languages owe little in their structure to Latin. The 1768 preface declares, "I own I am surprised to see so much of the distribution and technical terms of the Latin grammar retained in the grammar of our tongue; where they are exceedingly awkward, and absolutely superfluous" (vi–vii). The *Lectures* essentially concede the impossibility of a general description of verb tenses, because of the diversity of practice among different languages (107–15). The preface applies this, with specific reference to Priestley's favorite anti-Latin theme, to English:

> . . . We have no more business with a future tense in our language, than we have with the whole system of Latin moods and tenses; because we have no modification of our verbs to correspond to it; and if we had never heard of a future tense in some other language, we should no more have given a particular name to the combination of the verb with the auxiliary *shall* or *will*, than to those . . . with the auxiliaries *do, have, can, must,* or any other.
>
> The only natural rule for the use of technical terms to express time, &c. is to apply them to distinguish the different modifications of words; and it seems wrong to confound the account of *inflections*, either with the grammatical uses of the *combinations* of words, of the *order* in which they are placed, or of the words which express relations, and which are equivalent to inflections in other languages.[28]

Lecture 12, "Of the regular growth and corruption of Languages," maintains, on Hartleyan associationist grounds, that languages "have a kind of regular growth, improvement, and declension" (169). There are limits beyond which a language cannot improve itself, as ideas and the combinations and relations of them conveniently expressed by words cannot be infinite (176): ". . . the best forms of speech, the most commodious for use, and the most agreeable to the analogy of the language, will at length establish

---

28. Preface, vii–viii. This statement earned Priestley the praise of Ian Michael, *English Grammatical Categories*, 405–6, as one of the very small group of grammarians that could strictly apply formal criteria to analysis of English verb tenses because its members knew what they meant by tense.

themselves, and become universal, by their superior excellence," at which time that language may be said to have arrived at its maturity and perfection (178–79). That this had not yet happened to English, Priestley indicates in the 1768 preface: ". . . the best we can do . . . at present is to exhibit its actual structure, and the varieties with which it is used. When these are once distinctly pointed out . . . the best forms of speech, and those . . . most agreeable to the analogy of the language, will soon recommend themselves, and come into general use; and when, by this means, the language shall be written with sufficient uniformity, we may hope to see a complete grammar of it" (xv–xvi).

The relationship between the *Lectures* and the new preface to the *Rudiments* is, however, best seen in a declaration of the latter that "It must be allowed, that the custom of speaking, is the original, and only just standard of any language. We see in all grammars that this is sufficient to establish a rule, even contrary to the strongest analogies of the language with itself" (ix). The *Lectures* systematically portray languages as an evolutionary development from sounds to letters, to combinations of sounds (and letters) into words, to the functional combinations of words into simple and then complex sentences. In such a theory of language development, spoken language must be prior (and superior) to written, in determination of its forms. However "philosophical" the *Lectures* might be, it is clear from this and earlier examples, that Priestley's philosophy of language was social and utilitarian and that his universal grammar would not derive from theological or scientific absolutisms.

John Locke, in book 3, "Of Words," in the *Essay Concerning Human Understanding,* challenges the view that language was of divine origin and even rejects the notion that words are natural, as opposed to arbitrary, signifiers of things. Priestley's reading of Locke may well have been the origin of his attitude toward languages; certainly these opinions form the basis of the discussions of the *Lectures.*[29] Priestley would not deny the possibility that writing had been a gift of God, but the "imperfections of all alphabets . . . seems to argue them not to have been the product of divine skill, but the result of such a concurrence of accidental and gradual improvement as all human arts" (29). Though he accepts the Adamic story, which argues for a single origin of spoken language, and a comparison of

29. For Locke's views of language and their importance to eighteenth-century language study, see Hans Aarsleff, *The Study of Language in England, 1788–1860* (Princeton: Princeton University Press, 1967), esp. 10–14.

alphabets even supports the single origin of writing (possibly Hebrew or Samaritan!), Priestley prefers to believe in a gradual development of both, in a number of places, as classes of sounds, "labials, dentals, linguals, palatines, gutterals, and the like," produced different letters and words. This is a more natural explanation than a literal acceptance of the Babel miracle and "more agreeable to most other operations of the deity" (288).

Nor are languages to be regarded as immanent works of nature, in spite of such an authority as James Harris, for nature is thoroughly uniform while languages are not (115). Language is an art of men, and of ordinary men, not philosophers, "suggested by the necessities of beings in their first uncultivated state, and enlarged as their further occasions prompted and required" (113–14). But the arts are based, if unknowing, on science. To improve language arts, as to improve any other, the laws of science on which it is based must be understood; that science, for language, is the science of man and of man's societies.

Priestley's universal grammar is, then, not in the seventeenth-century mode of a philosophical and universal language, though he was sympathetic to such plans as Bishop Wilkins's *Essay towards a real Character and philosophical Language* (1668), when "language itself, as an abstract science . . . is sufficiently understood to succeed in so grand a scheme" (301). His *Lectures* are "the most substantial survey of the idea of language universals of the period," but his notion of universality is limited to functions common to most languages. The truly original and universal requirement is that of human communication. Languages differentiate by historical and local variations between groups of users, but the functions, the universals, of language, are the consequence of unalterable "rules of right reason."[30]

For Priestley, a study of language in general becomes a study of people, their societies, and their cultures. This meant a study of psychology, history (the variation of societies in time), and comparative "anthropology" (the variations of societies in space). The *Lectures* speculate on the psychological (sensationalist and associationist) origin and genetic development of language, and declare: "The little light that hath yet been stuck out upon the subject of language in general hath resulted from the comparison of properties of different languages actually subsisting"[31] and because Priestley as-

---

30. Cohen, *Sensible Words*, 101, 57, 122; Priestley, *Lectures*, 237–38.
31. *Lectures*, 296; he uses examples from Hebrew, Latin, Greek, French, Welsh, Italian, Arabic, Malay, German, modern Greek, Armenian, Saxon, Lapp, Chinese, Japanese, Ethiopian, Persian, Carib, Poconchi, Peruvian, Huron, Brazilian, and Algonquin—most obscure language references being taken from books such as the "miscellaneous dissertations" of Hadrian Reland.

sumes that the act of thinking is a linguistic activity (*homo sapiens* being much the same as *homo loquens*), he emphasizes variations of language that convey different expressions of thought. He is concerned with stylistic differences for sake of elegance, ease, harmony, variety, precision, or degree of emphasis (e.g., 148, 152–53, 156–57). It seems that his *Lectures on a Theory of Languages and Universal Grammar* not only refers to Priestley's reading of Locke and Hartley, and of contemporary grammarians, but anticipates the teaching he is shortly to do in history, oratory, and criticism.

Priestley's interest in languages was of long standing when he commenced teaching and continued long after he had left Warrington. His contributions to Croft's *Dictionary* are dated circa 1787–88 and in 1796–1800 he corresponded with Benjamin Smith Barton on comparative languages, lending him a copy of Peter Simon Pallas's *Vocabularia Comparativa*.[32] On the subjects of oratory and criticism, however, he had not made a formal study and he ceased systematic investigation of them on leaving Warrington. Yet elements of these subjects had necessarily entered books he read at Heckmondwike and courses he took at Daventry, for these were topics of major importance, not only to potential preachers but to anyone who contended publicly, in speaking or writing. Exercises at Cambridge and Oxford were still carried out through formal oral disputations and pamphleteering was, throughout the century, a standard means for advancing political, economic, or theological ideas. It was to answer the needs of such contentions that courses such as that at Warrington Academy were intended. For together, oratory and criticism comprised those subjects frequently then still described as rhetoric: the art of using language skillfully in argument and persuasion.

In his early reading of Watts's *Logic*, which he was again to read at Daventry, Priestley had been introduced to the subjects of argumentation, disposition, definition, and the use (or misuse) of formal topics for the organization of ideas—all elements in that part of rhetoric called oratory. At Daventry, where sermon-making was part of the curriculum, Priestley's speech impediment had discouraged emphasis on delivery, but his tutors had substituted sermon writing and the formal composition of essays, homilies, and commentaries, with considerable concern for English literary style—and these were the elements of rhetoric included in criticism. Al-

---

32. Priestley to Benjamin Smith Barton, 20 July 1796, 27 November 1800; nos. 154, 168, in *A Scientific Autobiography of Joseph Priestley: Selected Scientific Correspondence*, ed. Robert E. Schofield (Cambridge: MIT Press, 1966); hereafter cited as *SciAuto*. Pallas's *Linguarum totius orbis vocabularia comparativa* was published in parts at St. Petersburg in 1786, 1789.

though he may not have had systematic formal instruction in the theory of rhetoric, Priestley had been instructed in its practice.

When faced with the task of teaching its elements at Warrington, Priestley could not rest content with rhetorical practice alone. Although he instituted the custom of public exercises at the academy, where "every Saturday the tutors, all the students, and often strangers, were assembled to hear English and Latin compositions, and sometimes to hear the delivery of speeches, and the exhibition of scenes in plays" (W 1/1:54), he had also to provide lectures on theories of oratory and criticism. For this purpose, he seems to have done what teachers in like circumstances have always done. He acquired copies of texts in oratory, rhetoric, style, and criticism and melded them into a set of lectures of his own, it being "the business of a *Lecturer,* to bring into an easy comprehensive view whatever has been observed by others: and in this respect I hope it will be thought I have not acquitted myself ill; few works of criticism, of any value having escaped my attention, at the time I was engaged in those studies."[33]

He had not originally intended that these lectures be published, but inevitably, Priestley being Priestley, they were published and, equally inevitably, when published as *A Course of Lectures on Oratory and Criticism* (1777), they were not simply pastiches of his reading, they contained original contributions of his own.[34] The immediate occasion of their publication was the campaign Priestley had begun, in the 1770s, to bring David Hartley and the "association of ideas" to the attention of the public, emphasis on Hartleyan associationism being what most clearly distinguishes Priestley's approach to rhetoric from that of others. The originality of *Oratory and Criticism* was not simply in its associationism, however, for Priestley adapted the classical system of rhetoric to the new logic of Locke and, in

33. *Oratory and Criticism*, preface, ii: he acknowledges, however, that there is less account of later publications, as he had since been engaged in other pursuits. In a letter of 13 April 1777, he wrote that he had not much revised the lectures, as he could not "bring my mind to it," and had not wanted to study the subject afresh; see Priestley to Newcome Cappe, 13 April 1777; MSS 1257B, Burndy Library, Smithsonian Institution Libraries, Washington, D.C. 20560; quoted by permission. (part in W 1/1:298–99).

34. I have used Joseph Priestley, *A Course of Lectures on Oratory and Criticism* (London: J. Johnson, 1777). This edition was reprinted in the Landmark in Rhetoric and Public Address series with an introduction and index by the editors, Vincent M. Bevilacqua and Richard Murphy (Carbondale: Southern Illinois University Press, 1965). The work had but one English edition during Priestley's life, though it was pirated in Dublin (1781) and translated into German (Leipzig, 1779). In the Rutt edition of the *Theological and Miscellaneous Works* it was included in the same volume (23; 1824) as the reprinted *Rudiments* and *Lectures* on language and was, therefore, separately issued in 1826 and 1833.

doing so, produced lectures "of special interest to historians of rhetoric because they show what could be made of . . . [the new rhetoric] when it became the subject of a treatise by someone whose fame . . . was ultimately to be linked with the development of modern chemistry."[35]

As published, *Oratory and Criticism* contained thirty-five lectures in three parts, with part 3, "Of Style," taking the last twenty-five. A fourth part of the course of lectures, on elocution, was never written down. Priestley notes, in the preface, that the written lectures were outlines of what he delivered in class and that he had omitted, in publication, "a considerable part of what I had composed . . . in the first part of this work, which is, in its own nature, more trite than the rest" (iv–v). Probably, then, the lectures delivered at Warrington from 1762 to 1767, contained more than was published on the first "offices" of rhetoric: invention, memory, and disposition, but Priestley was clearly more interested in treating the "office" of style, while that of delivery (called elocution, by Priestley, in partial acceptance of the contemporary elocutionary movement of Burgh, Sheridan, Walker, and Austin) is only touched on, here and there and in the last three lectures, and then primarily to illustrate critical theory.

Considering Priestley's own problems of speaking—"sometimes manifested," according to one of his Warrington students, "in discoursing from his written Lectures"—it is scarcely surprising that he minimized the treatment of elocution, though "his observations on . . . defects in speaking, and his directions how to remedy them, were very judicious; and he had the advantage of being able to refer . . . to excellent practical models in Dr. Aikin and Mr. Seddon."[36] His published theories of rhetoric relate, then, only to four of the five classical offices, and these he changed in meaning and application.

Classical rhetorical theory had undergone a revival of interest early in the eighteenth century, with at least four editions of Aristotle on rhetoric

35. Wilbur Samuel Howell, *Eighteenth-Century British Logic and Rhetoric* (Princeton: Princeton University Press, 1971), 632–33. Again one may wonder whether subsequent achievements in science have not attracted from Howell, and Bevilacqua and Murphy—whose introduction to the Landmark reprint, xxxiv, speaks of Priestley's "commitment to new science" and of his ideas being shaped by "contemporary scientific-psychological theory"—and other modern rhetoricians, more attention to Priestley's rhetorical writings than they would otherwise have received. Perhaps because these lectures were published after Priestley's major scientific achievements, there seems more justification for bringing science into the discussion than there was earlier with the works on language. See also Michael G. Moran, ed., *Eighteenth-Century British and American Rhetorics and Rhetoricians: Critical Studies and Sources* (Westport, Conn.: Greenwood Press, 1994), 175–85, with a good recent bibliography.
36. Quoted in Turner, *Warrington Academy*, 25, 26.

appearing during the century and an abstract printed in John Lawson's *Lectures Concerning Oratory* (1758), eleven editions of Cicero's *De Oratore* and a Lawson abstract, and four editions of Quintilian's *Institutes of Oratory*. Moreover, the standard "modern" text of midcentury, John Ward's *System of Oratory* (published in 1759, from lectures delivered at Gresham College c. 1720), was a faithful representation of the classical view.[37] Priestley specifically cites Ward as one of his chief sources.

Superficially, the first two parts of *Oratory and Criticism* have the form of classical theory, but closer examination reveals substantial deviations from classical orthodoxy. His replacing of the three offices of rhetoric (invention, disposition, and memory) by two (recollection and method) was more than a renaming.

In classical rhetoric, invention was a process of investigation, discovery, and the formulation of ideas. For Priestley, there was no need to crowd the speaker with more branches of knowledge than was required. Oratory consists of "rules for the proper use of those materials . . . acquired from . . . study and observation. . . . In order to speak, or write, well upon any subject, . . . [it must already] be thoroughly understood." A speaker must acquire a perfect knowledge of it before he can expect any assistance from the art of oratory. He must also have acquired the principles of grammar ("a knowledge of the inflection of words and . . . structure of sentences"), of logic ("the rules . . . relating to arguments, their perspecuity or confusion, their fallacy or their force"), and of human nature ("knowing the passions, prejudices, interests, and views of those he hath to deal with") (2–4).

For the purpose of classical "invention," a set of formal "topics" or commonplaces were listed to lead to discovery. Isaac Watts, among others, had been against using "topics," as ready-made formulas suitable only for trite arguments and Priestley acknowledges that criticism. Nonetheless, he includes three lectures on "topics," for to him, they were only mnemonic devices to aid the mind in recalling knowledge already there. As "ideas are associated by means of their connection with, and relation to one another," recollection of ideas is aided by means of other ideas "with which they were previously associated" (22). Examples of topical ideas, by which "middle-term" concurrence between a subject and its attribute might be recalled,

---

37. For a discussion of eighteenth-century English rhetorical theory and practice, see William P. Sandford, *English Theories of Public Address 1530–1828* (Columbus, Ohio: H. L. Hedrick, 1931), W. S. Howell, *Rhetoric*, and Moran, *British and American Rhetorics*.

were definition, adjunct, antecedent, contraries, example, authority, cause, effect (8, 10). Priestley here created a scheme of topical analysis filling "the vacuum created by his exclusion of invention and discovery from the province of rhetoric."

> By grounding the theory of the middle term on Hartleian psychology, Priestley made a notable contribution to rhetoric, offering for the first time a psychological rationale for topical theory. More important, he made clear as part of his rhetorical and inventional theory that distinction in [Francis] Bacon between the investigatory and communicative function of rhetoric echoed in the works of Campbell, Blair, and Whately.[38]

In his five lectures on Method, he departs still further from classical rhetoric structure. Dividing all composition into narrative and argument (instead of the three kinds of classical rhetorical speeches: the assembly, the bar, and the ceremonial occasion), the arrangement of materials for narrative becomes, simply, the *order of nature: time* for events and *place* for subjects of natural history: " . . . a writer can never be blamed if he dispose the materials of his composition by an attention to the strongest and most usual *association of ideas* in the human mind" (35). Characteristically, Priestley's greater interest lies in argumentative composition. For this he replaces the six-part classical oration (introduction, narration, proposition, confirmation, refutation, conclusion) with a totally new oratorical structure: analysis and synthesis.

> The . . . processes, of *synthesis* ["beginning with more general and comprehensive propositions . . . descend to the particular propositions which are contained in them" (42)] and *analysis* ["proceed from particular observations to more general conclusions" (42)], are calculated either to demonstrate truth unknown to others, or to set one that is known in the strongest light; and when a person proposes to treat a subject fully . . . he cannot do better than to take one or other of those methods, according as the nature of the case will direct. (66)

---

38. Bevilacqua and Murphy, eds., introduction to Priestley, *Course of Lectures*, quoted by permission of the Southern Illinois University Press, xxi.

Although analysis is the method of discovery, while synthesis is that of exposition, the former is useful in argument "because . . . beginning with no principles or positions but what are common, and universally allowed, we may lead others sensibly, and without shocking their prejudices, to the right conclusion" (43).

Here Priestley's rhetorical theory may be supposed most influenced by his scientific interests. The division of philosophical argument into analysis and synthesis is commonly an approach of natural philosophy and the arguments and examples he uses to justify that division are typically derived from the sciences, although Locke, Hutcheson, Hume, Hartley, and Harris are the examples given prominence in the lectures. Throughout the first two parts of *Oratory and Criticism,* mathematics and science are used to exemplify terms and topics such as "universal propositions" (9), "consequents" (14), or "amplification":

> *Newton's Principia* is a remarkable instance to the present purpose. The demonstrations in that treatise are extremely concise, a great number of intermediate steps being omitted; and therefore but few, even of mathematicians, are capable of understanding it without comment. The commentary *amplifies,* by supplying the steps . . . suppressed by the author; and thus the book may be fitted for more general use. (27)

In the sections on Method, *all* proper arguments are shown to be analogous to scientific ones: "The form in which evidence is presented by *Euclid* . . . gains the readiest and most irresistible admission into the mind. . . . Such a successful method . . . certainly deserves the attention and imitation of all . . . desirous to promote . . . any kind of truth" (45). Newton's argumentative style, as illustrated in the *Opticks, Chronology,* and *Comment on Daniel,* offers many relevant examples, says Priestley quoting David Hartley, "and it is probable that his great abilities and practice in algebraic investigations led him to it insensibly" (57).

Possibly Priestley adopted this arrangement, especially the analogy between geometry and argument, from William Duncan's *Elements of Logic* (1748). Though he does not mention Duncan as a source, he repeats his arguments and might well have been introduced to his work by John Aikin, who taught logic at Warrington and had been an associate of Duncan while at Aberdeen. But Duncan was not responsible for Priestley's making analysis and synthesis stand for the whole of "disposition" in oratory and, doing so, went beyond all other "new rhetoricians" of the century.

> Thus he became the first rhetorician of his time to establish a new theory of rhetorical structure. And his *Lectures on Oratory* are the only eighteenth-century rhetorical work to which . . . [to] turn . . . to argue that the structure of . . . the Declaration of Independence, is authentically oratorical. . . . [This structural principal] became the official property of rhetoric when Priestley borrowed it from logic and made it stand where the doctrine of six-part oration had stood since the days of Cicero.[39]

The twenty-five lectures, "Of Style," are prefixed by an admonition: Let the first, and principal view of every orator, whether in writing or speaking, be to *inform the judgment*, and thereby *direct the practice;* and let him only attempt to please, or affect when it is subservient to that design.[40] This is not an indication that Priestley thought "Style" of less significance to rhetoric than "Recollection" or "Method," but it is an intimation of the practical role he expected it to play. He would not deny that style had a purely ornamental function, but it had also a utilitarian one. The "bare" matter of a speech or composition was "adapted to do little more than make an impression upon those persons who, of themselves, and from a regard to the nature and importance of the subject, will give their attention to it" (71). By style, one could attract attention and consent to an argument, independent of or supplemental to its matter, from people who might otherwise not attend.

Priestley's attitude is scarcely new. Watts's *Logic* would early have informed him that language was more than a communication of ideas. It treats such elements of style as metaphor and figures of speech as designed to represent ideas "with vivacity, Spirit, Affection, and Power . . . to move, and persuade."[41] Priestley's approach, however, was to emphasize the working of the mind to which the appeal is made rather than properties supposed innate in the stylistic elements of the appeal. Although the subjects he discusses are those of contemporary writers on aesthetics and style, his explanations, in terms of David Hartley's association of ideas, leads

39. Howell, *Rhetoric*, 661, 641–42.
40. Actually the last paragraph of the part of *Oratory and Criticism* immediately preceding the beginning of part 3, "Of Style," 69.
41. See Chapter I.

him to a quasi-relativism regarding standards of taste at odds with most contemporary thought and suggestive of Archibald Alison's later romanticism.[42]

John Aikin had taught belles lettres at Warrington before Priestley's arrival and his example would surely have had some effect on Priestley's teaching. Exactly how Aikin taught is unclear, but he was described as having "a rooted passion" for literature, particularly ancient literature, a familiarity with drama, and an "early taste for poetry." These characteristics, combined with the general description of his teaching at Warrington, suggest that he would treat criticism by means of literary appreciation, with emphasis upon the beauty of passages, in themselves and in relation to ancient models.[43] Priestley's treatment was more managerial, but surely some of his aptness in literary illustration—from Homer, Virgil, Shakespeare, Milton, the Bible, Addison, Pope, and so forth—was owing to Aikin's suggestions.

When Priestley addressed the composing of formal lectures on criticism, the principal authorities on style were Alexander Gerard's *Essay on Taste* (1759) and Henry Home, Lord Kames's *Elements of Criticism* (1762). They were the major influences on part 3 of his *Oratory and Criticism*. His concept of "vivid representation," for example, is drawn "virtually intact" from that of Lord Kames, while the discussion of the operations of the imagination "is little more than a patchwork of debts and amendments to the view of Gerard and Kames."[44] Gerard and Kames, however, were neoclassicists in their views of taste and drew their psychological justifications from the "faculties" psychology of Scottish Common Sense Philosophy, in which there were separate, irreducible, inherent attributes of the mind and sensations and values to appeal to each of them.

Priestley treats the common principles and stylistic devices of belles lettres: novelty, uniformity, sublimity, grandeur, comparison, contrast, meta-

42. See, for example, Samuel H. Monk, *The Sublime: A Study of Critical Theories in XVIIIth-Century England* (New York: Modern Language Association, 1935), 117–19; Andrew C. Smith, *Theories of the Nature and Standard of Taste in England, 1700–1790* (Chicago: University of Chicago Libraries, 1933), 130–51.

43. See Turner, *Warrington Academy*, 13, 16. That Aikin could communicate his taste and love of literature is seen in the accomplishments of his family. His daughter, Anna Laetitia (Aikin) Barbauld, son, Dr. John Aikin, and grandchildren, Lucy and Arthur Aikin, were together responsible for more than fifty publications, ranging from mineralogy and physic to politics and theology, but chiefly relating to literature: memoirs, poetry, songs, editions of British authors, and children's stories.

44. Bevilacqua and Murphy, eds., introduction to Priestley's *Course of Lectures*, xl, xlvi.

phor, allegory, antithesis, hyperbole, personification; but he does not address them only as traditional elements to be described and illustrated. Nor does he attribute their influence to separate senses: of fitness, morality, or awe, for example. He divides the elements of style into aspects that act on the passions (to win interest) on the judgment (to win assent) and on the imagination (to win admiration). Each of these aspects is described, illustrated, and explained in terms of the principles of Hartley's associationism. That is, he attempts to "lay open the sources of all the pleasures we receive from this most refined art, explaining what are the properties, or principles, in our frame which lay the mind open to its influences" (72).

The first two of these seem comparatively simple; illustrations are easy to find and practical precepts follow directly. The passions, for example, being blind and mechanical principles, are excited whenever there is the appearance of relevant circumstances. "*Vivid ideas and strong emotions . . .* having been . . . associated with reality, it is easy to imagine that, upon the perception of the proper feelings, the associated idea of reality will likewise recur" (89). Hence one might use the present tense to heighten belief in accounts of past events or describe particular circumstances in detail with sensible images, avoiding general and abstract terms. As "we are, in all cases, more disposed to give assent to any proposition, if we perceive that the person who contends for it . . . believes himself" (109), a discourse will gain readier assent if the manner of its presentation be natural to a person greatly in earnest; gestures, air of countenance, and whole manner should correspond to give the appearance of conviction, candor, spontaneity, and mastery of the subject (114–15).

When he comes to discuss "pleasures of the imagination," Priestley is in greater difficulties. It was one of the most debated aesthetic topics of the eighteenth century; yet no consensus had been reached. Its problems were subtle and for them he could offer little personal insight. More than half the lectures on style (more than 150 pages) are devoted to aspects of imaginative pleasures and their explanation, with Gerard and Kames providing most of the matter and Hartley the reason. In general, the results are unexceptionable: " . . . the principle sources of pleasure which enter into the works of genius and imagination . . . [are such as] must either *draw out and exercise our faculties* or else, by the principle of association, must *transfer from foreign objects ideas that tend to improve the sense*" (279). An example of the former, is "the transposition of words and clauses from their natural order, and occasional parenthesis" (143), which will sometime have a good effect as it produces moderate surprise and effort of the will. For the latter, the use of metonymy: " . . . in all cases, provided the sense be in no danger of being mistaken, the writer is at liberty to substitute, instead of a proper term, any word which, by its association can bring along with it ideas that can serve to heighten and improve the sentiment" (238).

There are some things in his discussions that seem characteristically Priestley. His discussion of sublimity, for example, is drawn almost entirely from Gerard, but the choice of science as an illustration has the Priestley touch: "The sublime of science consists in general and comprehensive theorems, which, by means of very great and extensive consequences, present the idea of *vastness* to the mind. A person of true taste may perceive instances of genuine sublime in geometry, and even in algebra; and the sciences of natural philosophy and astronomy, exhibit the noblest fields of the sublime that the mind was ever introduced to" (157).

And who else could write so seriously about humor and its usages, devoting some thirty pages to wit, the risible, ridiculous, burlesque, parody, mock-heroic, irony, riddles, and puns without an indication that he had ever found anything really funny and without an example bringing the reader to laughter? "When we are advanced in life, a variety of passions, and a regard to decorum, check the propensity to laughter; whereas many idiots continue to laugh upon the slightest occasion imaginable" (200).

It is also in the section on the imagination that Priestley makes what might, potentially, have been his greatest contribution to aesthetic theory. So long as beauty or excellence was thought to inhere in the admired objects or compositions, the philosophical problem of taste was the determination of those characteristics by which one object was naturally to be preferred to another. Even when the nature of good taste was, to some degree, transferred to the perceiving mind, if the mind were possessed of innate senses of beauty, proportion, and so forth, by which objects were judged, then standards of taste were immutable and transcended experience though some poor individuals might have less active senses of taste than others. Robert Boyle, John Locke, and Isaac Watts had each indicated belief that changes of taste were inevitable with changes of time or culture. David Hartley had been specific: " . . . there must be great Differences in the Tastes and Judgments of different Persons; and . . . no Age, Nation, Class of Men, &c. ought to be made the test of what is most excellent in Artificial Beauty; nor consequently of what is absurd."[45]

45. Hartley, *Observations on Man*, 1:442. Robert Boyle, in his *Some Considerations Touching the Style of the Holy Scriptures* (1661), which is likely to have been included in Priestley's reading sometime before the *Oratory and Criticism* was published, refers to changes

But none of these, including Hartley, accepted a complete relativism of taste and few of the other writers, involved in the eighteenth-century drift toward subjectivism, were as prepared as Priestley to declare that absolute and universal standards for taste were unknown.

> Had all minds the very same degree of sensibility, that is, were they equally affected by the same impressions, and were we all exposed to the same influences, through the whole course of our lives, there would be no room for the least diversity of taste among mankind. For in those circumstances, we should all have associated precisely the same ideas and sensations with the same objects, and the same properties of those objects; and we should feel those sentiments in the same degree. (133–34).

As this cannot be true, different persons at different times and places will have different associations with the same objects and conceive different sentiments of them. There is, today, enough similarity in situation, to afford a foundation for *similarity in taste*, but " . . . a *standard of taste* . . . cannot be applied to those persons whose education and manner of life have been very different." Moderns will, thus, not be sympathetic toward admired compositions of the past, what is admired in the East will not be so in Europe, what is thought admirable in France will not meet the same admiration in England (134).

Conforming to his generally idealist philosophy of theology, science, and language, this did not mean to Priestley that a perfect standard of taste was unattainable. It certainly did not mean that everyone's tastes were equally good. It did mean that one could not yet choose, on the basis of any available criteria, and particularly not on so vague and undetermined a business as ascertaining the opinion of some select body of connoisseurs, what that standard might be. The justness of taste *was* best determined by those most conversant with the objects of it, but there was no special selectivity of sensibility: " . . . all the principles of taste in works of genius . . . are within the reach of all persons whatsoever. . . . Nothing can be requisite to the acquisition of taste, but exposing the mind to a situation in which those associated ideas [of the work of genius] will be frequently presented to it" (74–75). Here, as in the establishment of good usage in language, the

---

in the standards of eloquence with differences in customs over time or region; cited by Howell, *Rhetoric*, 471. See also Martin Kallich, "The Association of Ideas and Critical Theory in XVIII-century England" (Ph.D. diss., Johns Hopkins University, 1945), passim.

test seems to be the marketplace of cultural competition: " . . . we may expect that, in consequence of the growing intercourse between all the nations of the earth, and all the *literati* of them, an uniform and perfect standard of taste will at length be established over the whole world" (135).

Had Priestley's *Course of Lectures on Oratory and Criticism* been published when the lectures were first delivered, it might well have had as substantial an influence as his works on language and grammar. The fifteen-year delay in its appearance allowed George Cambell's *Philosophy of Rhetoric* (1776) to take the lead in placing rhetoric firmly on a base of human nature. Of course there were also conservative elements to oppose the philosophical tendencies of Priestley's utilitarian and managerial aesthetic. When William Enfield, the editor of a manual of elocutionary exercises, *The Speaker* (1774) and tutor in belles lettres at Warrington following Priestley, reviewed *Oratory and Criticism* for the *Monthly Review*, he noted the industry and ingenuity of the performance, quotes at length from a chapter on imagination, illustrating the use of Hartley, then doubts the "propriety of resting . . . criticism, on . . . metaphysical speculation, upon the powers and principles of the human mind," and concluded that "the art of criticism will be more successfully as well as more agreeably taught, by lively exhibitions of the beauties and defects of compositions than by profound speculations on the powers and operations of the human mind."[46]

The most influential book on rhetoric of the period was Hugh Blair's *Lectures on Rhetoric and Belles Lettres* (1783), with numerous editions to as late as 1870. Blair was conservative in his approach and did not contribute significantly to rhetoric theory. Though he quotes Priestley, it was on grammar and not on oratory. There is some indication that *Oratory and Criticism* was known in the United States, and the third edition of John Walker's *Rhetorical Grammar* (London, 1801) excerpts nearly three pages from Priestley's Lecture 4, on invention, but Priestley's rhetorical views were, on the whole, essentially without influence.[47]

Even in the connection between association of ideas and imagination, it was Archibald Alison's *Essays on the Nature and Principles of Taste* (1790),

46. [William Enfield], review of "A Course of Lectures on Oratory and Criticism by Joseph Priestley," *Monthly Review* 57 (for August 1777): 89–98.

47. Contemporary copies of *Oratory and Criticism* are to be found in the libraries of Harvard, Yale, and the American Philosophical Society while Brown University ordered a copy from England in 1783; see Warren Guthrie, "The Development of Rhetorical Theory in America, 1635–1858," *Speech Monographs* 14 (1947): 38–54; also W. P. Sandford, *Public Address*, 117–36.

backed by the influence of Francis Jeffrey and the *Edinburgh Review,* that dominated attention. Though Alison's aesthetic theory was avowedly associationist, his book does not refer to Hartley nor to Priestley, whose *Oratory and Criticism* had developed many of the same themes at least a dozen years earlier.[48] William Wordsworth's debt to association theory for his concept of poetic invention is generally acknowledged and credited to the influence of Alison's *Essays.* One study of Wordsworth's preface to the *Lyrical Ballads* does, however, find frequent cause to refer to Priestley's *Oratory and Criticism* and another literary scholar imagines that Wordsworth read Hartley "in Priestley's second edition printed by J. Johnson, in whose shop Wordsworth foregathered with his malcontent friends."[49]

Nevertheless, so far as Warrington Academy was concerned, it can scarcely be argued that Priestley's lectures on oratory and criticism rendered less than full value to his students. In them, as in his lectures on English grammar and on language and universal grammar, he more than fulfilled the function for which he had been employed: to provide the students of this most utilitarian and middle-class of eighteenth-century English schools, two of the three parts of the *trivium,* the standard liberal arts curriculum of the schools since the Middle Ages. Academy trustees were not unappreciative of his efforts, nor of the unexpected public-relations treasure they had found in their young tutor. In 1764 they set about to reward him— and to get the fullest advantage from his penchant for scholarship. Hugh, Lord Willoughby of Parham, first president of Warrington's trustees, and Dr. Samuel Chandler, minister of Old Jewry, London, prepared a recommendation to the University of Edinburgh that Priestley was a "person

---

48. I have read Archibald Alison, *Essays . . . , with corrections and improvements* (New York: G. and C. and H. Carvill, 1830), to find that he treats his entire system as essentially new with him, owing something, perhaps, to such acceptable authorities as Hutcheson, Reid, and Beattie. Jeffrey reviewed the second edition of Alison's *Essays* for the *Edinburgh Review* 18 (1811): 1–46, and expanded his review for a long article in the *Encyclopedia Britannica.*

49. W. J. B. Owen, *Wordsworth's Preface to Lyrical Ballads* (Copenhagen: Rosenkilde and Bagger, vol. 9 of *Anglistica,* 1957) cites Priestley's *Oratory and Criticism* and the *Lectures* on Language and Universal Grammar; indeed there are more references to Priestley than to any other persons save Coleridge, Hartley (same number), Samuel Johnson, and Dorothy Wordsworth. Willard L. Sperry, *Wordsworth's Anti-Climax* (Cambridge: Harvard University Press, 1935), 126–27, speaks of Wordsworth's reading of Hartley and quotes his letter to Richard Sharp, in 1808, describing Hartley as among the men of real power, who go before their age, and exclaims of the *Observations on Man,* "How many years did it sleep in almost entire oblivion." Yet it was Priestley who brought Hartley to the critical attention of the late eighteenth century and rescued the *Observations* from their oblivion. See also, Kallich, "Association of Ideas," 223–24.

worthy of receiving a degree of Doctor of Laws." That recommendation was endorsed and forwarded to Principal Robertson of Edinburgh by Thomas Percival, son of a Warrington trustee and its first enrolled student, who had become Robertson's friend while studying medicine at Edinburgh. On 4 December 1764, the Faculty in Law having concurred, the *Senatus Academicus* responded favorably and Joseph Priestley became, henceforth, Doctor Priestley.[50]

One of Priestley's reasons for going to Warrington from Nantwich had been a desire to extend his "connexions"; now that extension had been formally recognized in his acquisition of the LL.D. His next, and more important accomplishment, for Warrington Academy, himself, and ultimately the history of education, was to transcend the limitations of his formal academy appointment in languages and belles lettres and, going beyond grammar, rhetoric, and logic as the basic school curriculum, redefine what would henceforth be meant by a liberal arts education.

50. This differs from Priestley's account, "After the publication of my 'Chart of Biography' Dr. Percival of Manchester, then a student at Edinburgh procured me the title of Doctor of Laws from that University" (*W* 1/1:58), but Priestley's memory was at fault. The *Chart of Biography* has a colophon indicating publication "according to Act of Parliament, Feb. 2, 1765"; although it had been engraved earlier, for its dedication to Lord Willoughby remained though his lordship had died on 21 January 1765. The date of Priestley's degree is given in MS Da 31.5, Minute Books, *Senatus Academicus*, 1:156–57, Edinburgh University Library; cited by permission, University Archivist. Willoughby's and Chandler's involvement is stated in Priestley's letter to Richard Price, 8 March 1766, no. 3, *SciAuto*, and Percival's role confirmed in Dr. William Robertson's letter to Dr. Percival, 8 March 1774, *The Works, Literary, Moral, and Medical of Thomas Percival, M.D.* (London: J. Johnson, 1807), 1:xl. Willoughby's recommendation was probably his only service to the academy since accepting presidency of its trustees in 1757, but he was, in addition to being England's only Presbyterian peer, also a trustee of the British Museum, first president of the Society of Antiquaries, vice president of the Royal Society, where he exerted influence to get Percival elected fellow. Samuel Chandler had been a friend of Doddridge, was a fellow of the Society of Antiquaries and the Royal Society and held D.D. degrees from Edinburgh and Glasgow.

V

WARRINGTON ACADEMY, 1761–1767

Liberal Education, History, Biography

The first proposals issued for prospective subscribers to Warrington Academy had declared: "It is now become a general and just Complaint that some Publick Provision is wanted for the education of young gentlemen designed either for the learned professions *or for Business*."[1] A principal attraction of John Holt, as tutor, had been his years of conducting a commercial school. Early annual reports of Warrington trustees emphasize availability of instruction in business: "Merchants-accounts are . . . taught; and those who are intended for Business will be properly exercised in Writing." Lectures on chemistry, "Principally of the Commercial Kind," are planned. A writing-master has been provided, "practised in book-keeping, and . . . well acquainted with the forms of business in merchants counting-houses."[2]

Then, suddenly, in 1766, without dropping references to commercial courses, the emphasis of the trustees' report changed. "The great and primary object of the institution was the education of youth in general; as well those who are *intended for civil, active and commercial life*, as of such as are designed for the learned professions." There is an added reference to a course of studies "distinct from that, which is adapted to the learned profes-

---

1. Quoted by McLaughlin, "Warrington Academy," 11; my emphasis.
2. *A Report of the State of the Warrington Academy, By the Trustees at their annual Meeting July 1st, MDCCLXII (June 30, MDCCLXIII)*, 3 (1).

sions . . . new articles of academical instruction, such as have a nearer and more evident connection with the business of active life."[3] The cause of this change of emphasis was Joseph Priestley; more explicitly it was his publication the previous year of *An Essay on a Course of Liberal Education for Civil and Active Life,* dedicated to the "President and . . . the Trustees of the Academy at Warrington."[4]

This was one of the most frequently published of Priestley's works. It was reprinted separately, appended to early editions of his *Miscellaneous Observations Relating to Education* and then moved to prefix *Lectures on History and General Policy,* for at least sixteen printings.[5] And it, along with the *Miscellaneous Observations,* has provided the basis of claims that Priestley was the most considerable English writer on educational philosophy in the period between Locke's *Some Thoughts Concerning Education* (1693) and Herbert Spencer's "What Knowledge is of the most Worth" (*Westminster Review* [1859]), reprinted as the introductory chapter of his *Education* (1861). It has even been claimed that Spencer derived his most striking and original observations from Priestley.[6] There seems no proof for the claim, but there are parallels in ideas and similarities in many of their basic opinions, such as the proper relationships between church, state, and education suggesting that Spencer read Priestley.

The major thrust of the first part of the *Essay* is critical: There is no "proper course of studies . . . provided for Gentlemen . . . designed to fill the principal stations of active life, distinct from those . . . adapted to the

---

3. *A Report . . . June 26th MDCCLXVI,* 1, 3; my emphasis.

4. Joseph Priestley, *An Essay on a Course of Liberal Education for Civil and Active Life. With Plans of Lectures on: I. The Study of History and general Policy. II. The History of England. III. The Constitution and Laws of England. To which are added, Remarks on a Code of Education, proposed by Dr. Brown, in a late Treatise, entitled, Thoughts on Civil Liberty, &c.* ([London]: C. Henderson, T. Becket and DeHondt, J. Johnson and Davenport, 1765).

5. Separately in 1765, 1768, 1778, 1793; appended to *Misc. Observ.* (minus *Remarks on . . . Brown*), London, 1778, Birmingham, 1778, Dublin, 1780; shorn also of *Plans of Lectures* and prefixed to *Lectures on History,* Birmingham, 1788, Dublin (1 and 2), 1788, Dublin, 1791, 1793, United States, 1803; Rutt separate, 1826, and translated into Dutch, 1793 and French, 1798.

6. H. G. Good, "The Sources of Spencer's Education," *Journal of Educational Research* 13 (1926): 325–35; H. M. Knox, "Joseph Priestley's Contribution to Educational Thought," *Studies in Education. The Journal of the Institute of Education* 1 (1949): 82–89. The value of this general estimate of Priestley's importance as an educator is reduced by the consideration that historians of education rarely mention *any* English writer between Locke and Spencer, but McLachlan, "Warrington Academy," 28, quotes Kenneth Lindsay's *English Education,* 24, as joining Priestley with T. H. Huxley, Thomas Arnold, and Albert Mansbridge in his praises of their contributions to English educational ideas.

learned professions" (1). This might once have been acceptable, but knowledge of the "true sources of wealth, power and happiness, in a nation" is now available and unless persons who have influence in national affairs are better educated, England will fall behind other nations (3, 4). The state and subjects of learning common in schools and universities is everywhere ridiculed: ". . . necessity . . . has . . . forced a change, and . . . increasing necessity will either force a greater and more general change, or we must not be surprised to find our schools, academies, and universities deserted; as wholly unfit to qualify men to appear with advantage in the present age" (23–24).

These criticisms were scarcely new (and, indeed, are repeated nearly every academic generation); Milton complained of the failure of schools and universities to meet new conditions of life, in his "Of Education" (1644), Locke's *Some Thoughts Concerning Education* recommends attention to "real knowledge" instead of the subjects of the schools, and criticism of school and university is repeated throughout the century.[7] This part of Priestley's *Essay* is a summary of liberal education criticism, bringing together what was being said, in scattered parts, elsewhere. The claim that Priestley "did more perhaps than any other man to modernize the curriculum of the dissenting academies" rests on his ability to make some immediate recommendations for educational practice.[8]

These recommendations appear in later sections of the *Essay* and in Priestley's *Miscellaneous Observations relating to Education*.[9] The latter was not published until Priestley was years away from Warrington and had seen more of the world, and even of education, in the loftier circles of Lord Shelburne's household. He notes that though it was written "at different times, as particular occasions suggested" (xi), liberal education had "been the business of a great part of my life to study and to conduct" (v–vi), whereas many other writers on the subject "appear never to have had much

7. See John W. Adamson, "Education," chap. 15 in *From Steele and Addison to Pope and Swift*, vol. 9 of *The Cambridge History of English Literature*, ed. A. W. Ward and A. R. Waller (Cambridge: Cambridge University Press, 1933), 425–62.

8. The quotation is that of Frederick Elby and Charles Flinn Arrowood, *The Development of Modern Education* (New York: Prentice Hall, 1934), 606, but the sentiment is a common one; see, for example, J. W. Ashley Smith, *Birth of Modern Education*, 152–59. As curriculum modernization at the universities lagged that of the Dissenting academies, often by decades, the claim is not unimportant.

9. Joseph Priestley, *Miscellaneous Observations Relating to Education. More Especially, as it respects the Conduct of the Mind. To which is added, An Essay on a Course of Liberal Education* . . . (Bath: for J. Johnson, 1778); reprint, Birmingham and London, 1778, Cork, 1780, Birmingham, 1788, and United States, 1796, as well as the Rutt collected ed., 1831.

to do with the *conduct* of it and to have given little attention to its real influence" (xii). This explicit connecting of his *Miscellaneous Observations* with his teaching experience, and its frequent association, in publication, with the *Essay,* justifies joining the two here in a single summary of Priestley's educational practice and curricular recommendations.

His general discussions of educational philosophy are marked by common sense, usefulness, and, inevitably, by the Hartleyan associationism that provided the frame for much of his thinking: the most decisive and lasting education is that which is "natural"; that is, directly derived from personal experience. Experience, however, is a hard and slow teacher; we try by "artificial" education to anticipate nature, communicating knowledge sooner than experience and more easily because in a more regular order (*Misc. Obs.* 2–3). By experience a man learns the best way of doing a particular task in a particular context, but a man used to only one way of doing things and given no idea of others, "will be wholly at a loss when it happens that that track can no longer be used," while a person given the principles of action can strike out differently (*Essay* 20–21).

At the age of sixteen or seventeen, students are as capable of study in any subject as they ever will be (*Essay* 14). All that can be expected of formal education is a smattering of a subject. Education is only finished by continued application after leaving school. But it is better to acquire early, partial knowledge of principles when they make their deepest impression, than to have no regular theory at all (*Essay* 15). "All real ability might have been applied *originally* with equal success to one pursuit as to another, and where two objects of pursuit have a great resemblance, the application to one of them may prepare the mind for applying to the other with advantage." But one must not expect training the mind in one area to transform to another. "In fact, ingenuity and address in one thing has very little proper connection with that in another" (*Misc. Obs.* 39).

For the teacher, Priestley has practical recommendations based upon experience: A lecturer should have an organized set of notes, including the principal arguments he will use and the facts that support them. These notes become the text for class meetings, of twenty to thirty students, for no longer than an hour. Sessions should begin with questions to and from students, to assure that they understood the subject of the preceding lecture before going on to another. Students should be able to read the text (and even to copy it). References should be made to principal authorities, especially those on both sides of any controversial topic. An account of these references, and perhaps a written abstract of them, should occasionally be required. Discussion is to be encouraged; queries, objections, and remarks welcomed. "A Tutor must be conscious of his having made very ridiculous pretensions, and having given himself improper airs, if it give him any pain to tell his class, that he will reconsider a subject; or even to acknowledge himself mistaken. . . . Every tutor ought to have considered the subjects on which he gives lectures with attention, but no man can be expected to be infallible" (*Essay* 31).

It was in curricular formation, however, that Priestley made his major contribution: "The chief object of education is not to form a shining and popular character but an useful one . . . to render a man happy in himself and useful to others" (*Misc. Obs.* xii–xiv). A first consideration, then, is a person's religious education and, second, the arts needed for subsistence. Priestley does not concern himself with the education of laborers, persons of "mechanical employment" who do not need a liberal education. He was concerned with the education of women, such as clergymen's daughters, "whose families must live genteely," with few prospects when their parents die except servile dependence upon others. "On this account, whatever [such] parents are able to provide should be disposed of in favour of daughters in preference to sons," but better, daughters should be "taught such things as women can maintain themselves by doing" (*Misc. Obs.* 137–38).

His primary focus was on the young men attending academies and universities, able to afford a liberal education and not intending to spend their lives in the mere mechanical parts of their businesses. Their preparation should begin with languages—English, of course, and then some knowledge of the learned languages is preferable, though not necessary. Of these, Latin is best, for it gives access to classical literature and enables one to recognize English words derived from it. French should then be studied and the student should become acquainted with the more useful branches of practical mathematics and, if possible, with some algebra and geometry (*Essay* 18–19): ". . . the learning of one language, and the comparing it with another, is a very useful exercise, and is an excellent introduction to that most important distinction of ideas which are expressed by words" (*Misc. Obs.* 44–45). "Great excellence in any of the *elegant arts* is an unfavourable circumstance to youth, and except they be intended to exercise those arts, as a profession, a mediocrity is more desireable" (*Misc. Obs.* 57).

In the *Essay*, Priestley directs his attention to persons likely to become magistrates and legislators, lawyers beyond practicing attorneys, military men at higher ranks of preferment, merchants "beyond the servile drudgery of the warehouse or counting house" (9). In *Miscellaneous Observations,* he considers also persons of landed property or fortune; for these, study of land cultivation "in the most perfect and ornamental manner" is the first object and to this is related "the study of nature, including . . . *natural history* and *natural philosophy*" (15). This is not merely a study of immediate and personal practical concern; it is the "most liberal, most honourable, the happiest, and what will probably be the most successful employment" for persons of fortune. It is also one which they owe to society. "Since others till the ground and do all the drudgery of life for them, they ought, in return, to employ their time and fortune for the common benefit" (19). But for these, as for those persons of the *Essay*, the primary recommendation is a study of modern history, civil policy, the constitution and laws of England—subjects no more beyond the capacity of students to learn than are logic, metaphysics, or algebra to which they are now directed (*Essay* 13–14).

As it was in his emphasis on modern history that Priestley most transformed academy curricula, it is essential to note that he did not intend, in the study of history, a tracing of evolutionary developments of society. In society and government, as in language and criticism, Priestley believed in progress, but this was not a dialectical process of society always in a state of becoming. Priestley's history was not the history or the historicism of the nineteenth century. Nor, however, was it simply a continuation of the "exemplary" form that had dominated historical theory from the days of Cicero, Livy, Quintilian, and Tacitus.[10] In the eighteenth-century battle of ancients and moderns, the combatants had agreed about the purposes of history, but there was profound disagreement about its method. Priestley was on the side of the "moderns"; he subscribed to the view that history was "philosophy teaching by example."[11] His *Lectures on History* declared: ". . . real history resembles the experiments made by the air pump, the condensing engine, or electrical machine, which exhibit the operations of nature, and the God of nature himself . . . and are the ground work and materials of the most extensive and useful theories."[12]

---

10. See George H. Nadel, "Philosophy of History before Historicism," *History and Theory* 3 (1963–64): 291–315.

11. The modern source of that definition, originally credited to Dionysius of Halicarnassus, appears to have been Lord Bolingbroke's *Letters on the Study and Use of History,* to which Priestley refers in his *Lectures on History and General Policy* (Birmingham: for J. Johnson, 1788), for example, 6, 7, 10, 253.

12. Joseph Priestley, *Lectures on History and General Policy,* 5.

As these *Lectures* were not published for nearly a quarter of a century, only Warrington's students would hear anything so specific from Priestley for many years, but the implications are strong in his *Essay* of 1765, where he commends the study of "whatever may be demonstrated from history to have contributed to the flourishing state of nations, to rendering a people happy and populous at home and formidable abroad" to aid the subjection of history "to the highest uses to which it can be applied; to contribute to its forming the able statesman, and the intelligent and useful citizen" (*Essay* 10, 11). History, that is, was to become a kind of natural philosophy, or science, of society. It was not to be confined to the "ancients'" moral didacticism, related to rhetoric and generally taught in the schools as a part of literature. Priestley was of the new, analytical, school of history involving, in its method, philology, antiquities, government and ecclesiastical records, anything that would further an investigation into the entire life of the past.[13] By collecting data of past lives and events, one could determine more accurately the laws of society and knowledge of those laws would as naturally lead to progress as did knowledge of physical nature.

It was in this spirit that Priestley included in his *Essay* syllabi of courses in modern history, English history, and law. He designed his syllabus for *Lectures on the Study of History* as an introduction to historical literature and method. His list of topics necessary for the study of English history included the arts, language, education, letters, food, dress, manners, sentiments and diversions, manufactures, trade, public works and transportation, cities and housing, as well as politics, religion, war and treaties. He recommended study of commercial geography and chemistry: "absolutely necessary to the extension of this useful branch of science [i.e., geography]." And his syllabus on the constitution and laws of England suggests a practical, but not professional, course organized with historical insight. The courses represented by Priestley's syllabi would clearly constitute a better "liberal" program for studies for civil and active life than the remnants of the trivium that it replaced; even recognizing that grammar, rhetoric and logic, in the medieval schools, might well encompass aspects of history, government, and law.

Priestley's new courses were adopted at Warrington and parts of them were retained after he left and until Warrington Academy folded in 1786.

---

13. For a perceptive treatment of history as viewed by "ancients and moderns," see Joseph M. Levine, "Ancients and Moderns Reconsidered," *Eighteenth-Century Studies* 15 (1981): 72–89.

The program seems also to have been installed at Hackney New College (1786–96), where Priestley lectured from 1791 to 1794. Some of it can also be found in the curricula of a few other Dissenting academies (e.g., Northampton, Manchester) where Priestley's liberal influence may be inferred, but its full implementation, prior to 1788, would have been difficult. The *Essay* had contained more or less complete syllabi for the three courses of the program: 63 lecture-topics for the "Study of History," 50 subject-divisions for each chronological section of the history of England, and 64 lecture-heads on the laws of England. Not until publication of the *Lectures on History and General Policy* (1788), however, was there a full text for each of the three, and only one of these was Priestley's.

In the preface to the *Lectures,* he explained that publication of William Blackstone's *Commentaries on the Laws of England* (1765) and Francis Sullivan's *Historical Treatise on the Feudal Law and the Constitution and Laws of England* (1772) had made unnecessary the publishing of his lectures on English law, while Robert Henry's *History of Great Britain* (1771–93) had done the same for the lectures on English history.[14] He does not say that he had never intended to publish either of those lectures, nor does he explain the long delay in appearance of the *Lectures on History* which he had intended for publication from the beginning.[15] When it finally was published, although the lectures on how to do history probably remained much the same as when originally delivered, those on the uses of history in defining general policy (roughly half the whole) had surely become substantially more sophisticated during the course of twenty-three additional years of active public life.

The *Lectures on History* was then the only published text that supplemented the syllabi of the *Essay* and that did not appear for years. While still at Warrington, Priestley did, however, provide a few aids to help in

14. Joseph Priestley, *Lectures on History and General Policy,* vi. Henry's *History of Great Britain, from the first invasion of it by the Romans under Julius Caesar, written on a New Plan* was published in five volumes from 1771 to 1785, and completed, posthumously, in a sixth volume in 1793. Thomas P. Peardon, *Transition in English Historical Writing, 1760–1830,* Studies in History, Economics and Public Law no. 390 (New York: Columbia University Press, 1933), 58–59, describes Henry's "new plan" as "almost exactly" that outlined in Priestley's 1765 description of his course on the history of England.

15. A Warrington student recalled reading manuscript lectures in history which he was not allowed to copy as he could others, for Priestley "intended them for publication" (*W* 1/ 1:50n). The *Syllabus of a Course of Lectures on the Study of History* was separately printed (Warrington: William Eyres, 1765). The preface to *An Essay on the First Principles of Government . . .* 2d ed. (London: J. Johnson, 1771), xvi, notes the intention to publish "in due time" his "Course of lectures on history and civil policy." Nothing was said about publishing the

the development of a "course of liberal education." One of these was an implicit demonstration of the uses of history in political argument, in his "Remarks on a Code of Education" appended to the *Essay.* Two others were the publishing of "mechanical methods . . . to facilitate the study of history": a *Chart of Biography* (1765) and a *New Chart of History* (1769), with their accompanying *Descriptions.*

Of these, the charts and descriptions were surely the more useful. Although the "Remarks" were thrice reprinted with the *Essay,* their obvious purpose was politico-theological, as their reprinting with the second edition of Priestley's *Essay on the First Principles of Government* (1771) makes clear. Few persons would have regarded the "Remarks" as aids to the construction or teaching of a history curriculum. The charts, on the other hand, were developments out of familiar practice.

Priestley mentions "Mechanical methods" as subjects for Lecture 17 of his *Syllabus on the Study of History,* specifically referring to chronological and genealogical tables. Isaac Watts had recommended such tables as more inclined to fix the subject in the mind than many rereadings, in his *Treatise on the Improvement of the Mind* (1741) and Philip Doddridge used the table of Francis Tallent's *A View of Universal History* (1681) at Northampton Academy. Probably, therefore, chronological tables continued to be used at Daventry and also, perhaps, a new "mechanical" aid, recently imported from France: *A Chart of Universal History* that replaced the tabular form by a format of vertical columns, divided horizontally by lines of various scales representing the passage of time.[16]

Certainly Priestley was aware of such a chart, and may well have used it when he commenced teaching at Nantwich, for he then began to design a similar chart, on biography, but with an improved plan. On his chart, time was represented by a single, horizontal scale, divided equally by perpendicular lines into centuries from 1200 B.C. to A.D. 1800. Five horizontal

other lectures, whose manuscripts were destroyed in the Birmingham Riots of 1791, see Priestley, *An Appeal to the Public on the Subject of the Riots in Birmingham* (Birmingham: by J. Thomson for J. Johnson, 1792), 38.

16. *A Chart of Universal History (Done from the French, with considerable improvements)* (London: Thomas Jeffreys, 1750). According to Samuel Miller, *A Brief Retrospect of the Eighteenth Century* (London: printed at New York, reprinted for J. Johnson, 1805), 2:359, the French designer of this chart was the Abbé Pierre Nicolas Lenglet du Fresnoy (1674–1755), who certainly wrote or compiled a number of aids to historical study, but is not listed as author of a chart in the holdings of major national libraries. The chart Priestley describes as "lately imported from France," *Lectures on History and General Policy,* Lecture 18, 153, in words clearly transcribed from the notes of 1764–65 may have been a later chart than that of London, 1750.

lines divided the width into sections representing classes of achievement (statesmen, divines, mathematicians, poets, and so forth). Within their appropriate classes were drawn horizontal lines representing the lengths of the lives of the most famous personages. The chart was incomplete when Priestley arrived in Warrington, but early in 1764 he enlisted the aid of academy trustees in obtaining information for it and by July 1764 it was being engraved for publication.[17]

About the same time a *Description* of the chart was printed and early in 1765 both were published: *A Chart of Biography* and *A Description of a Chart of Biography* to accompany the chart.[18] The *Chart* was roughly three feet long and two feet wide, impressed upon paper, backed and pasted onto linen. There were about two thousand persons represented on the *Chart*: famous people, but not necessarily meritorious; chosen with care, but not with singularity. "No two persons living would make the same choice. I will even venture to say . . . no one person would, at different times, make the same choice" (*Description* 21). Proportionally more Englishmen are represented, and more men of letters, not out of national or cultural prejudice, but to suit the needs of those most likely to use the chart. Space was left for purchasers to add other lines.

The sources used for dates were often unclear and frequently contradictory. Priestley usually selected a source that gave reasons for its dates and indicated, by dots or broken lines, when a date was uncertain. For his early

17. On 27 February 1764, Thomas Bentley, soon to be vice president of academy trustees, forwarded a letter from Priestley to the Reverend Samuel Pegge, an antiquary writing for the *Gentlemen's Magazine* under the pseudonym Paul Gemsage, asking for information on English antiquities. On 21 April 1764, Priestley wrote again to Pegge for specific dates for the *Chart*, about to be sent to the engraver. On 11 July 1764, he wrote to Thomas Birch, encouraged by Lord Willoughby, asking that some last few dates be supplied to the engraver, Mr. J. Mynde, and hoping Birch might get Lord Willoughby's acceptance of the chart's dedication; see Priestley to [Pegge], 26 February 1764; Bentley to Pegge, 27 February 1764; Priestley to Pegge, 21 April 1764, all MS Add. C 244, fols. 13 and 15, used by permission. Bodleian Library, Oxford; and Priestley to Birch, 11 July 1764, Birch Papers, Add. MSS, used by permission, British Library.

18. *A Chart of Biography . . . Published According to Act of Parliament, Feb. 2, 1765* by J. Johnson; *A Description of a Chart of Biography: with a Catalogue of all the Names inserted in it, and the Dates annexed to them* (Warrington: by William Eyres, 1765). I have used the second edition of the *Description* (Warrington: for the author, 1765). Both *Chart* and *Description* were, in first editions, dedicated to Ld. Willoughby. According to a note appended to a review (*Monthly Review* 32 [1765, misprinted 1764], 160) they had been printed some six months prior to their publication, too soon to delete the dedication at Willoughby's death. The title page of the *Description* was, however, printed after 4 December 1764, for it carries the LL.D. designation that Priestley was henceforth to use on every publication bearing his name.

dates, he adopted the chronology of Sir Isaac Newton, which was not accepted then (or now) by most chronologers and which cut about four hundred years from the accepted chronology of Greek history.[19]

The *Description of the Chart of Biography* was provided with each *Chart*, and also sold separately. It contained a general justification for the *Chart* and an alphabetical listing of all the names on it, with the dates more precisely indicated and more specific designation of the principal areas of achievement than could be given on a chart. Priestley had initially intended to give a short account of the actions of each of the persons represented, "as a reason for giving him a place in the chart," but the *Description* was already too large and he put it off, possibly to be published at a later time. That "Account of all the persons whose names are introduced into my Chart of Biography" was destroyed in the Birmingham Church-and-King Riots of 1791.[20]

The *Chart* and the *Description* filled a useful purpose, selling well throughout the rest of the century. This may partially be a result of Priestley's newly displayed talent for advertising. The *Essay on Liberal Education* mentions charts of biography and history, as subjects for discussion in lecture-topic 18 of the *Syllabus on the Study of History* and contains "A Specimen of a Chart of Biography" with a two-page "Short Account of a Chart of Biography." The specimen and short account were also appended, by the publisher, Joseph Johnson, to at least one book he published in 1766: David Jennings's *Jewish Antiquities*, while Priestley mentions the chart or includes samples of it in books as various as his *History of Electricity* (1767), *History of Optics* (1772), and *History of Early Opinions concerning Jesus Christ* (1786), as well, of course, as his *Lectures on History and General Policy* of 1788.

Surely, though, the reception given the *Chart* and *Description* indicates more than response to an advertising campaign, however persistent. The forthcoming chart was a manifest element in winning, for Priestley, his cherished LL.D. from Edinburgh and its publication clearly was a factor

---

19. For Newton's *Chronology of Ancient Kingdoms Amended* (1728), see Frank E. Manuel, *Isaac Newton, Historian* (Cambridge, MA: Harvard University Press, 1963), 166–93; and Richard S. Westfall, *Never at Rest: A Biography of Isaac Newton* (Cambridge: Cambridge University Press, 1980), esp. 805–8. Newton's chronology was part of a concealed argument for his Arian theology, but, though Priestley was, at this time, an Arian, there is no indication that he adopted Newton's dates for that reason. He says that Newton's was the "most rational system" and makes persons "contemporary upon . . . natural principles" who would have been separated by centuries by the "bulk of chronologers" (*Description*, 17).

20. See Priestley, *Appeal . . . on . . . the Riots in Birmingham*, 38.

in his nomination to become a Fellow of the Royal Society of London.[21] The original plates for the *Chart* were destroyed in a fire at Joseph Johnson's bookshop in 1769 and plates were promptly reengraved, with improvements, for a reimpression of the *Chart* by 1770. It was reprinted in 1805, with additions bringing it to the close of the eighteenth century, and there may well have been pirated printings and some versions in bound sheets, for the *Description* was printed many times in the United States and Europe as well as in England.

Indeed, the *Description of a Chart of Biography* went through at least nineteen editions, including four American, one Dutch, and one Italian, before its final appearance in 1840. The *Charts* of biography and of history were among the first acquisitions ordered for the new Library of Congress in 1800 and, as late as 1853 the form of biographical time-lines was used, with due acknowledgment to "the clear form which Priestley employed in his *Chart of Biography* nearly a century ago," by J. C. Poggendorff for a set of "Lebenslinien" charts for sixteenth-, seventeenth-, eighteenth-, and nineteenth-century scientists.[22]

Priestley's *New Chart of History* (1769) and *Description of a New Chart of History* (1769) were as successful as the biographical chart and its description. The *New Chart* was reprinted, with additions to the end of the eighteenth century, in 1805. The *New Description* went through at least twenty editions, including one Dutch, an American, an Italian, and an Austrian, by the time it was reprinted in the *Works*.[23]

Priestley had hoped that an improved form of the French chart might be "executed by the Proprietors of the former Chart." Failing in that hope,

21. See Priestley's Royal Society certificate, no. 4, *SciAuto*, 20, where his only specified publication is "a chart of Biography."

22. One American edition of the *Description* was *A Description of a Set of Charts of Biography* (Philadelphia: Matthew Carey, 1804). J. T. Rutt included the *Description* minus the catalog of names and dates, appended to volume 24 of the *Theological and Miscellaneous Works* (1826), separately published in 1840. For Library of Congress acquisitions, see Theodore Besterman, ed., *Publishing Firm of Cadell & Davies: Select Correspondence and Accounts, 1793–1836* (London: Oxford University Press, 1938), 123–29. J. C. Poggendorff, *Lebenslinien zur Geschichte der Exacten Wissenschaften seit Wiederherstellung Derselben* (Berlin: Alexander Duncker, 1855), v.

23. Joseph Priestley, *A New Chart of History . . . Engraved and published according to Act of Parliament, April 11th. 1769, by J. Johnson. A Description of a New Chart of History, containing a View of the principal Revolutions of Empire that have taken Place in the World* (London: J. Johnson and J. Payne, 1769); I have used the fifth edition (London: J. Johnson, 1781). Rutt's reprinting of the *Description of the New Chart*, from the 1816 ed., appears as an appendix, with the *Description of the Chart of Biography*, volume 24 of the *Works* (1826), separately issued in 1840.

and encouraged by the reception of the *Chart of Biography,* he was "at length . . . induced to undertake it himself" (*New Description* 5), though the task was not completed until after his leaving Warrington. The *Description of the New Chart* provides justification for it, in comparison with the French chart, whose general plan was excellent, but whose execution was faulty, with many mistakes both of composition and engraving. The *New Chart* was the same size as the *Chart of Biography* and, on it, time was again represented as "flowing laterally, like a river, and not falling in a perpendicular stream" (*New Description* 8) and the same equally divided scale of time was used as in the biographical chart.

Because this chart was to provide a "succinct view of all the revolutions of empire," the simple linear form of life-lines was inadequate. There were too many variables other than time to be represented. Horizontal bands now represent geographical areas or countries (not nations). Within these, strips of different widths and lengths represent kingdoms and empires of differing importances and durations. A horizontal line of dots indicated indistinct boundaries between contiguous empires; horizontal broken lines, a dependent state of one peoples on another.

A horizontal strip was broken off when one empire was absorbed by another, whose strip was then enlarged. Solid vertical lines between strips indicated land acquired by conquest, broken vertical lines, acquisition by peaceful means (i.e., marriage, voluntary cession, and so forth); dotted vertical lines indicated uncertainty of date. Because extensive empires could not always be represented by contiguous spaces, divided parts of empire were colored. Only four colors were used, but the same color could represent different empires when sufficiently separated in time.

Priestley intended to show, on his chart, "the history of all nations that have ever made any figure in the world, the exact dates of the rise and fall of every considerable state" (*New Description* 6). Some one hundred and five nations, peoples, or countries are, in fact, represented on the *New Chart* and cited in the "View of the Principal Revolutions of Empire" of its *Description.* Sequences of state are indicated, baldly and without comment, in the "View," but the introduction to the *New Description* contains Priestley's characteristic didactic religious moral.

On examining the *New Chart,* he says, one could not but be appalled at the record it shows of "torrents of human blood . . . restless ambition of mortals . . . depravity of human passions." Yet, it is possible to retain ". . . faith in the great and comfortable doctrine of an over-ruling Providence . . . all the evils that infest this mortal life are, in His hands, subservient to the most benevolent purposes." Indeed, ". . . It is even easy to show, in a sufficient number of instances, that wars, revolutions of empire, and the necessary consequences of them, have been, upon the whole, extremely favourable to the progress of knowledge, virtue and happiness" (*New Description* 18–20).

Even in the best of circumstances, it would have been difficult to determine, in the middle years of the eighteenth century, the "exact dates of the rise and fall of every considerable state," assuming that those events possessed exact dates. Although Priestley probably began his reform of the chart of history while at Warrington, where he would have the use of the academy and town libraries, the completion of his task when he went to Leeds was not a best circumstance. The Leeds circulating library did, however, soon acquire a number of historical works for him to use.[24]

He states that he took but two or three dates from the French chart and depended chiefly upon the data to be obtained from the "Universal History, ancient and modern parts." He had first read the ancient parts of *An Universal History, from the Earliest Account of Time* while a student at Daventry and the modern section, in sixteen volumes, had been completed the year of publication of the *Chart of Biography*. A cooperative work of a number of "scholars," it has been described by Harry Elmer Barnes as the first of efforts [in England?] to produce a universal history and as containing "an unparalleled body of materials on all peoples of all ages." Nonetheless, the *Universal History* presented a number of problems and deficiencies. The various authors were of vastly different abilities and one of them, George Psalmanazar, was a notorious literary imposter.[25]

Francis Baily, who prepared another chart of history and another *Epitome of Universal History* in 1813, claimed that the "chronology adopted by the several writers of that work was so various and the facts so differently related" that its use was ill-advised. Priestley also introduced more complication by again adopting Newton's chronology, dismissed by Baily as "built upon too slight a foundation to overturn the concurring testimony of an-

24. According to the *Laws for the Regulation of the Circulating Library in Leeds; and A Catalogue of the Books Belonging to it* (Leeds: Griffith Wright, 1768), that library possessed a forty-four-volume edition of the *Universal History, Modern* as well as some seventeen other historical works, mostly on ancient history.

25. John Campbell, George Sale, John Swinton, Archibald Bower, George Psalmanazar, et al., *An Universal History, &c.* (London: Bathurst, Rivington, Hamilton, et al., 1736–65). Harry Elmer Barnes, *A History of Historical Writing* (Norman: University of Oklahoma Press, 1937), 171.

cient writers."[26] Yet, despite his criticisms of Priestley's *New Chart* and *New Description*—that the arrangement was confusing, information carelessly and hastily drawn up, abounding in gross errors and misrepresentations varying with every edition—Baily concedes that the *New Chart* was excellent in plan and design and adapted both chart and description to his own chart, with new dates and information based upon nearly half-a-century of historical research since Priestley's compilation.

Nor was Baily's the only historical chart, inferentially based upon Priestley's, to be published during the nineteenth century. In the *New Chart of History* as in his other educational writings, Priestley was to influence, directly or indirectly, several generations of teachers. It is possible, however, to overemphasize the extent of that influence. Similarities in approach or subject matter taught obviously do not necessarily demonstrate dependence upon Priestley or his writings.

As the *New Chart* was not a totally original conception, neither were his other works on other aspects of teaching. Parts of Priestley's philosophy of education, teaching methods, and even of the liberal curriculum can be found in educational writings and institutions of the late eighteenth century, independent of his influence. Indeed, Priestley's general philosophy of education clearly derives from the writings of John Locke, Isaac Watts, and David Hartley, while his teaching methods reflect those developed by Philip Doddridge, under which Priestley and Aikin had been schooled and which, no doubt, they used at Warrington.

Even the designation of history as a practical study for men of affairs was scarcely new. King George I had endowed a professorship in modern languages and history at Oxford and another at Cambridge in 1724. Lord Bolingbroke's *Letters on the Study and Use of History* (1752) had recommended modern history, especially, for the man wishing to be of service to his country and Lord Chesterfield's letters to his son had proposed it as training for the diplomatic service as early as 1760 (though his *Letters* were not published before 1774). History had been one of the subjects taught at Daventry Academy when Priestley was there and Aikin had not only lectured on history at Warrington before Priestley arrived; he had also given

26. Francis Baily, *An Epitome of Universal History, Ancient and Modern, From the Earliest Authentic Records to the Commencement of the Present Year* (London: J. Johnson and Co., and J. Richardson, 1813), v, xxiii; see also ii, 770–85. Baily, better known as an astronomer and a founder of the Astronomical Society of London, claims he discussed the many errors of Priestley's chart with Priestley himself—probably during 1791–94, when Priestley was lecturing at Hackney New College; Baily would then have been between seventeen and twenty.

lectures on the "Elements of *Jurisprudence*, so far as it is derived from the same sources with the Doctrines of Natural Religion and Private Morality."[27]

All this having been said, however, it remains true that Priestley made systematic and operational what had been fragmentary, frequently impractical, and unused before him. It is far from clear that the course in history at Daventry had included any modern history, while that taught by Aikin at Warrington appears, from references to its contents, to have involved ancient and ecclesiastical writers only, in an "exemplary" mode. The Regius Professorships at Oxford and Cambridge had become sinecures by the end of their first decades. Not until John Symonds's appointment to the chair at Cambridge in 1771 were lectures in modern history regularly given there, while Oxford had to wait until Thomas Arnold was appointed in 1841.[28]

Moreover, the reception of Priestley's ideas as fresh and original, when reintroduced by Spencer nearly a century later, suggests that they were neither so commonplace nor so obvious as to make them merely a summary of the educational practice of Priestley's contemporaries. In this, as in so many other of the extraordinary range of subjects on which he wrote, Priestley made an original contribution because he saw the immediate systematic implications of what others suggested in parts and had the courage to develop their fragmentary remarks into an operative whole. On almost every subject on which he wrote, there were contemporaries who were more profound and analytic, but few persons made significant contributions to so many different subjects.

27. *Report of . . . Warrington Academy . . . MDCCLXII,* 3. Aikin, who is supposed once to have studied to be a lawyer, is described as possessing "a deep and extensive knowledge of the constitution and laws of England" and as having "made some considerable progress in an elementary work" on that subject until Blackstone published the *Analysis of the Laws of England* in 1754; see Turner, *Warrington Academy,* 13.

28. See C. H. Firth, "Modern History in Oxford, 1724–1841," *English Historical Review* 32 (1917): 1–21. Curiously, both Symonds and Arnold are reported to have read and referred to Priestley's *Lectures on History.* Symonds recommended the work at Cambridge; see McLaughlin, "Warrington Academy," 55. Emma Jane Warboise, *The Life of Dr. Arnold (Late Head Master of Rugby School)* (London: Burnet and Isbister, 1897), 4, reports that Arnold read Priestley's *History* in 1803, at the age of eight, as a student of a Dr. Griffiths in Warminster School, Wilts., and quoted it from memory thirty-eight years later when filling the chair of Regius Professor of Modern History at Oxford.

VI

WARRINGTON ACADEMY, 1761–1767

Electricity

Probably the most striking example of Priestley's talent for creative synthesis is shown in an area for which nothing in his previous experience would have suggested any capacity: his work at Warrington Academy, which led, in 1767, to the publication of *The History and Present State of Electricity.* Priestley may well have had more instruction, informal and formal, in the sciences than most contemporary teachers of languages and letters, but there had been little to show for it when he arrived at Warrington. He had, while teaching in his school at Nantwich, purchased some scientific apparatus for instructional purposes; ". . . I had no leisure, however, to make any original experiments until many years after this time" (*W* 1/1:43). He must have given Warrington Academy representatives, bringing him the invitation to be tutor of languages, an unpleasant surprise, if, as he remembered the episode nearly a quarter of a century later, he told them that he "should have preferred the office of teaching the mathematics and natural philosophy" (*W* 1/1:47).

Not that Warrington Academy was uninterested in the sciences, for it, like most other Dissenting academies, included more experimental science in its general curriculum than could easily be obtained at Oxford or Cambridge. But Warrington already had, in John Holt, its tutor in mathematics and natural philosophy. It appears that Holt was less than inspiring as a teacher and had, upon occasion, to be reminded to make his lessons less theoretical. Nonetheless, though his major attraction was his teaching of commercial mathematics, there was no criticism of his knowledge of the natural philosophy he also taught. He brought with him a respectable collection of scientific apparatus, and was active, from the beginning of his appointment, in ordering science books for the academy library and more apparatus for its classes.[1] Under the circumstances, academy trustees might well have regretted their selection of a tutor of languages who declared an unwarranted preference for a very different set of subjects.

As it happened, of course, Priestley threw himself vigorously into the studies entailed by his appointment. Yet, soon after his arrival at Warrington, he found himself also involved in some of the scientific activities of the academy. Sometime during his first two years he gave some twenty lectures on anatomy—(which he had studied considerably at Daventry). During 1762, his developing friendship with John Seddon was encouraged by a mutual interest in the sciences. Seddon corresponded with John Roebuck, an academy trustee from Birmingham, about experiments on the manufacture of graduated thermometers; he seems also to have been the prime mover in bringing chemical instruction to Warrington. Writing to Seddon in March 1762, on the plans for a course of chemistry lectures, Dr. Matthew Turner, of Liverpool, declared: "I greatly admire Mr. Priestley and am certain any school must flourish that has such teachers in it, he seems to possess all the Qualifications necessary for the encreasing and diffusing nowledge."[2]

When Seddon set forth from Warrington on his conciliatory and fundraising campaign during the spring of 1762, Priestley was responsible for completing arrangements with Turner. His correspondence in April and May refers to the chemical lectures, requests Seddon to procure the "glasses" (lenses) for his clergyman friend, Mr. Brereton—suggesting that

---

1. On Holt, see Turner, *Warrington Academy,* 5–6; and the MS Minutes of the Proceedings of the Trustees of Warrington Academy, 33, 52, 134–35. According to Alexander Gordon, article; "John Horsley (1685–1732)" *Dictionary of National Biography* 9 : 1276–77, Holt's scientific apparatus had descended from Horsley to Caleb Rotheram to Holt and Warrington Academy and would, in time, pass to Hackney New College and then to the Dr. Williams's Library, from which it has long since disappeared.

2. Matthew Turner to John Seddon, 16 March 1762; Library, Harris Manchester College, Oxford; quoted by permission. On Priestley's anatomy teaching, see McLachlan, "Warrington Academy," 57. On Matthew Turner, surgeon and chemical consultant to Josiah Wedgwood and Matthew Boulton, see *SciAuto,* 8–11. The letter from Roebuck to Seddon is quoted in Robert E. Schofield, *The Lunar Society of Birmingham* (Oxford: Clarendon Press, 1963), 29; the letter is not dated, but must be before 1764 and perhaps as early as 1760.

Mr. Canton might be asked to examine them—and reminds him that Mr. Holt wants a solar microscope. When the lectures were finally given Priestley attended them, for his *Philosophical Empiricism* of 1775 notes his having assisted there in the making of a quantity of spirit of niter.[3]

None of this requires an inference of a shift toward scientific work. Nor can that be surprising in a person who was, as Priestley wrote, spending "five hours every day in lectures, public or private" (W 1/1:56), during a period in which he was also revising his book on English grammar, writing that on language and universal grammar, preparing lectures on oratory and criticism, on history and general policy, on English history and English law, designing the chart of biography, and writing the *Essay on A Course of Liberal Education.* If there is any sense of a shift in the emphasis of his interests and his teaching, it was toward history.

That he had not lost his interest in teaching natural philosophy is shown by the introductory paragraphs to the *Description of a Chart of Biography,* written not later than the early months of 1765:

> The proper employment of men of letters is either making new discoveries . . . or facilitating the communication of the discoveries which have been made already . . . few are qualified to make new discoveries of importance: a considerable share of natural genius, opportunity of making experiments, and a favourable concurrence of circumstances are requisite to it. . . . But when discoveries have been made, and the principles of science have been ascertained, persons of inferior abilities . . . are sufficient to digest those principles into a convenient method, so as to make the knowledge of them much easier than it was to the inventors. (3–4)

It is true that Priestley goes on to adapt that declaration to the organization of the work of the "great Historians, Chronologers, and Biographers of all ages and Nations." Nonetheless, the principal application to natural sciences is clear, as it was in the statement of John Locke, from whom he must initially have derived his sentiment:

> The Commonwealth of Learning is not . . . without Master-Builders, whose mighty Designs, in advancing the Sciences, will

3. A printed source of substantial portions of Seddon-Priestley correspondence is cited; see note 3 to Chapter IV. The originals are Priestley to Seddon, 9 April 1762, used by permission, Boyd Lee Spahr Library, Dickinson College, Carlisle, Pa.; 19 May 1762, Wedgwood

leave lasting Monuments to the Admiration of Posterity: But every one must not hope to be a *Boyle,* or a *Sydenham;* and in an Age that produces such Masters, as the great—*Huygenius* and the incomparable Mr. *Newton,* with some other of that Strain; tis Ambition enough to be employed as an Under-Labourer in clearing Ground a little, and removing some of the Rubbish that lies in the way to Knowledge.[4]

Surely, when Priestley wrote his statement, he saw himself as one of those "persons of inferior abilities" with a mission to digest the principles of science into a convenient method. And it is not surprising that he should combine that ambition with his recent concerns for history. For history, to Priestley was a means to discovery—of the laws of society, for example—and a means of conveying information and of persuasion. That is, it was teaching. In his treatment of rhetorical methods in *Lectures on Oratory and Criticism,* Priestley described the narrative method of ordering events by time (i.e., historically), as in the "order of nature" in its accordance with the association of ideas. In the method of philosophical argument, analysis: proceeding from particular observations to general laws, was recommended as the best way of leading people insensibly to form right conclusions (35, 42).

Given the task (extension of the knowledge of science) and method (historical), there remains the question of why Priestley's first subject was a history of electricity? For an answer to this question, Priestley is of little help. The preface to his *History of Electricity* describes electricity as "my own favourite amusement" and a later research paper mentions electrical demonstrations performed "with a view to amuse myself and my friends" several years before making original experiments.[5] But the only experimental natural philosophy for which he had demonstrated any continued interest was chemistry, which he had studied in the published lectures of Boerhaave while at Daventry and more recently in the lectures and laboratory experiments of William Turner.

Papers, The John Rylands University Library of Manchester. Joseph Priestley, *Philosophical Empiricism* (London: J. Johnson, 1775), 45.

    4. John Locke, "Epistle to the Reader," in *An Essay concerning Human Understanding* (London: A. Churchill and A. Manship, 1733), 1:sig. b2v.

    5. Joseph Priestley, *A History and Present State of Electricity, with Original Experiments* (London: J. Dodsley, J. Johnson and B. Davenport, T. Cadell, 1767), vii; Joseph Priestley, "An Investigation of the Lateral Explosion . . .," *Philosophical Transactions* 60 (1770): 192.

The most *popular* area of natural philosophy in England was electricity, however, and had been since the discovery of the Leyden jar and the publication, in 1754, of Benjamin Franklin's *New Experiments and Observations on Electricity.* Appearance of a third edition of Franklin's *New Experiments* (1760–62) may have reminded Priestley that Franklin was just such a "Master-Builder" as Locke had mentioned. And Franklin was again in London. Not only was Franklin in London, so was John Canton, Franklin's good friend and the first Englishman to confirm Franklin's experiments, and Canton, Priestley knew from his correspondence during 1762, was at least an acquaintance of John Seddon.

No doubt Priestley was aware of the scholarly value of consulting Franklin personally, if he were to write a history of electricity, for his appreciation of primary sources was better than that shown by most historians for several generations: "I have carefully perused every original author, to which I could have recourse; and every quotation in the margin points to the authority that I myself consulted, and from which the account in the text was actually taken. Where I could not procure the original authors, I was obliged to quote them at second hand, but the reference will always show where that has been done" (*Hist. Elect.* ix).

Given Priestley's assiduous advertising of the *Chart of Biography*, it is clear that he would also be aware of the value to the sale of his work, if he could obtain for it the endorsement of Franklin. Late in December 1765, he set forth on a three-day journey to London, carrying a letter of introduction to Canton with an important postscript:

> As my friend, Dr. Priestley, Tutor in the Languages and Belles Lettres here, is going up to Town, and has a great desire of being introduced to you, I have undertaken to do him this kind office. . . . You will find him a benevolent, sensible man, with a considerable share of Learning; besides the studies which belong to his Profession, he has a taste for Natural Philosophy, which will not render him less agreeable to you. . . .
>
> P.S. If Dr. Franklin be in Town, I believe Dr. Priestley would be glad to be made known to him.[6]

Priestley's account of the episode, while true enough, is something less than candid: ". . . going to London, and being introduced to Dr. Price,

6. John Seddon to John Canton, 18 December 1765; *SciAuto*, no. 1.

Mr. Canton, Dr. Watson (the physician) and Dr. Franklin, I was led to attend to the subject of experimental philosophy more than I had done before, and having composed all the lectures I had occasion to deliver, and finding myself at liberty for any undertaking, I mentioned to Dr. Franklin an idea that had occured to me of writing the history of discoveries in electricity, which had been his favourite study" (W 1/1:55).

He must have carried with him, on that visit to London (his first since he had passed through on the way to Nantwich in 1758) more than a letter of introduction. His charm was such that he made warm personal friends of Canton and Franklin. Yet surely he did not count on charm in enlisting their support for that "idea that had occurred to me of writing the history of discoveries in electricity." A provincial schoolmaster of languages does not request such assistance from two of the foremost British authorities on electricity without something to demonstrate that assistance would be worthwhile.

On 14 February 1766, but recently returned to Warrington, Priestley wrote Canton that he had completed a portion of his history "to the year 1742, when Desaguliers wrote."[7] That "portion," when printed, filled sixty-three quarto pages in the first edition and contained some eighty references to twenty different books or journal articles. Many of these references—to the works of Theophrastus, William Gilbert, or Otto von Guericke, for example—are to books supplied to Priestley after the London trip, but the majority are to works of Francis Bacon, Robert Boyle, and Isaac Newton, or to articles reprinted in the *Philosophical Transactions, Abridged* to which he would have had access in the Warrington circulating library or in that at the academy.[8] It seems likely that when Priestley "mentioned to Dr. Franklin to the plan of a treatise I *am* writing on Electricity," he also showed him an early draft of some of that portion "to the year 1742."[9]

7. Priestley to Canton, 14 February 1766; *SciAuto*, no. 2.

8. For books in Warrington libraries, see MS Minutes of Warrington Circulating Library and MS Catalogue of Books in the Library of the Warrington Academy; Library, Harris Manchester College, Oxford, cited by permission. The latter was compiled after Priestley left, but may be assumed to have some resemblance to the pre-1767 collection. There are several versions of the *Philosophical Transactions, Abridged;* some in several editions. Although the edition Priestley used is not known, the version appears to be that begun by John Lowthorp in 1705 and continued by Henry Jones, John Eames, and John Martyn, covering *Philosophical Transactions* to 1750. Note that the articles are scarcely cut and frequently reproduce the words of the authors; abridgment is achieved by omitting book notes, astronomical, weather and tide tables, and catalogs of plants and animals.

9. "Mentioned to Dr. Franklin" is from the Priestley to Canton letter of 14 February 1766, loc. cit; my emphasis.

In any event, Canton and Franklin immediately began to help Priestley. They introduced him to William Watson, the third of the triumverate of premier British electricians of the mid-eighteenth century. Each of the three—Canton, Franklin, and Watson—had won the Copley Medal of the Royal Society for his researches in electricity. Priestley also met Richard Price, who took him as a guest to a meeting of the Royal Society of London on 9 January 1766. Price, Priestley discovered, was also a good friend of John Canton and Benjamin Franklin. Together, the four—Canton, Franklin, Price, and Watson—helped obtain the books Priestley needed for his history, they insisted that he repeat the experiments he was to write about, and they sent him letters of advice, instruction, and correction: ". . . my letters circulated among them all, as also every part of my History as it was transcribed. This correspondence would have made a considerable volume, and it took up much time; but it was of great use with respect to the accuracy of my experiments, and the perfection of my work" (W 1/1:57–58).[10]

When Priestley returned to Warrington, late in January 1766, with his mission successfully completed, his friends there joined in helping him with his project. Academy scientific apparatus was placed at his disposal for experiments and John Holt assisted him in making them. Additional apparatus was ordered for his use. In the 14 February letter to Canton, Priestley notes: "We have given orders for an electrical kite, and shall make the experiments you desire," while brother Timothy Priestley, now an evangelical Calvinist minister, wrote:

> . . . finding I could work in either brass or wood, and turn his work in a lathe . . . he [Joseph] proposed a partnership, for making electrifying machines. . . . Many electrifying machines were made, and sold to his friends. . . . I made him a kite of fine silk oil case, six feet four inches wide; he could put the whole in his pocket, the frame would take to pieces . . . the string was composed of thirty-six threads, and one wire, this was to bring electrical fire from the clouds.[11]

10. Priestley's letters to Watson have not been found, but many of those to Canton, Franklin, and Price have been preserved and are printed in *SciAuto*. Except for an occasional draft, also printed in *SciAuto*, their letters to him have not survived. Priestley's attendance at the meeting of the Royal Society is recorded in the MS Journal Book Copy, vol. 25 (1763–66): 707, cited by kind permission of the President and Council of the Royal Society.

11. Timothy Priestley, *Funeral Sermon*, 42–43.

But it was his new London friends who were in a position really to be useful. When he wrote Canton: "My friends here imagine it would be a great advantage to the publication, if I were fellow of the Royal Society, and have persuaded me to be a candidate for that honour," the London friends rallied. A nomination certificate was prepared, signed by seven nominators, including Franklin, Canton, Price, and Watson, and posted on 13 March 1766. After much anxious waiting, Priestley was informed by Emanuel DaCosta, Clerk to the Society, of his election on 12 June 1766 and he could (and would always) henceforth identify himself as Joseph Priestley LL.D. F.R.S.[12]

The months from mid-January 1766 to early April 1767 were, for Priestley, crowded with activity. He continued his lecturing at Warrington, abstracted the books sent him from London, wrote his *History*, and carried on an increasingly sophisticated series of electrical experiments. By mid-September 1766, William Eyres at Warrington had begun to print the first parts of the *History of Electricity*, which Priestley then estimated would be complete in four to five hundred quarto pages and be printed off just after Christmas. In late January, the whole was not quite done, the preface is dated March 1767, and by April he could write asking his friends how they liked the finished work—now 700 quarto pages including the description of his own experiments.

The general response was surely all that he could have asked for. William Bewley wrote a three-part, forty-page review for the *Monthly Review*, filled with such phrases as: "excellent performance," "judicious and well-informed," "accurate and faithful," "talent for distribution and management," and ending with a summary; design and plan are excellent, execution

12. Priestley to Canton, 14 February 1766, loc. cit.; Priestley to Emanuel DaCosta, 18 May and 21 June 1766, DaCosta to Priestley, 14 June 1766 (W 1/1:65–67). Priestley's account, in his *Memoirs*, is again misleading: "my new experiments in electricity were the means of introducing me into the Royal Society, with the recommendation of Dr. Franklin, Dr. Watson, Mr. Canton, and Dr. Price" (W 1/1:58). His Royal Society certificate (see *SciAuto*, no. 4.) describes him only as "Doctor of laws, Author of a chart of Biography and several other valuable works . . . and very well versed in Mathematical and philosophical Enquiries." Nothing is said of electrical experiments; he had as yet sent no papers to the Society; the *History of Electricity* had not yet been published; and there is no reason to suppose that the majority of fellows voting would have cared whether or not there were experiments to confirm his being "versed in . . . philosophical enquiries." Possibly his principal nominators would not have signed their names without the letters describing his experiments, but the three other signers included Samuel Chandler, who had been one of his supporters for the Edinburgh LL.D. and was not one of his electrical correspondents.

masterly, the "joint product of labour and genius."[13] The work sold well and was read widely. It went through five English editions (1767, 1769, 1775 [2 vols. octavo; reprint, 1966], 1775, and 1795) with substantial additions and some corrections in the second and third. There was a French translation in 1771 and a German in 1772. Copies went everywhere electrical research was performed; Volta read it in Italy, Torbern Bergman in Sweden, Jan Ingenhousz in Vienna, Martinus von Marum in Holland. Franklin purchased several copies for distribution in the colonies; the American Philosophical Society in Philadelphia got one, Harvard another, and Yale, which also received a copy, adopted it in 1788 as a supplemental text in natural philosophy. John Winthrop and Andrew Oliver wrote to Priestley on electrical topics from America, Henry Cavendish read the *History* and was inspired by it to some of his own electrical researches. William Herschel based his earliest scientific papers for the Bath Philosophical Society on Queries taken from it.

More than half a century after its first appearance, Priestley's *History* was still providing instruction and information to scientists. Thomas Young cited it in his *Course of Lectures on Natural Philosophy* of 1807; Alexander Dallas Bache read the work in the United States in 1840; during the 1840s, Michael Faraday was clearly aware of it; and Peter Theophil Reiss refers to material in the *History* for a paper in the Transactions of the Royal Academy of Sciences, Berlin in 1849. By that time it had already become something of a classic. As other histories of electricity were written, material was taken (sometimes without acknowledgment) from Priestley's *History* and it is still used as a source of information for the history of eighteenth-century science in general and eighteenth-century electricity in particular.[14]

13. [Wm. Bewley], Review of "History and Present State of Electricity," *Monthly Review; or Literary Journal* 37 (1767): 93–105, 241–54, 449–65.

14. See, for example, Sir Edmund Whitaker, *A History of the Theories of Aether and Electricity* (New York: Philosophical Library, 1951), vol. 1; I. Bernard Cohen, *Franklin and Newton,* Memoirs 43 (Philadelphia: American Philosophical Society, 1956), and J. L. Heilbron, *Electricity in the 17th and 18th Centuries: A Study of Early Modern Physics* (Berkeley and Los Angeles: University of California Press, 1979). Thomas Young, *A Course of Lectures on Natural Philosophy and the Mechanical Arts* (London: J. Johnson, 1807). An examination of Faraday's work (e.g., his "A Speculation touching Electric Conduction and the Nature of Matter," *Philosophical Magazine* 24, ser. 3 [1844]: 13–44), shows so many echoes of ideas of Priestley that it seems obvious he read Priestley's scientific work. Certainly he knew of that in electricity, for he was referred to it in the 1840s by three people: Richard Phillips, William Snow Harris, and Emil du Bois-Reymond. See Thomas Martin, ed., *Faraday's Diary: Being the Various Philosophical Notes of Experimental Investigation* (London: G. Bell and Sons, 1932), 2:177; and L. Pearce Williams, ed., *Selected Correspondence of Michael Faraday* (Cambridge: Cambridge University Press, 1971), 405, 465–66. See also Robert E. Schofield, ed.,

There was considerable justification for the intellectual, as well as commercial, satisfaction with which Priestley made his acknowledgments, in the preface to the *History of Electricity:*

> With several of these principal actors it has been my singular honour and happiness to be acquainted; and it was their approbation of my plan, and their generous encouragement that induced me to undertake the work. With gratitude I acknowledge my obligations to Dr. Watson, Dr. Franklin, and Mr. Canton for the books, and other materials with which they have supplied me, and the readiness with which they have given me any information in their power to procure. In a more especial manner am I obliged to Mr. Canton, for those original communications of his, which will be found in this work, and which cannot fail to give a value to it, in the esteem of all lovers of electricity. My grateful acknowledgments are also due to the Rev. Mr. Price, F.R.S. and to the Rev. Mr. Holt, our professor of Natural Philosophy at Warrington, for the attention they have given to the work, and for the many important services they have rendered me with respect to it. (vii)

The assistance he describes must have contributed substantially to the quality as well as to the sales of the work. But an examination of his correspondence during the period of its preparation shows that the end product was Priestley's own. In fact, the "masterly execution" of the "excellent design and plan" is an example of Priestley's characteristic intellectual qualities.

Typically, it was a peculiarly appropriate time for writing a history of electricity, "when," as Priestley writes, "the materials were neither too few, nor too many to make a history; and when they were so scattered, as to make the undertaking highly desirable" (vii–viii). Electricity, in 1766, was still static electricity and its discharges. Descriptions of its phenomena were inevitably confused and confusing. Static voltages are high, but discharge currents are generally low and minor variations in humidity, textures or compositions of substances used, even configuration of objects or numbers of viewers around an experiment, can change results apparently without reason.

---

introduction to the reprinted 3d ed. of Priestley, *The History and Present State of Electricity, With Original Experiments,* Sources of Science, no. 18 (New York: Johnson Reprint, 1966); and Robert E. Schofield, "Electrical Researches of Joseph Priestley," *Archives Internationale*

It is typical of the difficulties involved in understanding static electrical phenomena, that conductivity should first have been noted in hempen string and wooden rods, which are now classed as nonconductors. Only with the work of William Watson and Benjamin Franklin, dating from the mid-1740s, had some theoretical unity begun to enter the subject. Though there were a few general surveys of electrical studies available in 1766, none was in English and none seriously tried to organize the subject as an educational endeavor, as Priestley was to do.

In the first part of the *History*, approximately half the whole, Priestley summarizes, in ten chronological sections, the experiments and observations on electricity from 600 B.C. to 1766. He prided himself on keeping "clear of any mean partiality toward my own countrymen and even my own acquaintance" (x). No doubt he tried, but in effect it was not entirely true.

English sources were easier to obtain than foreign ones and German sources the hardest to find and read. The *Philosophical Transactions* of the Royal Society, *Histoires et Mémoires* of the Royal Academy of Sciences, Paris; treatises on electricity in English, French, Italian, and Latin, reported on and sometimes published accounts of German electrical researches. Nevertheless, the first edition of Priestley's *History* is inadequate in its coverage of German work.

The French translator complained that Priestley was unjust to the French. With some reason, Priestley indignantly explained the complaint as the accusation of a follower of the Abbé Nollet, angry because he had favored Franklin's electrical theory over that of Nollet. There is some validity to the complaint. Priestley discussed Nollet's experiments fully and with praise, but when he described various electrical theories, those of the British are given the most attention. That of Nollet was called a "strange hypothesis" and Nollet's arguments in support of it were described as "very unsatisfactory . . . more ingenious than solid (452–53).

Priestley attempted to keep discussion of theory out of the chronological part of his *History,* proposing to give there "only known facts." In condensing his sources, he used words of the authors themselves as much as possible. Nonetheless, a work intended to lead the reader, by stages, through a detailed account of "what has really been done by others, and where the science stands at present" (483), almost inevitably will have positivist and whiggish overtones. Priestley, that is, interpreted the work of the past in terms of the accepted view of his own day and could not resist, occasionally, providing a "correct" explanation of phenomena "wrongly" interpreted or not explained at all. As an Englishman, and the grateful protégé of Benjamin Franklin, he naturally tended to adopt the British and Franklinian view of electricity. There was no real doubt in Priestley's mind, at the time of writing the first edition of the *History*, that electricity was a substance sui generis, probably a fluid (or perhaps two fluids); now and then, even in the chronological part of the work, he slips into "electric fluid" terminology.

These, however, are comparatively minor flaws in a Baconian history that does, as it intended, bring together more observations on electrical phenomena than had ever before been collected in one place. By the time he had completed the revisions for the third edition of the *History*, Priestley had collected and used roughly forty books and at least three sets of scientific journals, covering the period to 1766. A comparison of Priestley's work with modern historical bibliographies of electricity will show that, by the third edition at least, Priestley mentions every significant figure in electricity's history of whom he could have known.[15] And almost as a virtue of its positivist failings, the *History of Electricity* also became an authoritative statement on eighteenth-century British electrical research and a primary document for interpreting the work of his patrons.

The second half of the *History* (parts 2 through 8) illustrates Priestley's conception of how history is to be used. As he had recommended in his history lectures, so in the *History of Electricity*, after he describes events in chronological sequence, he attempts to extract meaning and practice from them. Probably more of this material than he quite realized came from his personal philosophy and experience (he described his own experiments in part 8) but the greater part is a distillation of his historical research. He summarized (still, so far as he could, without theory) propositions "comprizing the general properties of electricity," discussed the major contending theories, gave some hints, queries, and "Desiderata" for further investigations, described electrical apparatus, offered "practical maxims" and descriptions of "entertaining experiments" for young electricians. For his contemporaries, this half of Priestley's book would comprise the best available text for understanding electricity and directing electrical research,

15. See, for example, Paul Fleury Mottelay, *Bibliographical History of Electricity and Magnetism chronologically arranged* (London: C. Griffen & Co., 1922); William D. Weaver, ed., *Catalogue of the Wheeler Gift of Books, Pamphlets and Periodicals in the Library of the American Institute of Electrical Engineers* (New York: American Institute of Electrical Engineers, 1909).

while even a modern reader might find it informative and would surely discover delight in those "entertaining experiments."

It is also possible to discern, in this half of the *History*, those elements that combine in Priestley's career as a scientist. The *History of Electricity* was after all, written when he was thirty-three; he was a beginner in science, but already assured as a scholar with techniques and attitudes to transfer to his work in the sciences. The Baconian chronicle had demonstrated his skill in history; many of the remaining parts would illustrate other qualities that he would bring to the task of scientific research.

That on "theories of Electricity," for example, shows Priestley's independence of mind and concern for fair play. The accepted theory of electricity in 1766 and the influence of his advisers combined to recommend Franklin's single-fluid explanation of phenomena. There was, however, a two-fluid theory that, he believed, had merit, though its major English proponents had done the theory less than justice. Although he thought it probable that Canton, at least, would "not approve of the light in which I have represented . . . the theory of two electric fluids," Priestley "followed my best judgment" and elaborated upon the two-fluid theory, making it a viable alternative to Franklin's theory.[16]

The part on theories also reveals a deficiency in Priestley that was to weaken his scientific accomplishments all his life. He had no appreciation of the essential values of mathematical and quantitative studies in science. His inclinations were always toward qualitative experimentation and this is shown in his failure to comprehend the significance of a mathematical elaboration of Franklin's theory. That theory had encouraged the neglect of both mathematics and quantity, for its explanation of phenomena in terms of a single fluid of self-repelling particles emphasized the restoration of a material balance between bodies with an excess of fluid and those with a deficiency of it. The achievement of this null situation was described qualitatively; there was no definition for a quantity of fluid, and no occasion for computations.

There was one phenomenon—the repulsion of negatively charged bodies—more easily explained by two self-repelling fluids, for why should bodies lacking fluid repel one another? Two-fluids, however, lacked the

16. Priestley to Canton, 21 April 1767, *SciAuto*, no. 14. His treatment of a two-fluid theory involved Priestley in a charge of plagiarism from an irascible and eccentric Anglo-Irishman, Major Henry Eeles, though Priestley seems never to have learned of the accusation. For an account of the episode, see Robert E. Schofield, ed., introduction to reprint of the 3d. ed. of *History of Electricity*, xxiv–xxvii.

economy of means of one-fluid, and were no more quantified than it was. Many eighteenth-century electricians (including Franklin) adopted an alternate suggestion by the German scientist, Franz Ulrich Theodor AEpinus (1724–1802).

AEpinus had arrived at his explanation of negative repulsion, within Franklinian terms, in his *Tentamen Theoria Electricitatis et Magnetismi* (St. Petersburg, 1759). He addressed the subject of electricity in a mathematical, Newtonian fashion, divesting Franklin's theory of its clumsy electrical-atmospheres instrumentality by adoption of action-at-a-distance forces. In consequence, he found it necessary to assume that particles of ordinary matter, were, at short distances, self-repelling. Their repulsion would normally be counteracted by their attraction for particles of electric fluids, when bodies possess their natural amount of electricity. When bodies lost fluid (were negatively charged), their natural repulsion appeared. When they were positively charged, there was repulsion between the excess amounts of fluid.

This theory clearly introduces some new problems: It raises again the question of action-at-a-distance forces, which mid-century scientists liked to ignore; it proposes that particles of ordinary matter be at once gravitationally attractive and electrically repulsive. Its real failing, however, for most of AEpinus's contemporaries, was in its elaborate form. The *Tentamen* is not quantitative, for AEpinus does not assign numbers to his repulsive and attractive forces, but it is mathematical, with numerous equations.[17]

Priestley claimed that he "greatly admired AEpinus," and he described his experiments and the qualitative consequences of his theory at length. He stated, in a section of the *History*, on "Branches of knowledge peculiarly useful to an electrician": "AEpinus has lately given us an excellent specimen of what use MATHEMATICS, and especially algebraical calculations may be to an electrician; and their use, will probably, in time, be found still more extensive" (503). But after favorable treatment of this extension of Franklin's theory, with admiration for the replacing of Franklinian atmospheres by "spheres of influence," he wrote:

> He that reads the first chapter, as well as many other parts of his [AEpinus's] elaborate treatise . . . may save a good deal of time and

---

17. On AEpinus, see John L. Heilbron, article: "AEpinus," in the *Dictionary of Scientific Biography*, 1:66–68, and Heilbron, *Electricity in the 17th and 18th Centuries*, 388–402. The *Tentamen* has been printed in a modern English translation by P. J. Connor, R. W. Home, ed., *Aepinus's Essay on the Theory of Electricity and Magnetism* (Princeton: Princeton University Press, 1979).

trouble by considering, that the result of many of his reasonings and mathematical calculations cannot be depended upon; because he supposes the repulsion or elasticity [of] the electric fluid to be in proportion to its condensation; which is not true, unless the particles repel one another in the simple reciprocal ratio of their distances, as Sir Isaac Newton had demonstrated, in the second book of his Principia. (463)

As it happened, Priestley knew the law of repulsion between electrical particles. He ended his discussion of his own experiments with a reference to Franklin's observation that cork-balls were "wholly unaffected by the electricity of a metal cup, within which they were held" and, after verifying the observation, says; "May we not infer from this experiment, that the attraction of electricity is subject to the same laws with that of gravitation, and is therefore according to the squares of the distance; since it is easily demonstrated, that were the earth in the form of a shell, a body in the inside of it would not be attracted to one side more than another" (732). The situation is not quite that simple. The demonstration Priestley mentioned is based on a sphere not a cylinder, on fixed not fluid particles. It is not clear that the laws governing self-attraction are applicable to that between different substances, nor that laws of attraction need be the same as for repulsion. But, once Priestley's statement was made, these are questions that could be discussed and argued. Priestley did not do so.

Priestley the nonmathematician, has provided a mathematical quasi-demonstration of the inverse-square force law for electrical charges. It was the first respectable claim for that law, out of which came the development of a mathematical theory of static electricity. And Priestley played no role in that development. His major claim to fame in electrical studies is his "discovery" of the inverse-square law. Yet he does not refer to that discovery in his brief reference to the algebraic work of AEpinus. His failure to elaborate on that work or to see how it might be improved by introduction of the inverse-square relationship constitutes a major failure in his *History of Electricity*.[18]

Priestley was not the only British electrician with no appreciation for mathematical studies. Early in January 1767, John Canton wrote to Priest-

---

18. Priestley wrote of his admiration of AEpinus to Franklin, 13 April 1766, *SciAuto*, no. 7. John Robison, *System of Mechanical Philosophy* (Edinburgh: J. Murray, 1822), 4:2–3, makes his neglect of AEpinus's significance a major criticism of Priestley's *History*. Robison was not unbiased; he was politically reactionary and disliked Priestley's radicalism; he also had claims

ley, "I have lately been making experiments on the electrified cup of Dr. Franklin which I find does not contain any mystery in electricity." The following year, Benjamin Franklin recommended Priestley for the Copley Medal of the Royal Society; none of the experiments he cited as deserving of that honor has mathematical or quantitative implications. Probably only Henry Cavendish, of Priestley's British contemporaries, understood and effectively employed mathematics in their application to electricity and Cavendish's best work, including his own development of the inverse-square law discovery, remained unpublished for lack of encouragement.[19]

This general attitude perhaps explains Priestley's neglect of mathematics, but cannot justify his further neglect, in describing his own research, of the quantitative relationships in electrical experiments he had mentioned elsewhere in the *History*. Twice he noted that the "power" of a charged Leyden jar (condenser) is proportional to the quantity of coated surface and the thinness of the glass (439, 542), but when describing his battery of Leyden jars or discussing the power of an electrical discharge, though he meticulously recorded the area of coated surface, he never mentioned glass thickness. Later, performing experiments on "the force of the electrical explosion" in wires, he observed: ". . . I found a very remarkable difference occasioned by the length of the circuit in wires of the same thickness . . . which, I own, surprised me very much." Yet, under "practical maxims," in the *History*, he had earlier written: ". . . in wires of the same thickness, the forces that melt them will be in proportion of the lengths; and in wires of the same length in proportion of the squares of their diameters."[20] Moreover, he assumed the "maxim" in the design of a set of elaborate experiments on the comparative conductivities of different metals, but it took the French translator to remind him that relative melting points of the metals must be considered in drawing conclusions from his experiments. In the end, a promising line of inquiry was not carried through. Experiments from which Priestley might have determined coefficients of resistivity were not pushed to their quantitative conclusions.[21]

---

to discovery of the inverse-square law. His criticism has, however, frequently been repeated; see Heilbron, *Electricity*, 402, where Priestley is said to have mistaken AEpinus's meaning on the repulsion. Robison, however, does have AEpinus using a 1/R law of attraction and repulsion.

19. Canton to Priestley, 10 January 1767, *SciAuto*, no. 11. Franklin to the Royal Society, 10 March 1768, *SciAuto*, no. 21. See also Russell McCormmach, "Henry Cavendish: A Study of Rational Empiricism in 18th-Century Natural Philosophy," *Isis* 60 (1969): 293–306.

20. Joseph Priestley, "Various Experiments on the Force of Electrical Explosions," *Philosophical Transactions* 59 (1769): 65; *History of Electricity*, 545.

21. Priestley, *History of Electricity*, 1st ed., 728–30; 3d ed., 2:368–71.

When it came to qualitative experiments, however, Priestley demonstrated in the *History of Electricity,* that fertility of experimental imagination that was to characterize his entire research career. He had not intended to make original experiments, but having begun by attempting to verify the experiments of others, he found himself making observations of his own. Some of these, on the conductivity of ice and hot glass, for example, were not original as others had reported them earlier. Priestley's independent observations of "a current of real air" coming from the electrified points, however, though previously noted by Nollet and especially by Benjamin Wilson, were repeatedly disputed by Canton, Franklin, and Watson. Persisting in his claim for the reality of the phenomenon, Priestley required his critics to try the experiments for themselves and eventually won their acceptance of it. Franklin included this "discovery" as one of those for which Priestley deserved the Copley Medal; and what is now described as "the well-known electric wind" attracted the attention of scientists such as Michael Faraday, James Clerk Maxwell, and Svante Arrhenius in the nineteenth century and is still being studied today.[22]

Priestley decided that this kind of work did not, after all, require "a considerable share of natural genius," concluding that the progress of electrical studies "might be quickened, if studious and modest persons . . . could be brought to entertain the idea, that it was possible to make discoveries themselves" (577). Supposing that, with such a "prodigious number of electrical machines . . . in the hands of so many ingenious men in different parts of the world, . . . all that could be done in little, had been tried" (659–60), he imagined that the best chance of a real discovery was in the use of greater electrical "force" than had previously been tried. He constructed large arrays of Leyden jars ("batteries") and some of his observations, such as those partial rankings of comparative conductivities, could not have been made without them.

He was aware, however, that meaningful experiments are seldom the results of random manipulation of apparatus, however much bigger or better than that of others. Experiments must be designed and "every experiment in which there is any design, is made to ascertain some *hypothesis*" (444). What he did not realize, nor did most of his peers throughout his scientific career, was the idiosyncratic nature of his hypotheses. This was not primarily in the nature of epistemology, on which his views, though perhaps a bit

22. Priestley, *History of Electricity,* 591–97; Myron Robinson, "A History of the Electric Wind," *American Journal of Physics* 30 (1962): 366–72.

old-fashioned, were much the same as those of his English contemporaries. Priestley was not a philosopher of science, in the sense of his having constructed a logical and coherent view of nature and of scientific activity; few practicing scientists are. He did, however, set forth a few general principles on the nature of science, derived primarily from Bacon, Locke, Watts, and Franklin. One of these sounds almost modern in its empiricist nominalism: ". . . nature exhibits nothing but particulars; . . . all general propositions, as well as general terms, are artificial things; being contrived for the ease of our conception and memory; in order to comprehend things clearly, and to comprise as much knowledge as possible in the smallest compass" (442).

But Priestley was not a disciple of logical positivism, misplaced in time. Immediately following that statement is another: ". . . we actually see in nature a vast variety of effects, proceeding from the same general principles." In fact, for Priestley, the object of science was the discovery of its "ultimate and most general principles" (480). "The more we know of any science, the greater number of particular propositions are we able to resolve into general ones; and consequently, within narrower bounds shall we be able to reduce its principles" (434). Priestley's apparent nominalism in science was essentially the same as his rejection of fixed rules in language and aesthetics:

> . . . there is a *ne plus ultra* in everything, and therefore in electricity, . . . but what reason is there to think we have arrived at it? (482). . . . our business is still chiefly with *facts*, and the *analogy of facts* . . . far too few of these have been discovered to ascertain a perfect general theory . . . all the present hypotheses can do for us must consist in suggesting further experiments. (480–81).

When beginning the description of his original experiments, he pledged himself to report the hypothesis suggesting each of them, "false and imperfect as they often were" (573). That he did not always entirely do so is hardly surprising. It is comparatively easy to specify the immediate occasion of an experiment. To place the experiment in the context of a large research design is more difficult, especially when the ramifications of that design are unclear. Priestley's research was generally consistent with the Newtonian natural philosophy of eighteenth-century Britain. He would scarcely feel it necessary to declare his Newtonianism (everyone, except a few eccentrics, was a Newtonian), nor would he think to define its ontological implications.

Perhaps, at this early date, he could not even have distinguished these himself nor recognize the difference between his old-fashioned, physico-theological Newtonianism, learned at Daventry, and the secular, pragmatic Newtonianism of the mid-century readers of his work. Their object was the use of Newtonian principles to achieve understanding of the sciences; his was the use of the sciences, through Newtonian principles, to achieve understanding of the universe in relationship to God. This program is revealed primarily in the unity imposed on his research; it is not easily manifest in the design of particular experiments.

One of the earliest sets of these led to the "discovery," reported with much enthusiasm to his correspondents, of the conductivity of mephitic air (usually carbon dioxide). He began those experiments, having noted that a candle would not burn in mephitic air, reasoning: ". . . that all metallic bodies which are the most perfect conductors we know, consist of a vitrifiable earth, and what the chymists call *phylogiston*, or probably nothing more than this same mephitic air in a fixed state" (600).

He shortly discovered that mephitic air was not, as he had thought, a conductor, moisture in the air had misled him; but by that time he had been led to discover the conductivity of charcoal "from which mephitic air was procured in the greatest quantity" (607). And this discovery led to the question: ". . . since the calces of metals, which are electric bodies [i.e., non-conductors], become metals, and conductors, by being fused in contact with charcoal; are not metals themselves conductors of electricity, in consequence of something they get from the charcoal? Is not this the mephitic air, as the modern chymists suppose, that this is all that the metallic calces require to their revivication?" (607–8). The experiments were electrical, but the observation is a chemical one and the initial motivation to the experiments seems to have been chemical as well.

It is not strange that Priestley should have chemical questions on his mind, considering his recent involvement in chemical studies. What is surprising is the extent to which explicit chemical considerations enter so many of the experiments he performed in electricity. In his first letter to Canton, Priestley mentions having made some of Canton's artificial "phosphorus"; refers to discussions of such phosphorescent substances by the German chemist Caspar Neumann, in "his chemistry, published by Lewis"; and recommends that Canton read Neumann. Did Priestley subject this "phosphorus" to electrical experimentation? In the *History of Electricity*, he notes speculations on the relationship between "electric light" and phosphorescent light, by Francis Haukesbee and Dr. Hall. He suggested that "electric light [is] a real vapour ignited, similar to that of phosphorus."[23]

He tested the conductivity of inflammable air, comparing it with that of common and fixed airs. He proposed a "course of chymical electricity," testing conductivity of chemical substances: salts, acids, earths; and took electrical "explosions" through ores, noting especially the effect of sulphur in lead ore and arsenic in tin (616–20, 666). Next only to experiments done with larger electrical "force," those done with chemical associations contributed most in swelling the number and variety of Priestley's original electrical experiments.

His previous studies in chemistry were not solely responsible for his thus relating chemistry and electricity. In David Hartley's *Observations on Man* is a statement which Priestley found so important that he twice paraphrased it in the *History of Electricity*. In the preface:

> Hitherto philosophy has been chiefly conversant about the more sensible properties of bodies; electricity, together with chymistry, and the doctrine of light and colors, seems to be giving us an inlet into their internal structure on which all their sensible properties depend.

Further on, Priestley wrote:

> . . . chymistry and electricity are both conversant about the latent and less obvious properties of bodies; and yet their relation to one another has been little considered. . . . Among other branches of Natural Philosophy, let the doctrine of LIGHT AND COLOURS be also particularly attended to. It was this that Newton thought would be the key to other, at present occult properties of bodies.[24]

Now Hartley had recommended the combined study of chemistry and light; he had elsewhere related light and the aether and, in a third place, electricity

23. Priestley to Canton, 14 February 1766, *SciAuto*, no. 2. *History of Electricity*, 10, 17, 487. The book he referred to was William Lewis, ed., *The Chemical Works of Caspar Neumann* (London: W. Johnston, G. Keith, A. Linde, P. Davey and B. Law, T. Field, T. Caslon, and E. Dilly, 1759); a copy was to be found in the Academy library. The editor declares (preface sig. A4v), that Neumann is "biased by no theory, and attached to no opinions," but this work was a major avenue for the introduction of phlogiston ideas into British chemistry. It is probable that Priestley's notions about mephitic air and metallic calces were obtained from Neumann's work; see 53, 463, 473.

24. *History of Electricity*, xiii, 502–3. For the Hartley statement, see Chapter II.

and the aether. These paraphrases are typical examples of Priestley's skill at textual conflation; they also define the larger design of his researches.

Not only does the joint consideration of chemistry, electricity, and optics provide structure for Priestley's entire scientific career—his next major scientific work was a history of discoveries in vision, light, and colors, and he then commenced researches in chemistry—it also is the clue to the questions his electrical research aimed at answering. The very first "Query" or "Hint for further research" in the *History of Electricity* refers to the spectra of electric lights (spark discharges or those taken in vacuo) and suggests observing whether the electric fluid from different bodies had different properties with respect to light (487). Describing his own experiments, he wrote: "As the flames of different bodies yield very different proportions of the prismatic colours, I have often thought of attempting to ascertain the proportions of these colours in electric light, and compare it with the proportion of colours from light procured in various other ways . . . but I have not had leisure to pursue the inquiry" (726).

In another of the Queries, he asked: ". . . is the refractive power of glass the same when it is charged or excited?" (494). Priestley did not declare the investigation of the "latent and less obvious properties of bodies" as the motivation for his electrical research, but most of his experiments can be seen as contributing to such an investigation. One of the Queries demands: "Is there only one electric fluid, or are there two? Or is there any electric fluid *sui generis* at all, distinct from the ether of Sir Isaac Newton?" (488). Taken as a coherent whole, the rationale for his research can be understood as an attempt to answer that query and such questions as, What is the nature of the electric fluid? How does it differ from that of other bodies? "How are the properties of bodies influenced by the possession of electric fluid? How do bodies that differ in their reception of electric fluid differ from one another?

As he discovered that bodies cannot be differentiated into conductors and nonconductors: ". . . there is a gradation in all substances, from the most perfect conductors of electricity" (435), he attempted to classify substances by their ease of conduction and other properties. Examining Franklin's notion that "electrics" (relative nonconductors) naturally contain larger quantities of electric fluid than conductors do, he allowed a piece of red-hot glass (a conductor) to cool in contact with an insulated piece of copper until it became cold, and therefore a nonconductor. Priestley argued that, were Franklin correct, the cold glass must take its necessary additional fluid from the copper during cooling. The copper should then have a deficiency of fluid (i.e., be negatively charged). It was not, Franklin was wrong, and the essential difference between electrics and conductors remained undiscovered (716).

A research program integrating the physical sciences in an attempt to discover the internal structure of bodies, "on which all their sensible properties depend," was a very different undertaking from that contemplated by most of Priestley's contemporaries. It was also a program that could provide problems sufficient to keep an experimental natural philosopher fully occupied. But Priestley was a natural philosopher only in his spare time and even that was soon to be decreased, as he took on a new range of duties. In April 1767, shortly after the publication of the *History of Electricity*, he wrote to John Canton that he was leaving Warrington for Leeds. He did not intend to give up scientific interests, for he asked Canton's help in obtaining apparatus, such as a microscope and telescope, to replace that belonging to the academy, which he had been using.[25] Nevertheless, he would certainly be doing science from a completely different social context.

The *Leeds Intelligencer* gave the details, in a brief announcement of 9 June 1767: "The Rev. Dr. Priestley of Warrington is appointed minister of the dissenting meeting-house at Mill-Hill in this town."[26] Priestley explains the change in his *Memoirs*: ". . . my wife having very bad health, on her account chiefly I wished for a removal. . . . It was not possible even living with the greatest frugality to make any provisions for a family. . . . I therefore listened to an invitation to take the charge of the congregation of Mill-hill chapel, at Leeds, where I was pretty well known, and thither I removed in September, 1767" (W 1/1:62).

Once again, his recollection is but a partial truth. Certainly Mary Priestley was overworked (and underpaid at £15 per annum per student) as "house-mother" to academy students living in the Priestley home; possibly she found the Warrington climate unhealthy. The Leeds salary of one hundred guineas a year plus a house was little better than Warrington Academy's one hundred pounds and house, but there may have been more certainty in a minister's salary than in that of an academy tutor, anxiously dependent upon increasingly dilatory annual pledges. Besides, a Dissenting minister might hope members of his congregation would find future provision for members of his family.

---

25. Priestley to Canton, 21 April 1767, *SciAuto*, no. 14.
26. "Extracts from the *Leeds Intelligencer*, 1763–1767; 1768," *Publications of the Thoresby Society, Miscellanea*, 33, no. 725, p. 198.

These reasons, however, ignore what surely was of greatest importance to Priestley. He had trained as a minister and had essentially failed in his services to two small and unimportant congregations. Even brother Timothy, whose education Priestley thought was deficient, had been more successful and was now a minister in the more challenging arena of Cannon Street Chapel, Manchester. Soon after his ordination in 1762, Priestley had begun occasional preaching in the Warrington area and, as his scholarly reputation increased, so did his opportunities to preach. Now he was to have a chance to return, full-time, to the calling that he was always to regard as the most important a man could have.

That opportunity came, not simply from a respectable congregation, but from the one, of all others, that he could not have refused. He was given the chance to return "home" in triumph! His boyhood congregation of Heckmondwike had rejected him. Now he was to be minister at Mill Hill Chapel, the oldest and largest Dissenting meeting-house of Leeds, the city that dominated, commercially, culturally, intellectually, and religiously, the area in which he had grown up.

The minutes of the Trustees of Warrington Academy, for 25 June 1767 note the resignation of the Reverend Dr. Joseph Priestley from his position as tutor of languages. It was to take three men to fill the vacancy left by his departure. John Seddon took over the lectures on grammar, oratory, and history; John Reinhold Forster (temporarily) those on modern languages, and John Aikin those on the classical languages. Priestley's success as a teacher was indisputable. It was time for him to go on to new challenges and new opportunities.

VII

LEEDS, 1767–1773

Theology, Natural Religion

It was during an exciting, if not entirely propitious, period for England, and for Leeds, that the Priestleys moved from Warrington. On the eve of the Industrial Revolution and of great economic and population growth, it was also the time of rapid growth of the evangelical religious movement, the Wilkes exclusion crisis, and of Grenville's Stamp Act and the Townshend Acts. In the educational sanctuary of Warrington Academy, Joseph Priestley had had little occasion to involve himself in any of these matters. Now, thrust into the turbulence of public life, he could scarcely avoid involvement, for both the Leeds Dissenting congregations and the Leeds merchants were vigorously active in their responses to these events.

The town was entering a period of expansion from a population of roughly 16,000 in 1767 to nearly 31,000 within a generation. Most of that growth occurred during the latter part of the century, as larger manufacturing replaced the merchant–small clothier economy, but elements of the shift were taking shape during the years of Priestley's stay in Leeds. Much of the communications and transportation system was already in place, organized and financed by leading cloth merchants. There were turnpikes to Halifax and to York; "flying coaches," leaving Leeds three times a week, reached London in two-and-one-half days. In 1770, a Leeds-Liverpool Canal was authorized and construction begun. Soon afterward the Aire and Calder River Navigation extended a cut to Selby, and there connected with the Ouse, the Humber, and the sea.[1]

---

1. The Leeds-Liverpool Canal was undercapitalized and its construction further delayed by the wars in America and Europe; it was not completed until 1816. Proprietors of the Aire and Calder Navigation successfully opposed a canal to Selby by opening the Selby-cut in 1774.

Economically the period was one of wide variations. Over the century Yorkshire's share of Britain's wool and worsted industry rose from 20 to 60 percent and the cloth-finishing and -merchandising side of the business centered in Leeds. But during the 1760s and early 1770s, a substantial part of Leeds's trade was with the American colonies whose nonimportation agreements, following the Stamp Act and again after the Townshend Acts, severely hurt Leeds's exports. On 7 March 1769, the *Leeds Intelligencer* reported that Leeds's merchants: ". . . have received orders from their correspondents at Boston and New-York, to the amount of many thousand pounds, yet not a single piece is to be shipt till the duties on paper, glass, &c. laid upon the colonists by the last parliament be repealed."

In August the *Leeds Mercury* kept feelings high, reporting: "It has been calculated, that the Government have actually lost more from the Relax of Duties paid on American exportation, in one month, since coercive Measures have been adopted, than the whole Yearly Produce of their intended Tax upon the Colonies would have amounted to."[2] When these acts were repealed, buoyant trade followed when America took as much as 30 percent of Leeds's exports. Leeds's merchants were, of course, earnest petitioners for repeal of the Stamp and Townshend Acts.

For all the political excitement and economic shifts, Leeds could be a pleasant place to live between 1767 and 1773. Urban amenities had been developed for established cloth merchants, whose mode of life was inferior only to that of country gentlemen. Waterworks supplied running water to subscribers; some of the streets were lit by oil lamps; there was a concert hall in one of the public houses and a "New Concert Hall," in which the "Messiah," other Handel oratorios, and occasional musical and dramatic entertainments were periodically presented. There was also a set of Assembly Rooms, and, after 1771, even a theater. Two newspapers, the "radical" *Mercury,* and the more conservative *Leeds Intelligencer,* supplied weekly news (mostly secondhand, from the London papers); bookshops and a printing press provided more substantial reading. There was a bank, a Grammar School (dating from the sixteenth century), a Free or Charity School, private boarding academies for boys and for girls, and, perhaps, a Dissenting grammar school as well.[3] Three establishment churches, Meth-

2. *Leeds Intelligencer*, no. 821, Tuesday, 7 March 1769; *Leeds Mercury*, no. 136, Tuesday, 8 August 1769.

3. The Dissenting grammar school was certainly there in 1730, but there is no reference to it in Priestley's correspondence nor in any Leeds local histories that I have seen. See W[illiam] Lawrence Schroeder, *Mill Hill Chapel Leeds, 1674–1924* (Leeds: Mill Hill Chapel,

odist and Quaker meeting-houses, English Presbyterian, Baptist, and Independent chapels answered the spiritual needs of the population, while a general infirmary and a proprietary circulating library took care of their physical and intellectual needs.[4]

Priestley probably had little to do with these civic amenities. When he and his family arrived in Leeds, the house found for them was on Meadow Lane, across the River Aire from the main part of town, next to Jacques Brewery. The next year he moved to a house on Basinghall Street, more convenient to Mill Hill Chapel but not in a fashionable part of Leeds, where he might be caught up in town activities.[5]

It is wholly unlikely that the social activities of a provincial Assembly would appeal to him; his interest in music, except as an amateur performer on the flute, was slight, though a sacred oratorio might have drawn him. His opinion of the theater, given the amateur theatricals at Daventry and Warrington, may have been more liberal than that of the majority of Dissenters of his day. Still, his reply to William Enfield's plea that Dissenting ministers cultivate the "lively fictitious representation of human characters" suggests that he would not sacrifice useful and important time and activities for attendance at a playhouse.[6]

For Priestley was a serious person, with little time for frivolity. His amusements were private, with family or friends. Arriving in Leeds, he

---

1924), 38, for reference to the school under Joseph Cappe, and John Cossins's 1725 map reprinted in "Printed Maps and Plans of Leeds, 1711–1900," ed. K. J. Bonser and H. Nicholls, *Publications of the Thoresby Society* 47 (1958): 1–148.

4. These introductory general remarks about Leeds in the period 1767 to 1773 are derived from a number of sources. Chief among these were: R. G. Wilson, *Gentlemen Merchants: The Merchant Community in Leeds, 1700–1830* (Manchester: Manchester University Press, 1971); N. W. Beresford and G. R. J. Jones, eds., *Leeds and its Region* (Leeds: for the British Association for the Advancement of Science, 1967); and Edward Parsons, *Civil, Ecclesiastical, Literary, Commercial, and Miscellaneous History of Leeds, Bradford, Wakefield, Dewsbury, Otley, and the District within Ten Miles of Leeds* (Leeds: Frederick Hobson, and Simkin and Marshall, 1834). I have also seen microfilm of the *Leeds Intelligencer, 1767–1773* (by courtesy of Leeds City Libraries) and the *Leeds Mercury*, though all numbers between January 1767 and January 1769, and January 1771 to January 1773 are missing. Extracts from *Intelligencer* and *Mercury* issues during Priestley's years there, edited by G. D. Lamp and J. B. Place, are in *Publications of the Thoresby Society* 33 (1930–32): 156–208, 209–27; 40 (1955): 1–247.

5. Toward the end of Priestley's ministry, the congregation built a parsonage next to the chapel, but it was not ready for occupancy until his successor arrived; see Schroeder, *Mill Hill Chapel*, 49. This probably accounts for the tradition that Priestley Hall of Mill Hill Chapel was built on the site where Priestley had lived.

6. Joseph Priestley, *Letters to the Author* [William Enfield] *of "Remarks on Several late Publications relative to the Dissenters, In a Letter to Dr. Priestley"* (London: J. Johnson, 1770), 62–63.

really knew no one but his Priestley relatives. Not all the Priestleys in the region were of Dr. Joseph's family, but his father, Jonas, still lived in Fieldhead. Amelia (who took over a girls' school in Fieldhead in 1778), John Sr. and John Jr. (who advertised for a lost dog, 28 March 1769), Jonathan (merchant of Leeds), Mr. Joseph (who died 1 April 1770), Joseph Jr., Samuel, Thomas, William Sr. and Jr., and a Mrs. Priestley (who died in February 1771) are noticed in the *Leeds Mercury* and identified with Dr. Joseph's birthplace of Birstall Fieldhead. There were at least a dozen relatives: father, sisters, half-sisters, brothers, aunts, uncles, and cousins, within fifteen miles of Priestley when he returned to live in Leeds. There is no evidence that he visited, or was visited by, any of them during his six years there.[7]

He had, however, many old friends elsewhere and soon acquired new ones. His departure from Warrington having been an amicable one, he corresponded with John Seddon and was visited by Seddon and his wife shortly before Seddon's fatal accident in January 1770. Previously Priestley had visited Warrington to oversee printing of a second edition of the *History of Electricity,* staying with the Aikins. He also corresponded with Anna Laetitia Aikin in 1769 about her poem "Corsica." She seems to have visited the Priestleys in Leeds, as her poem "An Inventory of the Furniture in Dr. Priestley's Study" appears to describe that in the house on Basinghall Street.[8]

Apparently he did not become friendly with any of the churchmen or Dissenting ministers of Leeds, though Thomas Whitaker of Call-Lane Independent Chapel, baptized Joseph Jr., born 24 July 1768. Priestley did, however, become friendly with the Dissenting minister, William Turner of Wakefield. The two exchanged pulpits, frequently traveled together visiting other Dissenting ministers, such as Newcome Cappe in York, and on 19 June 1769, they visited Archdeacon Francis Blackburne, rector of Rich-

7. The verso to the half-title of Priestley's *Familiar Introduction to the Study of Electricity* (London: J. Dodsley, T. Cadell, and J. Johnson, 1768), refers to a venture, with brother Timothy, then minister of an evangelical Independent Chapel in Manchester, in the manufacture and sale of electrical machines; the venture is again mentioned in a letter to John Canton, 28 February 1770, *SciAuto*, no. 26. Lack of evidence is notoriously weak ground for an argument but lack of any reference in his correspondence or published writings, to Priestley's relations in Fieldhead (while there is a note on visiting his in-laws in Wales) suggests that they continued ostracizing their heretical relative. This would be consistent with the behavior of fervent eighteenth-century evangelicals.

8. See [Aspland,] "Brief Memoir of . . . Seddon," *Christian Reformer* 11, n.s. (1855): 365–74. A copy of the letter to Miss Aikin, 13 June 1769, is in the Yates Priestleyana Collection, no. 15, used by kind permission of the president and council of the Royal Society of London. See appendix to this chapter, for a copy of the poem.

mond, Yorkshire, where they met Theophilus Lindsey, rector of Catterick, Yorkshire, and husband of Blackburne's stepdaughter. This was to prove a momentous meeting, for Lindsey became Priestley's closest friend. Soon the two were exchanging frequent letters and Priestley rarely wrote for publication without first sending for Lindsey's criticism.[9]

There were few people with scientific interests in Leeds with whom Priestley might associate. He performed some experiments with the Leeds surgeon, William Hey, whom he nominated for fellowship in the Royal Society in 1775, and it is said he "worked together on gases" with the Dissenting minister, William Dawson of Idle, about seven miles away.[10] The Reverend John Michell, clergyman of nearby Thornhill, geologist, astronomer and physical scientist, and friend of Henry Cavendish, was often consulted during the writing of Priestley's *History and Present State of Discoveries relating to Vision, Light, and Colours* (1772). The engineer, instrument-maker, and experimentalist John Smeaton had a home near Leeds, in Austhorpe, and was occasionally visited by Priestley.

As Priestley's scientific reputation increased, with the appearance of successive editions of the *History of Electricity,* his correspondence with other "electricians" in Britain and on the Continent increased. So also did visits from them. It is probable that the visit to Leeds in June of the Danish Count Holstein was prompted by Priestley's international recognition as a science writer. For direct contact with science and scientists, however, he was primarily dependent upon visits to London.

Priestley began, during these years at Leeds, his practice of an annual visit to London to consult with his publisher, Joseph Johnson; attend meetings of the Royal Society (where he usually read a paper); and visit scientific and theological friends. And they occasionally reciprocated. James Douglas, Earl of Morton and president of the Royal Society when Priestley became a Fellow, passed through Leeds in April 1768; no doubt he visited Priestley on that occasion. Benjamin Franklin and Morton's successor as President of the Society, Sir John Pringle, certainly visited Priestley in Leeds in June 1772 and there witnessed some of his experiments. Clearly he did not need

9. The greater part of Priestley's surviving correspondence consists of letters to Lindsey and more of Priestley's private thoughts are there noted than in any other place. These are the letters most quoted (frequently with unnoted excisions and editing) in John Towill Rutt's edition of Priestley's life and correspondence; see W 1/1 and 2. The originals are preserved in the Dr. Williams's Library, London.

10. J. H. N., *O₂: The Bicentenary of the Discovery of Oxygen by Joseph Priestley* (Leeds: Leisure Services Department, Leeds City Museums, 11 September–26 October 1974, Exhibition Hall, City Square), 20.

to seek Leeds activities and Leeds associates to occupy time not devoted to professional duties at Mill Hill Chapel.

Priestley did, however, accept some civic responsibilities not immediately required by his ministry. He began the custom for Dissenting ministers to join the Leeds clergy in preaching an annual charity sermon for the benefit of the charity school—including, ironically (for he was becoming an Unitarian), one on Trinity Sunday 1770. He also preached charity sermons for the General Infirmary, which had been organized in May 1767, shortly before his arrival in Leeds.

His major civic interest, though, was in the establishment of the Leeds Library. There is a tradition that the library was his idea and that he personally compiled its first catalogue, but this cannot be confirmed. On this, as on so many other of his activities, his personal correspondence is silent. That he was deeply involved in the founding and first operation of the library is, however, quite clear. He had been a member of the Warrington Library committee and could have known details of the circulating library in Liverpool through his friendship with the Warrington Academy trustee, Thomas Bentley. In August 1768, at about the time he was moving from Meadow Lane to Basinghall Street, advertisements appeared in the Leeds paper calling a meeting of persons interested in the "scheme of a circulating library" and referring to the Catalogue and rules of the Liverpool Library as models. At the first business meeting, on 7 September 1768, Priestley was chosen secretary. The next year he was elected president of the Library Committee, a position he held until his departure from Leeds in 1773.

Operation of the Leeds Library, in its early years, showed considerable skill and planning, suggesting experience in initial organization. It also exhibited rather more political sophistication than one might expect of Priestley. The first president of the Committee was a prominent cloth merchant. Other members of the Committee included the vicar of Leeds parish, medical men, prominent members of Mill Hill and Call-Lane Chapels, a member of the local gentry, and the proprietors of each of the Leeds newspapers.

The "Catalogue of the Books belonging to the Circulating Library in Leeds," published in 1768, listed 503 volumes; by 1770, the library contained more than eight hundred volumes. Most of these were histories, travel books, and polite literature; there was a scattering of science and medical works, some volumes of reviews, and very little theology—none of it controversial.[11] The Leeds Library continued after Priestley left Leeds and remains today a substantial reminder of his service to the community.

---

11. [Joseph Priestley?], *Laws for the Regulation of the Circulating Library in Leeds; and a Catalogue of the Books Belonging to it: To which are prefixed, the Names of the Subscribers*

Of course Priestley's primary concern during his years at Leeds was with the activities of Mill Hill Chapel. Once again, after an hiatus of six years, he was a Dissenting minister and it was particularly important this time that he not fail. It was not only that he needed the salary now that he was married and had a child. True, the salary at Mill Hill was nearly equal that of the average superior clergy and most legal professionals, but if money alone were a deciding factor, he would probably have remained a teacher. The one hundred guineas a year was not much superior to the hundred pounds a year he had received upon going to Warrington, and, in either case, he would need to supplement his salary with income derived from his writing—most of which had been on educational subjects.[12]

But, in fact, he had given up teaching, where he was a conspicuous success, to return to the ministry, where he had twice foundered, because he regarded the Dissenting ministry as the most important position a man could hold. With two small and comparatively unimportant congregations, he had essentially failed. Now he was attempting his return with one of the oldest and most opulent of Dissenting congregations in the north of England, a congregation he had known of since his childhood. And he had, further, to justify by his success there his departure from the theological opinions of the Priestleys at nearby Fieldhead, at just the time when their evangelical views were increasingly accepted in Leeds and the surrounding region.

The congregation at Mill Hill had been formally organized in May 1672, within three months of the Act of Indulgence by King Charles II, which freed "erring and dissenting persons" from the penalties of the Conventicle and Five-Mile Acts of 1664 and 1665 if they applied for a license from a magistrate. The chapel was completed in 1674 at a cost of £400. Described at the time as the "first and grandest meeting-house in the North," it was a "proprietary chapel," built, paid for, owned and maintained by members of the congregation to whom the property would revert if, as in 1682, "conventicles" were again suppressed (see Fig. 6). The chapel was publicly

---

(Leeds: by Griffith Wright, 1768); Frank Beckwith, "The Beginning of the Leeds Library," *Miscellanea of the Thoresby Society* 37 (1941). I am also personally indebted to Mr. Beckwith, Librarian of the Leeds Library, for his courtesy in showing me about the library in the summer of 1978 and for the opportunity to examine copies of the 1768 and 1778 catalogs. The latter contains considerably more theology than was listed in that of 1768, but represents some five years of purchases after Priestley's departure.

12. The figures on salaries of clergy and lawyers are taken from Wilson, *Gentleman Merchants*, 82.

Fig. 6. Interior of Mill Hill Chapel, Leeds, where Joseph Priestley preached. Courtesy of the Department of Special Collections, Van Pelt–Dietrich Library Center, University Libraries, University of Pennsylvania.

reopened in 1687, at the Act of Indulgence of King James II and was permanently settled with the Toleration Act of William and Mary in 1689.[13]

The first minister, Mr. Thomas Sharp, was an orthodox Calvinist as, incidentally, was Mr. Nathaniel Priestley invited (unsuccessfully) to succeed Sharp in 1693. The only difference between the Independents and the Presbyterians in those early years of dissent had been disagreement over forms of church government. From 1660 to as late as 1689, however, the English Presbyterians had hoped for the establishment of a national church that would comprehend Anglicans and Presbyterians—and exclude Independents. While waiting for comprehension, they failed to establish the apparatus of presbyterian governance; and they moved, in an effort of doctrinal compromise, toward the Arminian position of the established church.[14]

13. See Schroeder, *Mill Hill Chapel*, and Charles Wellbeloved, "Presbyterian Nonconformity at Leeds," *Christian Reformer* 3, n.s. (1847): 392–400, 477–87, 522–37. I am also personally indebted to the correspondence of the Reverend Maurice Bonner, minister at Mill Hill Chapel, who searched the scanty pre-1773 records of the chapel for me in 1978.

14. As the elder Pitt was to say in a speech before the House of Lords in 1774, the established church of the eighteenth century had Calvinist articles, a Popish liturgy, and an Arminian clergy; see, for example, T. C. Holland, "Brief History of the Dissenters from the

When it finally became clear in 1689 that the established church would not comprehend the Presbyterians, there was an attempt to form a union of major Dissenting groups: Independents, Presbyterians, and Baptists, dropping distinctive names and joining a common fund to support Dissenting congregations, ministers, and students. Presbyterians, however, would not function within the discipline represented by such a union. The union began to dissolve, first in London and gradually over the country. The "conservative" Independents established their own fund for supporters of rigorous Calvinistic doctrine; the Presbyterian Fund supported the moderates and the theological "liberals." And, as congregations resumed distinctive names, they took that defined by the fund from which they received support, not by their historical connections.

At first, doctrinal differences were variations within a continuum. The decisive break occurred at the Salter's Hall Synod of 1719. There, at a joint convocation of ministers in the London Dissenters' union, the majority of Presbyterian ministers refused to subscribe to a creed to be imposed by the Synod, a creed that included a declaration of belief in the Trinity. Probably most of the nonsubscribers (which included Independents and Baptists) were Trinitarians, but thought it morally wrong to require subscription to a human expression of doctrine when the Scriptures provided a sufficient guide to Christian belief.

As the English Presbyterians had not established sessions of presbyters, their ministers were dependent only on the proprietors or trustees of the individual chapels that appointed them. These were likely to be the best educated, wealthiest, and most socially aware of their congregations. A consequence of the emphasis on the Bible as the sole theological authority in their chapels came to mean, therefore, the Bible as understood by reason, rather than through traditional interpretation. English Presbyterians increasingly became what Priestley was to call "rational Christians." This, in turn, frequently came to mean some form of anti-Trinitarianism: Arianism or Unitarianism.[15]

Revolution," *Monthly Repository* 12 (1817): 457. Arminians essentially rejected the Calvinist emphasis on original sin and predestination, arguing that salvation required faith in God, obedience to His will, repentance of sin, and the performance of good works.

15. See James C. Spalding, "The Demise of English Presbyterianism: 1660–1760," *Church History* 28 (1959): 63–83; J. Johansen-Berg, "Arian or Arminian," *Journal of the Presbyterian Historical Society* 14 (1968–72): 33–58. Arians made a distinction between the three persons of the Trinity; Son and Holy Spirit being subordinate to the Father, though created prior to the creation of the universe. Unitarianism, particularly the humanitarian Unitarianism increasingly adopted by Priestley, held that the Father only is God, that the Holy Spirit represents the

That is what happened at Mill Hill Chapel, Leeds. As late as 1730 to 1748, the minister, Joseph Cappe, was Arminian, but still Trinitarian. His congregation, said to number one thousand hearers, included an uncle of Joseph Priestley, who led members of the congregation in opposition to the Pretender's invasion of England in 1745. Cappe's successor, however, was Thomas Walker, an Arian, whom Priestley described as "one of the most heretical ministers of the neighbourhood" when he visited the Keighly home at Heckmondwike in Priestley's youth. Many of the members of Mill Hill Chapel objected to Walker's views and left to join the Independent Chapel on Call Lane, but Mill Hill's proprietors must have approved, for Walker's successor, Nathaniel White, was also Arian—and thus prepared the way for his successor, Joseph Priestley and his Unitarianism.

Unfortunately for those who wanted undiluted Calvinism, Thomas Whitaker, the minister at Call Lane Chapel from 1727, was a liberal in theology. By the time he retired in 1776 he had probably also become an Arian. Furthermore, the general drift toward conforming, which by mid-century was shrinking the size of Dissenting congregations in most areas, had, in Leeds, been halted by a delay between 1745 and 1751 in appointing the parish vicar and by subsequent public disputes over patronage rights. For a short time there was a lack of spiritual leadership in Leeds, except for the comparatively few persons satisfied by the intellectualism of "rational Christianity."

That leadership was, however, quickly supplied by the Methodist movement, or movements, for there were at least three aspects to mid-eighteenth-century Methodism. There were the Arminian, extra-Anglican followers of John Wesley; the Anglican evangelicals; and the Calvinist Methodists of George Whitefield and the Lady Huntingdon Connection. These three groups appealed to somewhat different constituencies, but each emphasized emotional appeal, original sin, and spiritual rebirth through a conversion experience. Together the three were responsible for the rapid growth of what turned into a new Dissenting movement during the second half of the century and for a new enthusiasm in parts of the established church.[16]

---

exerted will of God, and that Christ was entirely human, divinely selected to represent the benevolence of God, His power and will, and assurance of resurrection to believers.

16. See Alan D. Gilbert, *Religion and Society in Industrial England: Church, Chapel and Social Change, 1740–1914* (London: Longmans, 1976), passim; Bryan Dale, "Early Congregationalism in Leeds," *Transactions of the Congregational Historical Society* 2 (1905–6): 247–61, 311–25, esp. 259; and W. G. Rimmer, "William Hey of Leeds, surgeon (1736–1819): a reappraisal," *Proceedings of the Leeds Philosophical and Literary Society* 9 (part 8, 1961): 187–217, esp. 194.

By the time Joseph Priestley arrived in Leeds, all three forms of Methodism were well established. There was a Wesleyan meeting-house in Leeds and many in surrounding parishes. John Wesley made several visits to the area starting in 1742, regularly preaching at Birstal. By 1753, he had determined to hold his annual conference of Methodist preachers in Leeds once every three years. That held there early in August 1769 was said, by the *Leeds Mercury*, to be attended by several hundred ministers! The Reverend Henry Crooke, perpetual curate of Hunslet, in Leeds parish, was evangelical and an old friend of the Wesleys; Miles Atkinson, lecturer at St. Peter's, the parish church, was evangelical.

Some of the prominent wool merchants of Leeds were Methodists, including James Armitage whose family owned the Birstal Fieldhead estate where Priestley's relatives lived. The minister of Heckmondwike Chapel, the Reverend Mr. Scott, was a Calvinist-Methodist, as was his London patron, the Reverend Edward Hitchin, brother-in-law to Dr. Joseph's Fieldhead cousin, Mr. Joseph Priestley. The diary of the Reverend Henry Crooke for 1758 and 1759 notes visits from Mr. William Priestley and his son, the Mr. Priestleys of Fieldhead (wife, two sons, and daughter), and even a visit to the Priestleys of Birstal Fieldhead, where he joined them, Mr. and Mrs. Hitchin, and some other persons in dinner, conversation, a prayer and some hymn singing. The Priestleys of Fieldhead were clearly Methodist and brother Timothy was preaching as a minister in Lady Huntingdon's Connection.[17]

A principal problem that Priestley faced as minister of Mill Hill Chapel was, therefore, to maintain his congregation in size and loyalty to "liberal dissent" in the face of Methodist enthusiasm on the one hand and the social prestige and advantages of establishment conformity on the other. He addressed this problem with a tripart program in religious education, administration, and scholarship. The educational activities are summarized in "An Essay on the best method of communicating religious knowledge to the members of Christian societies," prefixed to one volume of his *Institutes of Natural and Revealed Religion*, first published in Leeds in 1772.[18]

17. See L. Tyerman, *The Life and Times of the Rev. John Wesley, M. A.* (New York: Burt Franklin, 1973), particularly vols. 2 and 3, passim; Wilson, *Gentlemen Merchants;* and the MS diary of the Reverend Dr. Henry Crooke, Clark MSS, Leeds District Archives, used by permission of John and David Clark.

18. The *Institutes* was first published in a 16 mo. format, three vols, with vol. 2 in 1773 and vol. 3 in 1774; there was another edition during Priestley's ministry in Birmingham, 8 vo. in 2 vols., 1782, reprinted in 1794 and in 1808. Joseph Priestley, *Institutes of Natural and Revealed Religion* (Birmingham: For J. Johnson, 1782), 1:xix–xliii.

The breaking up of traditional churchly structure and discipline has, said Priestley, placed the entire burden of religious counsel and instruction on the minister. But his time is necessarily insufficient to do everything that is needful: to warn the unruly, comfort the feeble, instruct the ignorant, confirm the doubtful, and save those in danger of being lost (xi–xli). Of course he must do everything he can, lest he violate the instructions of "the great shepherd and bishop of souls, Christ Jesus . . . to feed his lambs and his sheep" (xlii). Some things are possible "in the country" (as opposed to London), where the congregation and minister still exchange respect and affectionate concern.

But most of a congregation never see their minister except in the pulpit and that is a poor place from which to communicate religious knowledge in the regular and systematical way it ought to be taught. Certainly the minister should preach regularly each Sunday and his discourses on immediate religious practice may do something to reform the conduct of his listeners. There is little prospect, however, of making a good and lasting impression on persons after they have arrived at thirty or forty years of age. It is best to bear with the aged and even "those who are advanced to middle life and not attempt the arduous and almost hopeless task of rectifying their errors." The fruitful endeavor still left open to the minister is the instruction of youth.

And clearly, the "generality of youth in the present age" needs instruction, being extremely ignorant of religion. Parents, reacting from the strict maxims of their forefathers, have gone the opposite extreme, neglecting even daily family prayer and reading of Scripture. For want of being well established in the principles of "rational religion," many young people fall prey to enthusiasm (i.e., Methodism) or infidelity, while ignorance of the sentiment of religious liberty may lead to abandonment of the Dissenting interest. The minister should, therefore, establish special classes of religious instruction for young persons.

Priestley recommends three such classes: one for young men (and perhaps women, separately or together) from age eighteen or twenty to about thirty, to teach the elements of natural and revealed religion; another of children under fourteen, to teach the first elements of religious knowledge by way of a short catechism in the plainest and most familiar language possible; and a third, intermediate class, to teach knowledge of the Scriptures only.[19]

---

19. He used the same class divisions later in Birmingham, adding instruction in scriptural geography to the younger classes and a short course of Jewish antiquities, ecclesiastical history

Typically, Priestley also made, of his classes, a further opportunity for publication, continuing the policy he developed at Warrington. For each class he taught, he produced a text: in 1767, *A Catechism for Children and Young Persons,* with seven additional English editions, three American, and one in Welsh; in 1772, *A Scripture Catechism, consisting of a Series of Questions, with References to the Scriptures instead of Answers,* with five additional editions by 1817; and the *Institutes of Natural and Revealed Religion.* Except for the *Institutes,* which was intended to teach the principles of rational dissent, these texts were supposed to convey the fundamentals of Christian religion, without concern for the doctrinal disputes of theological systems: "I have, in imitation of the simplicity of Dr. Watts, studied to make this Catechism . . . very plain, and have not introduced into it, the technical terms of any particular system of religion . . . [and] inserted nothing but what will be acknowledged to belong to common Christianity" (*Catec.* vii).

This was the stance of most ministers of "rational dissent," who attempted to separate religion from theological argument. In the preface to his Leeds resignation sermon, Priestley declared that he had "kept the pulpit almost entirely sacred to the . . . more important business of inculcating just maxims of conduct, and recommending a life and conversation becoming the purity of the gospel." Perhaps inevitably this was a hopeless effort. Theological debate was a commonplace between the evangelicals and rationalists and the former found more reason for discontent in what the latter left out of their preaching and writing than in what they included.

An examination of Priestley's *Catechism for Children,* compared with that of the Westminster Assembly, which Priestley had learned as a child, illustrates their complaint. Except for three questions and answers (nos. 16, 18, and 56) out of sixty, each of which deals with the personhood of Christ, there is little positive reason for orthodox offense. In none of the questions and answers, however, is there any reference to the sin of Adam, reprobation, or atonement. The emphasis throughout is on the benevolence of God and on His forgiving of repented sins.

The *Scripture Catechism* is less open to obvious negative criticism, consisting as it does of various questions about persons or events in the Old and New Testaments, with answers to be found in the Scriptures themselves. The emphasis upon historical events might, however, be regarded as an avoidance of doctrinal issues, though Priestley explains that the historical

---

and so forth, to enable a more profitable reading of Scriptures; see "Method of Communicating religious Knowledge," xxxv–xliii.

parts of the Bible are the most interesting to young persons and those most conveniently taught in a catechial manner. A more orthodox selector of passages might well have chosen a different set of questions, leading, for example, to the "proof texts" on the divinity of Christ or the Old Testament passages taken as prophecies of His coming.[20]

The three volumes of the *Institutes of Natural and Revealed Religion* represent a different kind of enterprise. Comprising more than eight hundred pages (in the 12 mo. format, compared with the 86 of the two catechisms), the *Institutes* was primarily a summary of a half-century of the writing of liberal theologians on a number of issues and was to become a standard exposition of beliefs for generations of Unitarians.[21] Priestley had begun writing the *Institutes* while still a student at Daventry Academy. During his years at Warrington, he had continued his religious studies, but not intensively amid the pressures of other duties. In Leeds, he returned to serious theological reading and there, in consequence of "reading with care Dr. Lardner's *Letter on the Logos*" (W 1/1:69), he became a Unitarian.

Lardner's *Logos* had first been published anonymously in 1759, but its author was not a well-kept secret. Priestley must have known it as the work of someone who had been his friend since Lardner's advice on the publication of the *Doctrine of Remission* in 1761. Surely he had read *Logos* before; perhaps, as Alexander Gordon suggests, he was stirred to read it again by Lardner's death in 1769.[22] The arguments are hardly new—though they are not derived from any Socinian writers, being taken from Lardner's reasoning and interpretation of the Scriptures, supported by references to Grotius and early church fathers.

Why Priestley should now find them convincing is far from clear. Perhaps the insistence that, in the Scriptures, the "word body is not to be understood exclusively of the soul" (9), was already taking on special meaning for him, a monist meaning quite different from Lardner's intent. In any

20. Joseph Priestley, *A Catechism for Children and Young Persons* (Leeds: for J. Johnson, 1767); *A Scripture Catechism* . . . (London: J. Johnson, 1772).

21. See Russell E. Richey, "Joseph Priestley: Worship and Theology," *Transactions of the Unitarian Historical Society* 15 (1972): 41–53, 98–103.

22. Alexander Gordon, *Heads of . . . Unitarian History*, 109. Priestley notes (W 1/1:37) that he had visited Lardner in London in 1767, to find the old man's memory failing, but his judgment still sound in commending the sentiments of the *Doctrine of Remission*. The *Letter on the Logos* was several times reprinted; I have used an edition published as one of several Unitarian tracts: Nathaniel Lardner, *A Letter to Lord Viscount Barrington (written in the year 1730) concerning the Question, Whether the Logos supplied the Place of a Human Soul in the Person of Jesus Christ* (London: for the Unitarian Association, and R. Hunter, 1833).

event, at this time Priestley abandoned the Arian position he had held since his days at Daventry (roughly fifteen years) to be replaced by what was (incorrectly) called "Socinianism" for the rest of his life. He also adopted Lardner's argument that the notion of a preexisting inferior deity had been the creation of learned converts from heathenism (46), and Lardner's method of citing the opinions of early church fathers for his own theological polemics.

Priestley being Priestley, he announced his new theological convictions in a series of pamphlets over the next several years in Leeds. And his classes at Mill Hill Chapel provided the opportunity to summarize these, and his continuing studies, in the three-volume *Institutes*, as a systematic statement of advanced liberal Dissent. Publication of the three volumes was not completed until after he left Leeds, but one may reasonably assume that together they represent the sum of his religious teaching to the most advanced of his classes there.

He was to write many more clearly polemical works for philosophers and men of letters, but he did not value their opinion in matters of religion. "Love of knowledge or learning, like that of riches and power, was absurd when pursued as an end in itself."[23] Men of letters have their peculiar prejudices, which have been responsible for many corruptions of Christianity, a religion revealed to common workmen, intended equally for all mankind. Conscience might dictate differently in different nations and with different levels of education, but there was universal concurrence in the sanction given the same rules of conduct (1:67). By the use of the "light of nature," right reason and the Scriptures any person could achieve sufficient understanding of the nature of God, and of the duties and prospects of mankind.

Only rarely does the *Institutes* explicitly address differences of opinion among Protestant Christians, though there are, inevitably, specific references to "popish corruptions." The dedication enjoins young readers to have "nothing to do with a *parliamentary religion* or a *parliamentary God*" (vii). Priestley later urges them to avoid both *enthusiasm* and *superstition* in their devotions (1:121). Mostly, however, doctrines are conveyed in choice of subjects and the implications of the development of them; both of these are clearly rationalist and Unitarian.

The first part of the *Institutes* is a discussion of natural religion; that is, "how far unassisted reason might have been able to carry us in our inquiries concerning the being, perfections, and providence of God, though these

23. Priestley, *Institutes* (1782), 1:100; references in parentheses are to this edition.

conclusions were perhaps not drawn from these premises prior to Revelation" (1 : 173). Part 2 presents argument supporting the truths of Revelation. There follows a curious interpolated section in which Priestley argues that Revelation has followed a historical pattern paralleling his favorite scheme of associationism; the gradual development of religious knowledge through Revelation in Jewish history to Christianity being like the individual's coalescing, by association, of the ideas of difficulty, pain, misery, and so forth into the final larger picture of the genuine pleasures of life.

Finally, part 3 treats of the religious truths acquired through Revelation. As nothing can be admitted as contained in Revelation that is contrary to the plain principles of natural religion (1 : 272), most of these truths support what was learned from the first section, but there are some things (e.g., the propriety of prayer) that cannot be proved by the "light of nature."

The arguments from natural religion are mostly those familiar to eighteenth-century deism: proof of the existence of God on evidence of design in the universe. From this follow a priori characteristics of an uncaused intelligent first cause. One of these is divine unity. One cause is necessary; more than one would be divisive and complicate the smooth running of things.

This deduction is supported by evidence of comparative religions. The most important doctrines of natural religion must also have been revealed to the "first parents" of mankind. Except with the Jews (as seen in Revelation), these have been subsequently corrupted, particularly by philosophers, as the most primitive religion appears to have been unitarian (1: 180). Here is the clearest example of Priestley's technique of writing doctrine by omission. He declared his intention of avoiding controversial subjects—or, at least, not discussing them at large—by exhibiting the truths of Revelation in the words of the Scriptures themselves. However, in his selection of those words, he carefully skirted passages that other writers had used to prove the divinity of Christ.

He also indirectly attacked the fundamentals of Calvinism. The manifest tendency of things in the world show that God's design is to produce happiness. For any persons to be necessarily miserable in the whole of their existence would, therefore, be unpleasing to their benevolent creator. Sometimes it may be necessary that "the interests of a few give place to that of a greater number . . . [as] a greater sum of happiness can exist in a greater number than in a smaller" (1 : 50, 43).

Perfect moral behavior is not possible to imperfect humanity, but it would be useless to give men laws it was not possible for them to obey.

Man can, therefore, choose and pursue a life pleasing to God. God does not reject man's sacrifices as inadequate to their sins, but because He requires repentance and perseverance of effort to live a good life. Deathbed repentance is a miserable fallacy, for there must be a change of conduct (2:334, 303).

When questioning the evidences of Revelation, one must give the same value to scriptural accounts (including those of miracles and prophecies) as one gives to other historical accounts of fact. Much of the skepticism of revelation is based upon ignorance of the manners and customs, climate and geography, language and phraseology peculiar to Jews and other Oriental nations. As travelers report their experiences in the Near East, these doubts are increasingly removed. Other problems are caused by demanding too much from biblical accounts. Their truth is founded on leading facts, not on exactness in detail. One may, for example, allow fable or allegory in the account of the Fall of man without that affecting the truth of Mosaic history, or admit that, in some cases the apostles and evangelists might mistakenly imagine an Old Testament reference to Christ, without supposing all such prophecy incorrect.

> Christian divines having maintained the absolute inspiration of every word . . . of scripture, has been attended with very bad consequences, by laying . . . revelation open to . . . insignificant, but plausible objections; and this kind of inspiration is as needless, as it is impossible to be maintained.
>
> . . . Let us consider them as the production of honest and faithful men, well informed concerning all the great things of which they write, but not . . . with respect to every *puntilio* they mention. Let us consider the great truths which they deliver, as from God, to be divine . . . but when . . . advancing opinions which are plainly their own . . . let us consider them as . . . fallible, and treat them accordingly. (2:47–49)

Having demonstrated when, how, and why Revelation is to be taken as authoritative, Priestley then addressed the religious truths to be obtained from it. Most of these are seen to support natural religion, but one cannot expect clearness or consistency of moral principles from the light of nature alone. The Ten Commandments and rules of conduct laid down by Jesus and the apostles are of infinitely more service (1:243).

Neither consistency in support of natural religion nor agreement on clarity of moral principles carry over, however, to doctrines some have derived from Revelation. Infant baptism is, for example, *not* a support to the doctrine of original sin. Scriptural accounts of the baptism of whole families (including servants) show that to be part of a parent's profession of Christianity.

New Testament descriptions of the Lord's Supper show it to be a cheerful social occasion in memory of Christ's service to man, his death and resurrection. There is no suggestion of sacramental significance, no support of superstitious fear, and certainly no evidence for the "monstrous doctrine of transubstantiation" (2:352). The Lord's Supper is an affirmation of commitment to Christianity and every Christian should participate on all proper occasions.

Organization of the early church was independent of divine inspiration, but is worthy of our attention as the practice of Christians not corrupted by power and self-interest. The Scriptures show that bishops were the Greek equivalent of Jewish elders, "ordained" by *imposition of hands* in an ordinary ceremony of communal prayer. There is nothing sacred in a tradition of temporal passage of church office. Belief in the existence of the soul separate from the body is based on the "false philosophy of the east," and is wholly without scriptural authority, which rather promises resurrection of the body at the second coming of Christ. The concept of a separable spirit has led to idolatrous worship of Jesus Christ, the Virgin Mary and innumerable saints and angels (2:166, 216, 389).

The "light of nature" is unclear on the character of everlasting life or the duration of the punishment of the wicked, and the Scriptures are here of little additional help. It is clear that punishment must be long and severe if it is to have penitential value, but our reason makes it doubtful that God, with perfect foreknowledge, should select a system involving the everlasting, inexpressible misery of the greater part of his creatures. Priestley therefore argues for acceptance of Hartley's universalist concept of salvation.

During the period in which Priestley was teaching and writing this summary of his religious position, he also wrote pamphlets on each of its more controversial aspects and responses to the criticisms that these inevitably drew. The first set of these were *A Free Address to Protestant Dissenters on the Subject of the Lord's Supper* (first published in 1768 with second and third editions of 1769 and 1774); *Additions to the* [Free] *Address . . . on . . . the Lord's Supper, . . . and a Letter to the Author of the Protestant Dissenter's Answer to It* (1770); *Considerations on Differences of Opinion* among *Christians; with a Letter to the Rev. Mr. Venn, in Answer to his Free and Full Examination of the Address . . . on . . . the Lord's Supper* (1769); and *An Address to Protestant Dissenters on Giving the Lord's Supper to Children* (1773).[24]

The first and last of these pamphlets were admittedly derived from Bishop Hoadley's 1734 *Plain Account of the Nature and End of the Sacrament of the Lord's Supper* and James Pierce's *Essay in favour of the ancient Practice of giving the Eucharist to Children* of 1728. Priestley hoped he had made a contribution by bringing all the arguments into one view and into shorter compass. He also believed his manner of treating the subject of the Lord's Supper—by exhibiting the genuine Scripture doctrine and then tracing the corruptions of it historically—might have some advantage in correcting long-standing religious prejudices.

Following this procedure, it becomes clear that the Lord's Supper was originally instituted as an emblem of the cheerful and benevolent Christian religion, parallel to the Jewish celebration of Passover. All Christians therefore should celebrate the rite regularly and include their children, to establish them in early habits of worship. The notion of magic in the ceremony, suggested in the articles of the Church of England and in the Westminster Catechism, was an unreformed remnant of Popish superstition that Dissenters could easily discard, acting individually by congregations that have not subscribed to articles of faith. A rational Dissenter should consider the significance thought by others to inhere in the rite and not offend by intruding where he was not wanted, but, like "every person, who has a just regard for the honour of religion," he should bear "his testimony against so base a prostitution of its rites as making them a qualification for a civil office" (*Free Add.* 33).

These arguments had all been stated by earlier writers; indeed the immorality of encouraging occasional conformity and the belief that a necessary continuation of religious reform was more likely to occur in Dissenting congregations had recently been affirmed in *The Confessional* by Priestley's new friend, Archdeacon Francis Blackburne.[25] Nevertheless, the *Free Ad-*

---

24. Joseph Priestley, *A Free Address . . . on . . . the Lord's Supper* (London: J. Johnson, 1769), 2d ed.; *Additions to the [Free] Address* (London: J. Johnson, 1770); *Considerations on Differences* (London: J. Johnson and J. Payne, 1769); and *Lord's Supper to Children* (London: J. Johnson, 1773).

25. [Francis Blackburne], *The Confessional: or, a Full and Free Inquiry into the Right, Utility, Edification, and Success of Establishing Systematical Confessions of Faith and Doctrine in Protestant Churches* (London: S. Bladon, 1767), 2d ed., xlvi; x–xvii of preface to first ed.

dress at once elicited critical responses to which Priestley as quickly wrote answers. This was not entirely because he was contentious by nature. In spite of repeated experiences to the contrary, Priestley was always to believe that truth could be obtained through the competition of opposing views. He had early stated this opinion in his work on grammar and again in that on aesthetics; now, in the *Free Address,* he wrote: "Whenever any opinion is freely canvassed in open daylight, it will be easy to see on which side the truth lies" (vii). Perhaps it was on this principle that he entered so readily upon nearly every dispute offered him. On the evidence of his published writings, however, it would be hard to argue that he personally disliked controversy and only joined one on principle.

Appendix

AN INVENTORY OF THE FURNITURE IN DR. PRIESTLEY'S STUDY

(I)

A Map of every country known,
With not a foot of land his own.
A list of folks that kicked the dust
On this poor globe, from Ptol. the First;
He hopes—indeed it is but fair—
Some day to get a corner there.
A Group of all the British kings—
Fair Emblem!—on a pack-thread swings,
The Fathers, ranged in goodly row,
A decent, venerable show,
Writ a great while ago, they tell us,
And many an inch o'ertop their fellows.
A Juvenal to hunt for mottoes;
And Ovid's tales of nymphs and grottos.

(II)

The Meek-robed lawyers, all in white,
Pure as the Lamb,—at least to sight.
A shelf of bottles, jar and phial,
By which the rogues he can defy all,—
All filled with lightning, keen and genuine,
And many a little imp he'll pen you in;
Which, like Le Sage's spite, let out,
Among the neighbors makes a rout;
Brings down the lightning on their houses,
And kills their geese, and frights their spouses.
A rare thermometer, by which
He settles to the nicest pitch
The just degree of heat, to raise
Sermons, or politics, or plays.

(III)

Papers and books, a strange mixed olio;
From shilling touch to pompous folio;
Answer, remark, reply, rejoinder,
Fresh from the mint, all stamped and coined here;
Like new-made glass, set by to cool,
Before it bears the workman's tool.
A blotted proof-sheet, wet from Bowling,—
"How can a man his anger hold in?"—
Forgotten rhymes and college themes,
Worm-eaten plans, and embryo schemes:—

(IV)

A mass of heterogeneous matter,
A chaos dark, nor land nor water;
New books, like new-born infants, stand,
Waiting the printer's clothing hand:—
. . . .
And all, like controversial writing,
Were born with teeth, and sprung up fighting.
"But what is this," I hear you cry,
"Which saucily provokes my eye?"—
A thing unknown, without a name,
Born of the air and doomed to flame.

          —Anna Laetitia Aikin, c. 1769

The poem can be found in *A Selection from the Poems and Prose Writings of Mrs. Anna Laetitia Barbauld* (Boston: James R. Osgood and Co., 1874), 32–34.

VIII

LEEDS, 1767–1773

Religious Polemics, Theology

People who met Priestley personally, even people who disliked his opinions, found him quiet, pleasant, well-mannered, and even charming. And, of course, he cannot be held alone responsible for the pamphlet wars he entered.[1] A controversy requires at least two antagonists and Priestley had usually sufficient provocation when he concurred in one. Nonetheless, a substantial part of his writing is argumentative and the style of that part is generally free of the stiffness that mars much of the rest. He entered each controversy with a cheerful conviction that he was right, while most of his opponents were convinced, from the outset, that he was willfully and maliciously wrong. He was able, then, to contrast his sweet reasonableness to their personal rancor while responding, in kind to irony and sarcasm.

Although he claimed to be open to change of opinion and did, frequently, change his mind, he rarely did so in response to argument. Moreover, his readiness in responding to attack and persistence in argument could be infuriating to graduates of Oxford and Cambridge. They often felt that this provincial schoolmaster and Dissenting minister was ignorantly renewing a debate long since won in favor of orthodoxy.

---

1. See, for example, the anecdote reported of the Baptist minister of Philadelphia who met Priestley by accident, expected to dislike him, and ended by declaring him a "great and good man," Edgar F. Smith, *Priestley in America, 1794–1804* (Philadelphia: P. Blakiston's Son & Co., 1920), 75–77. Priestley remarks in his *Memoirs* that his speech impediment perhaps kept him from being disputatious in company (W 1/1:28).

The characteristics of Priestley's polemic style are all present in his first explicitly religious controversy: on the Lord's Supper. They are not obvious in the *Letter to the Author of the Protestant Dissenter's Answer to the Free Address* (1770).[2] Not "uniformly candid"—it declared that *Free Address* "designedly calculated to convey false and injurious ideas" (41)—the *Dissenter's Answer* was generally a respectful response and Priestley treated it gently. He was even led to some corrections and additions for the third edition of the *Free Address.* The author did not admit, however, that Calvinists took the ceremony to be more than a rite of remembrance and Priestley demanded that he answer some questions on the subject if the debate was to continue. An earlier response found in Priestley's *Considerations on Differences of Opinion among Christians with a Letter to the Rev. Mr. Venn* (1769), was more acerbic.[3]

Priestley would prefer, he declared, to debate with a man having some command of temper and of his subject, but Venn's answer has afforded him an occasion to expose the "gross misrepresentation and unworthy cavilling" that were used too frequently against merely speculative differences. Was it, for example, really relevant to the argument that Priestley prefixed his work by a motto from a pagan author [Persius]? If the origin of a motto has significance, does one infer a belief in transubstantiation from Venn's motto from a Popish cardinal? Was it true that Priestley's rational Christians—called "new schemers" by Venn—were arrogant and self-satisfied? More fairness, more candor, more meekness, and more generosity should, agreed Priestley, be expected from Christians than from men of the world and where, in Venn's writing, is there evidence of these virtues?

Leading Methodists were, no doubt, originally influenced by concern for men's souls, but they seem since seduced by their natural love of power (7). They appear unable to distinguish between differences in judgment and willful evil: ". . . let not our party prejudices blind us so far as to make us condemn and reject what is good in any set of men, merely because they hold it" (29). Venn declared that Priestley's method of quoting everything in Scriptures relating to the Lord's Supper was absurd and had been fully

---

2. This *Letter* was first published as an appendix to *Additions to the* [Free] *Address . . .* (London: for Joseph Johnson, 1770) and was then included in the third edition (1774) of the *Free Address.*

3. The Reverend Henry Venn (1725–97), vicar of Huddersfield, Yorkshire, to 1771, was a leader in the evangelical revival of the Church of England and intimate friend of Charles Whitefield and the Wesley brothers.

exposed. Priestley thought Venn should spend more time reading the Bible and less on books of controversy.

The "new schemers," according to Venn, insisted on the right to reject, as of suspicious authority, what in the Scriptures appeared to them to be unreasonable. Priestley evaded this valid accusation with an equally valid insistence that rational Christians believe in scriptural inspiration (56–57). Venn accused Priestley, variously, of being a deist, a pagan, and an atheist. Priestley observed that this had him, simultaneously, believing in one God, many Gods, and no God. He pointed to the hypocrisy in Venn's publishing a declaration that, abhorring every sort of persecution on religion's account, he would not inform the civil magistrate that Priestley was in violation of the law.

Finally, Priestley demanded that Venn, believing in a special religious efficacy for the Lord's Supper, defend compelling conscientious ministers to prostitute this "solemn religious rite" by administering communion to known infidels and persons of notoriously profligate lives, in order that they may be admitted to civil office (85). He closed by returning Venn's good wishes, with the addition "of as much modesty, candour, and meekness, as may be wanting to complete the Christian part of . . . [his] character" (88). It is not perhaps surprising, after such treatment, that Venn should shortly retire to a less demanding living, nor that Priestley should rapidly become known as a vigorous champion of rational Christianity.

Certainly Priestley seems to have gained confidence by the exchange, for the next in his series of controversial religious pamphlets was an explicit attack on the major elements of Calvinist doctrine. The topics are spelled out in the characteristically long eighteenth-century title: *An Appeal to the Serious and Candid Professors of Christianity. On the Following Subjects, viz. i. The Use of Reason in Matters of Religion. ii. The Power of Men to do the Will of God. iii. Original Sin. iv. Election and Reprobation. v. The Divinity of Christ. and, vi. Atonement for Sin by the Death of Christ. To Which are added, a Concise History of the Rise of those Doctrines; and the Triumph of Truth, being an Account of the Trial of Mr. Elwall, for Heresy and Blasphemy* (Leeds: J. Binns, 1770).[4] Possibly the reprinted account of Elwall's trial was an indirect response to Venn's scarcely veiled notice that Priestley's views were illegal. According to Elwall, he had been charged by some Staffordshire clergy with heresy and blasphemy, tried in the summer

---

4. I have used the third edition of the *Appeal* (London: J. Johnson et al., 1771) and a 1771 edition (Leeds: John Binns) of Priestley's republishing of Elwall's *Triumph of Truth* (1726).

assize of 1726, and released after an ingenuous defense of his anti-Trinitarian views. For Priestley, Elwall's most significant declaration, in what was otherwise a fairly standard argument, was probably "For as to the Holy Ghost, (their third God) it is evidently no distinct person from God, any more than a man's spirit is a distinct person from the man. . . . So the word of a man, and the spirit of a man, are not distinct persons from the man, but the man himself" (12). There are materialist implications here which Priestley was later to elaborate in his *Disquisitions Relating to Matter and Spirit* (1777).

Given the differences in circumstances—Elwall had been a vigorous supporter of the Hanoverian succession at a time when local clergy were still Jacobite—the potential parallel that Priestley attempted to draw was not a good one.[5] One should note, however, that Priestley was never charged for his beliefs, which remained illegal in England until 1813, when provisions of the Toleration Act were extended to Unitarians.

Priestley's *Appeal,* with or without the reprinted *Triumph of Truth,* was the most successful of his publications. It was published anonymously, by "A Lover of the Gospel," lest his name turn potential readers away, and at a cost of one pence so that copies might be distributed as tracts. It was written, Priestley says in his *Memoirs,* to combat the influence of the Methodists who were particularly active in Leeds at that time. But Wesley's Methodists were Arminian and the *Appeal* is anti-Calvinistic. Hence the Methodists it was combating were those of George Whitefield, such as the Priestley clan of Birstal Fieldhead. The work sold widely; five thousand copies within a few months of initial publication; 30,000 copies had been dispersed by 1787 according to the *Memoirs.* At least twelve editions (including one American) had been called for by Priestley's death, and a total of twenty English and three Welsh editions had been published by 1863.

The message of the *Appeal* is indicated by its title and was summarized in the *Institutes.* Man's God-given right reason is required for the understanding of God-given Revelation; they cannot be contradictory. No pride is involved in supposing man able to do what God commands; power to

5. Edward Elwall (1676–1744) was a mercer and grocer of Wolverhampton when he became a Unitarian. Though he adopted some of the thinking and turns of expression of the Quakers and Priestley believed him to be a Quaker (he was lent a copy of the *Triumph of Truth* by a Leeds Quaker), Elwall never became a Friend and ended his life a Sabbatarian Baptist. It appears that the case against Elwall, on the instructions of Hanoverian judges, did not go to jury on a technicality: the prosecution had failed to serve Elwall with a copy of the indictment. See [Robert B. Aspland], "Some Account of Edward Elwall and his Writings," *Christian Reformer* 11, n.s. (1855): 329–45.

do so comes from He who commands. Though we may be harmed by another's behavior, we cannot repent of it, for we were not responsible for it. Absolute election and reprobation is a doctrine of licentiousness, yet no sinner (and all men have sinned) can be alone justified by works. "We all stand in need of . . . free grace and mercy" (18). The "Practical Consequences of the Above Doctrines" (item vii, not listed in the title) is a desire to do the will of a loving and merciful God. This is simple Gospel doctrine, written for men of plain understanding; not the esoteric doctrine of theologians requiring acuteness, subtlety, and refinement.

There were immediate replies to the *Appeal*, to which Priestley responded with *Letters and Queries addressed to the anonymous Answerer of an Appeal . . .; to the Rev. Mr. Thomas Morgan, and to Mr. Cornelius Caley* (Leeds: by J. Binns, 1771).[6] Priestley was sad that his intent to restore Christianity to its primitive purity and efficacy should be so misunderstood and combated. The "anonymous Answerer" employed abusive terms: hypocrisy, Jesuitry, low pitiful cunning, artifice and fraud, falsehood, infidelity, blasphemy, and so forth. Priestley responded with: "Blessed are ye when men shall revile you." He posed a set of queries, asking for example, whether the Answerer really believed that nature, which was as much a gift of God as was grace, was an evil and malignant principle?

To the claim that his variant translation of Philippians 2:6 was a bold and manifest perversion of Scriptures, Priestley answered that his was better Greek and agreed with the version approved by the Christian fathers of the first centuries; namely, Origen, Novatian, Clemens Romanus, Justin Martyr, Lactantius and others, as Samuel Clarke proved in his *Scripture Doctrine of the Trinity* (1712).[7] And, finally, Priestley cited the Athanasian Creed to substantiate the heresy involved in the Answerer's interpretation of the Trinity.

To the Reverend Mr. Morgan, who appears to have been a friend of one of Priestley's friends, the response was less full, as "want of explicitness and precision . . . leaves little room for particular animadversions" (20–21). Priestley did object to the implied threat in Morgan's statement that his

6. I have used the reprint of the one edition of *Letters and Queries*, in *Works*, 21:3–28, where Morgan is identified as minister of Morley Chapel near Leeds and Caley as a preacher among the Wesleyan Methodists.

7. The "Authorized King James Version" of the passage reads: "Who [Christ Jesus], being in the form of God, thought it not robbery to be equal with God." Priestley's version read; "Who being in the form of God, did not think that being equal to God was a thing to be seized." Modern versions of the "Authorized," with variant readings in the margin, offer: "counted not equality with God a thing to be grasped at."

sentiments are not those of a "legal dissenter," claiming, however, that he "would rather glory in a religion which had no countenance but in the Kingdom of God and of Christ" (22). The reply to the Arminian, Caley, could be "in perfect good humour" as the two agreed on "Christian moderation, candour, and benevolence" and Caley acknowledged the sincerity of intention and moderation of spirit of the *Appeal*, though it was wrong in the fundamentals of Christianity (26).

In dealing with Caley's attempt to compromise with Calvinist doctrine, Priestley concluded, however, that the "genuine *Supralapsarian system*," wrong as it was, was nonetheless striking and consistent. Then acceptance of objections to any part, as in modern, qualified, intermediate systems, must necessarily bring down the whole. That left only the simple belief, that the merciful Parent of the Universe, intending only the happiness of creatures, "sent his well-beloved Son, *the man Christ Jesus*, to reclaim men from their wickedness . . . promising a life of endless happiness" for their repentance. That is "the essence of what is called *Socinianism*" (27–28).

The continued (perhaps unexpected) success of the *Appeal* produced other answers; Priestley saw or heard of at least seven by 1772, not counting adverse notices in other publications. To all of these, he wrote a single reply: *A Familiar Illustration of Certain Passages of Scripture relating to the Power of Man to do the Will of God, Original Sin, Election and Reprobation, the Divinity of Christ, and Atonement for Sin by the Death of Christ* (London: J. Johnson, 1772). Like the *Appeal* and *Letters and Queries*, *Familiar Illustrations* appeared anonymously, by "A Lover of the Gospel," and the designation was particularly appropriate this time. Priestley here ranged over the whole of the Old and New Testaments, taking each passage that had been cited against his views, interpreting it, in context, to show that, properly understood, there was no disagreement and then adding a selection of additional passages supporting those views. This is the most theological of his early religious controversial pamphlets.[8] Perhaps that is why the *Familiar Illustrations*, at four pence a copy, was less widely distributed than the *Appeal*, though it was one of the more successful of his works. At least eleven editions were printed by 1836.

As most of the biblical passages treated in the *Familiar Illustrations* had previously been referred to, if only in passing, and most of the doctrines

---

8. That is, in the *Familiar Illustrations*, Priestley explicitly addresses the systematic problem and methods of interpretation by which scholars were attempting to determine, from Scriptures, what religious faith should be. My references are to the first (1772) edition.

illustrated are already familiar from Priestley's earlier writing, the major attraction of this work today is the evidence it provides of Priestley's theological scholarship. There is, for example, his emphasis on determining the beliefs and practices of the primitive church, which he had previously mentioned in *Letters and Queries*. It seems an anomaly that Priestley, who is generally counted among the prophets of progress in the eighteenth century, should care so greatly for the beliefs and practices of early Christians.[9] But progress, for Priestley, was to be achieved through the gradual acquisition of ultimate truth, which, in nature and in society, was acquired by experiment, observation, and reason. In religion, truth had been *revealed*, gradually to the ancient Hebrews and finally to the first Christians. That revelation, to the common people, had since been corrupted by philosophers and priests. What was needed was a recovery of original, revealed truth, a recovery that had begun with the reformers of the sixteenth century and for which progress was continuing in the liberal Dissenting movement of his own day.

Unfortunately, one could not now go simply to the Scriptures for the recovery of revelation. What was clear to the people of the time when the writing of the prophets, apostles, and evangelists was "published" is "extremely difficult to us, who use a very different language, and whose manners and customs are so exceedingly unlike those of the Jews" (i). Moreover, the English translation of the Scriptures was made when the translators had "just emerged from the darkness of popery," from which they had kept and passed along many of the opinions which Priestley was combatting.

A new translation is needed, taking into account the highly figurative style of Oriental writers and the progress of Protestant scholars, but meantime, alternative translations and interpretations of particular passages can be compared, even by the ignorant and unlearned, and those preferred that are "most agreeable to the general strain of scriptures and to common sense" (ii). As he was later to say: ". . . it will be an unanswerable argument, *a priori*, against any particular doctrine being contained in the scriptures, that it was never understood to be so by those persons for whose immediate use the scriptures were written, and who must have been much better qualified to understand them, in that respect at least, than we can pretend to be

9. See, for example, J. B. Bury, *The Idea of Progress: An Inquiry into its Origin and Growth* (New York: Dover Publications, 1955), 221–22, 230; Lois Whitney, *Primitivism and the Idea of Progress* (Baltimore: Johns Hopkins University Press, 1934), 177–83.

at this day."[10] Priestley has here placed himself in the tradition of "modern" rationalist biblical criticism, started at least as early as Spinoza and Locke and continuing past Priestley to the German critics of the nineteenth century.

Indeed, as he was later systematically to apply his critical method, Priestley became a notable precursor of the Germanic "higher criticism," but, at this state, he was primarily repeating the judgments, good and bad, of others. He accepted, for example, the traditional view that the Gospel according to St. John was written by the author of Revelation. His attempt to deal with the mystical, Hellenistic strain of John is, perhaps inevitably given his prosaic rationalism, unconvincing. The whole tenor of the New Testament, said Priestley, is that the Father is the only true God, exclusive of the Son. The powers communicated to Christ, by God, especially after the resurrection and ascension, explain the expressions about Christ's supremacy in the New Testament, but that does not mean that Christ is God. Nonetheless, "it is certainly right that a very high degree of respect should be paid to him" (27).

Priestley cited Isaac Newton "and others" as proving the "three witnesses" verse of 1 John 5:7 to be a late interpolation, not found in any ancient manuscript. He also cited Newton respecting a change in the reading of 1 Timothy 3:16, adding: "The oldest manuscript in the world which I have examined myself, has been manifestly altered . . . as appears by the difference in the colour of the ink."[11] And, as a final example of his dependence on earlier scholars, Priestley again defended his translation of Philippians 2:5–6 by quoting a similar rendering by the French critic, Father Robert Simon. Priestley ended the *Familiar Illustration* with an anti-Calvinistic prayer and a "Conclusion" wondering that Arminians should so vehemently oppose Arianism and Socinianism and hoping that his writings might, to those who can reason, be conducive to the reformation of ancient error.

At the same time that he was developing the religious education part of his program for Mill Hill Chapel—and responding to the attacks this

10. Joseph Priestley, *An History of Early Opinions concerning Jesus Christ, compiled from Original Writers, proving that the Christian Church was at first Unitarian* (Birmingham: for the Author, and J. Johnson, 1786), 1:xv.

11. [Joseph Priestley], *Familiar Illustration*, 38. No doubt he is here referring to the *Codex Alexandrinus*, a uncial MS of the fifth century, which had been deposited, with the Royal Library, in the British Museum. The issue was whether the reading should not rather be "Who was manifest in the flesh" than "God was manifest in the flesh." Newton's criticisms are contained in his *Historical Account of Two Notable Corruptions of the Scriptures* (1690).

elicited—Priestley also worked to restructure the congregation there. It made little sense to instruct the young people in religious principles on a Sunday if they were to live in an irreligious environment for the rest of the week. As he made clear in his "Essay on . . . Religious Knowledge," Priestley found a paucity of religious activity in the homes of his congregation. To encourage a return to the family piety of earlier Dissenters, he wrote *A Serious Address to Masters of Families, with Forms of Family-Prayer*, first published in 1769 with two subsequent editions in Priestley's lifetime and a fourth edition in 1832.[12]

Heads of families have, according to Priestley, an authority more absolute than is possible in any civil government and their power to do good is equivalent to a necessity of doing so. While children's minds are pliable, they should be led to moral habits and toward performing their duties to God and their fellow creatures. There are many influences on them that may secretly undermine parents' efforts, but if they have tried the parents have, at least, the satisfaction of having done their duty. Family heads also have a duty to servants in a household, beyond the fulfilling of a legal contract, to see to their moral and religious instruction.

The *Serious Address* recommends a structure of practical religious exercises, free of superstitious notions and practices. Each day the family (including servants) should assemble for the reading of a portion of the Scriptures—from the four Gospels, Book of Acts, and the practical parts of the apostolic epistles—and join in rational devotion or prayer, which may be patterned after forms offered in the *Address*. These forms include a set of long prayers, each clearly exhibiting the rationalist, Unitarian doctrine, to which might be added any of a collection of twenty-four "Forms for Particular Occasions, To be used at the Discretion of the Person who conducts the Service" (46). The twenty-four range from morning and evening "petitions," those for family, sickness, journeys, and death, to those for the world, for country (king and royal family), Parliament, and king's ministers, and includes one for Sundays, in which thanks is given to God, for the opportunity of worshiping according to the dictates of conscience, without fear of molestation.

The "Essay on . . . Religious Knowledge" also noted the impossibility of the minister's doing everything, or, indeed, anything effective in main-

---

12. Joseph Priestley, *A Serious Address to Masters of Families* . . . (London: for J. Johnson and J. Payne, [1769]); 2d ed. (London: J. Johnson, [1771]), from which my references are taken; 3d ed. (1794).

taining a proper religious temper throughout the community he served. In his *Free Address to Protestant Dissenters on the Subject of Church Discipline* of 1770, Priestley described his solution to that problem.[13] Knowledge of Priestley's experiences in the Independent Chapel of his youth and in that at Needham Market, lends to *Church Discipline* autobiographical overtones. Clearly frustrated at the lack of moral authority in congregations of liberal Dissenters, he was also personally aware of the difficulties entailed in his proposed solution.

Genuine Christianity produced a serene and cheerful disposition to live a moral life according to the commandments of a benevolent creator. The success and zeal of primitive Christians in achieving this state was largely a consequence of their disciplinary organization. Being essentially without regular ministers, early Christian societies had organized themselves (or been so organized by apostles) with presbyters, elders, or "bishops," and deacons for admonishing, reproving, exhorting, and providing for the group. In time, this true system of Christianity became debased—partly by its adulteration with *Oriental* or more properly *Indian* philosophy (4–5), but also because the spread of the church led to factions, increased size, loss of internal discipline, creation of competing hierarchies, and use of civil authority to enforce conformity.

The result was a church guilty of unequaled cruelties, abasements, mortification, and fraud. The Reformation had improved things, but produced a church demanding gravity and austerity and that still enforced conformity of belief. *Independents,* for example,

> judge by rules exceedingly obscure and doubtful; their proceedings have a dangerous tendency . . . both to the candidates for church-communion, and those who sit as judges. . . . What can be more precarious than to judge . . . by certain *internal feelings,* which are incapable of being described, except by strong metaphors; by a kind of *faith* that is different from believing, and a *new birth,* that is something else than a change of affections and conduct . . . What room is not here left for self-delusion . . . to the candidate, and imposition . . . to the judges. . . . Men who act upon this system

---

13. Joseph Priestley, *A Free Address to Protestant Dissenters, on the Subject of Church Discipline: with a Preliminary Discourse, concerning the Spirit of Christianity, and the Corruptions of it by False Notions of Religion* (London: J. Johnson, 1770), 137 pp., 8 vo. [my reference]; 2d ed. (1774), 3d ed. (1788).

will be liable to be imposed upon by . . . the *fools* and the *knaves.* . . . I am very willing to think that the most corrupt species of christianity is better than none; but . . . in my opinion, it is better to have no church discipline . . . than that of the Independents.[14]

Yet some discipline is necessary in societies of rational believers, to oversee the spiritual and physical needs of members, but also to involve them in the activities of the church. In the established Church, members have at least to make responses and may not, without offending, habitually omit communion. They have something regularly to do in the functioning of their religious community. In congregations of liberal Dissenters, too often religious life consisted in going to chapel once a week to hear the "haranguing" of the minister.

Preaching was not unimportant and the minister who did it made a contribution to the religious understanding of his hearers. This was especially true if he was trained to expound the true meaning of Scriptures and answer the objections of unbelievers. For this he would need thorough acquaintance with the language of the Scriptures, the history of primitive times, the learning, languages, customs, and revolutions of the Jews, and other ancient eastern nations, and the Greeks and Romans. That minister should especially know the antiquities of the Christian Church, and the remarkable changes that have taken place since the time of Christ. He should "also be well acquainted with the writings and reasonings of the *modern literati,* on moral and metaphysical subjects" (75–76).

Nonetheless, the important parts of a Christian service were the reading of Scriptures, prayers, and singing of psalms, and maintaining the practical spiritual discipline of the congregation. A minister was not needed for the first part, and the last could not be done by him. The heroic days of Old Dissent were gone and with them the almost superstitious reverence for the minister. The newer, and younger, ministers could not assume the same spiritual authority. They hadn't the physical maturity to overawe older members of the group, and they were too dependent for their subsistence upon the good opinion of the congregation to risk offending them. "In every thing relating to society, we must go upon the supposition that we are *men,* as well as Christians" (87).

14. Priestley, *Church Discipline,* 37–40. He adds that the perceived necessity of a new birth drives many to despair at their abandonment by God, "In this Observation, I doubt not that I speak to the painful experience of many of my readers" (14).

Under these circumstances, it were well if the religious activities of the community were not dependent upon the minister, particularly in periods of doctrinal growth or development:

> It happens . . . that the minister, being the person who reads and thinks the most in the society . . . [is] first enlightened. He begins to call in question the truth . . . which he has hitherto preached. . . . A man of integrity . . . will always preach . . . consistently with his real sentiments . . . now considerably different from those of his hearers. . . . A young man will be apt . . . to obtrude his favourite opinions on his hearers; perhaps even preach on those subjects; and . . . not always take the most prudent methods of propagating truth.
> The usual consequence . . . is, that the congregation and especially the elder part . . . take alarm; while some . . . of the younger part, adhere to the minister, and adopt his sentiments. . . . If the society be large enough . . . they often divide in two; or the minister . . . if he be not absolutely dismissed, finds himself at length . . . relinquishing his office, and often without any maintenance. . . . In a society governed by elders, single persons, or their opinions, do not make so great a figure, or occasion so great alarm (109–12).

Furthermore, a minister educated as a learned expositor of Scriptures, a leader in achieving new understanding of religious truths, lacks the talents and knowledge of society and the world to be effective in admonishing and advising. What is needed is the joint labor of many persons of different talents, combined in a disciplinary organization. In this regard, rational dissent should follow the guide of the Presbyterians of Scotland, ". . . how despicable soever their principles or conduct may appear . . . in other respects" (424).

From their numbers, the congregation should choose a selection of the more serious members, wealthy enough to have freedom and authority to act without risk of displeasure at their censure. These elders with the minister form a consistory meeting regularly to consult on the state of the church and congregation. Elders could admonish careless or immoral behavior, but never differences of opinion. Notorious offenders might be referred to the congregation for public censure or exclusion from fellowship, but they should not be barred from worship.

Priestley was not sanguine about the introduction of this kind of discipline into the chapels of rational Christians. Indeed, he wrote to Theophilus

Lindsey: ". . . I know I shall be severely ridiculed by many of our free-thinking Dissenters for *molesting* the subject. . . . If I have ever been severe upon the Church of England, I have been here as much so upon ourselves, so that I expect soon to be in the case of Ishmael, 'My hand will be against every man, and every man's hand against me'; . . . I heartily wish I had done with controversy."[15] Yet a "true Christian Church" must, he believed, in time appear, as a true grammar, true aesthetic, and true rendering of Scriptures would in time be achieved. Meantime, serious members of Christian communities must do all they could to achieve the ends of religious discipline.

If Priestley really wished to be "done with controversy," he was clearly going about it the wrong way. In his public search for truth, he was never to understand that he incited opponents who knew they already had it. What Priestley hoped for was argument, or debate, carried on with civility between two equally earnest and seeking scholars and, for his favorite study, theology, he was to create a medium that permitted the kind of debate he wanted.[16] During 1768, as a third part of his role as Christian minister at Leeds, Priestley established a journal, *The Theological Repository*, to publish "Original Essays, Hints, Queries, &c. calculated to promote Religious Knowledge."[17]

In the editor's introduction to the first volume, he noted that advancement of knowledge requires assistance, the best being a general communication between workers in any field. In time, as more nations are added to the learned world and write in their own languages, "philosophers will be absolutely forced" to accept a "universal and philosophical language," such as that proposed by Bishop Wilkins (viii). Until that happened, national forms of communication must be encouraged. One might be a narrative of

15. Joseph Priestley to Theophilus Lindsey, c. mid-May, 1770 (W1/1:115). He also wrote S. Merivale, 8 December 1770: "I expect to be much laughed at for my piece on 'Church Discipline'" (W 1/1:125).

16. Priestley specifically called theology his favorite study in a letter to Theophilus Lindsey, 1 November 1770 (W 1/1:121) and repeats the general sentiment in his *Memoirs*, correspondence, and published works.

17. *The Theological Repository* was printed for J. Johnson, London; vol. 1 (1769), vol. 2 (1770), vol. 3 (1771) and then ceased publication until the 1780s, when vols. 4 (1784), 5 (1786), and 6 (1788) appeared. A publication history of the *Repository* is complicated; there seems to have been second and third editions of volume 1 and 2 in 1773 and 1795 and a second edition of volume 3 in 1795. The contents have never been thoroughly analyzed for the writing of Priestley; Rutt reprinted those articles by Priestley which, he felt, had not been republished in a different form, *Works*, 7:203–539; 20:524–28, but missed a few and confused others. I have used vol. 1 (1773), 2d ed.; vol. 2 (1770), and vol. 3 (1771).

the progress and present state of an art or science, such as his own *History of Electricity*. Another, popular, form was the periodical into which people could "throw every new thought" without waiting to complete a book or submerging the new in a mass of lesser or older materials.

There already were such journals for mathematics, agriculture, literature, and the natural sciences, but more were needed. As knowledge increased, journals could be "subdivided and multiplied ad-infinitum" (xii). He was, therefore, establishing the *Theological Repository* as a common channel of communication "open for the reception of all new observations that relate to theology; such as *illustrations of the Scriptures*, the *evidences of revealed religion*, with *objections of all kinds*, &c. &c. &c." (ix).

There were other religious magazines in England, but the *Theological Repository* was probably its first scholarly journal for speculative theology. Certainly it was the first for Unitarianism.[18] Priestley declared its pages open to all denominations of Christians, to nonbelievers, and even to deists ("The only way to have difficulties cleared up is to conceal nothing . . . expose every thing to publick inspection and examination [xii]") but few persons other than Arians or Socinians sent in contributions. Priestley and his friends wrote most of the papers. This was not Priestley's fault. As he wrote in the advertisement ending volume 1, the greater part of it had been written by persons of nearly the same sentiments, but no article of different sentiments had been rejected. Authors of other sentiments had themselves to blame if they failed to communicate them.

Nonetheless, Priestley was far from passive as an editor. He encouraged authors to write under pseudonyms and pledged that no attempt would be made to trace them. He excluded only indecent, abusive, or irritating language and he published contributions that argued strongly against his own views, including one reply to Lardner's *Letter on the Logos* that had converted him from Arianism to "Socinianism." He rarely responded in the *Repository* to anything written against his opinions. He actively solicited papers from friends and sympathetic acquaintances, published some responses "merely to prove my impartiality" and to encourage answers and, on at least one occasion, only accepted a piece because he thought he had been promised an answer to it.[19]

18. See Herbert McLachlan, *The Unitarian Movement in the Religious Life of England: Its Contributions to Thought and Learning, 1700–1900* (London: George Allen and Unwin, [1934]), "Chap. I. Sec. III. Journals and Periodical Literature," 168, for a discussion of the role of the *Theological Repository* in English Unitarianism and in biblical research.

19. Priestley to Theophilus Lindsey, 27 January 1771 (*W* 1/1:131). Most of his correspondence to Lindsey and to others during the period 1768–73, printed in Rutt's edition of the *Memoirs*, relates to contributions for the *Repository*.

Before commencing the *Repository,* Priestley had solicited the support of a number of Dissenting-minister friends: Newcome Cappe of York, Samuel Clark of Birmingham (his old teacher at Daventry, who died before the *Repository* had really begun), Andrew Kippis of Westminister, Samuel Merivale of Exeter, Thomas Scott of Ipswich (a neighboring colleague of Priestley's days at Needham Market), and William Turner of Wakefield. John Aikin of Warrington, and Richard Price of Newington Green had approved and encouraged the project, but not promised support. Most of these wrote some articles, but only Turner assisted in the editing and Priestley paid the costs of publication.

After three years, there was a "pause in the publication," because there had been little demand for the work and his costs had not been returned. With help from friends, who purchased multiple copies, his losses were held to something less than £30 by the last volume, but this was too much for a minister whose salary was only one hundred guineas a year.[20] Besides, by late 1771, Priestley was deeply involved in political activity and in scientific research. He concluded volume 3 with an expression of his conviction that the three volumes had contained "many truly original, and exceedingly valuable tracts" and a hope that the work might later be resumed (478).

Roughly one-third of the *Theological Repository* was written by Priestley, himself, as editor and, in the first three volumes, under the names Clemens, Paulinus, and Liberius. Some of what he wrote there was destined to reappear in a different format. For example, "Paulinus" on Judas and on the Lord's Supper in volume 1 and ten papers, signed "Liberius," in volumes 2 and 3, related to a "harmonizing" of differences in the four Gospels. These would be rewritten for the "Observations on the Harmony of the Evangelists" prefaced to Priestley's *Harmony of the Evangelists, in Greek* (1777) and . . . *in English* (1780).

The greater part of his first contribution, as "Clemens," "An Essay on the One Great End of the Life and Death of Christ, intended more especially to refute the commonly-received Doctrine of Atonement," was a rewritten version of his anonymous *Scriptural Doctrine of Remission* (1761).[21] He added a justification of differences of opinion among Christian scholars by comparing them to equivalent differences "concerning the opin-

---

20. He would have lost £20 from volume 2 and £30 from volume 3 without this help; see W 1/1:142, 181.

21. [Joseph Priestley], "Clemens," "An Essay on the One Great End of the Life and Death of Christ . . .," *Theological Repository* 1 (1773): 17–45, 121–36, 195–218, 247–67, 327–53, 400–430.

ions of Aristotle, Plato, Mohammed, and every other person who has set up for an instructor of others" (17–18). Unbelievers should take their notions of Christianity directly from the New Testament, not from reading critics and commentators.

Again Priestley emphasized the figurative, allegorical, and metaphysical language in which the Scriptures are written. He counseled the use of common sense in distinguishing the true meanings of disputed passages. He also now had the help of textual critics on variant forms of manuscript traditions (some of these derived from articles in the *Repository*). He had more (and better) descriptions of comparative religions: of primitive people, of Persians, Greeks, and Romans, citing Maimonides, Abrabanel, Buxtorf, Theophilus Gale, ancient Persian, modern Parsee, and Hindoo literature.

The conclusion of "Clemens," in this set of papers was, not surprisingly, the same as that of the *Doctrine of Remission:* Christ was not sent as a holy sacrifice for the sins of mankind. He was to be a pattern for man's behavior, showing that a man, like us, but passing his life and going to his death in obedience to the will of God, was invested with extraordinary powers and finally demonstrated the resurrection and future life open to all. The *Doctrine of Remission,* however, had been an exposition of Scripture in the framework of Locke, Watts, and Hartley. Now, in 1769 and some dozen years after writing that work, Priestley was more sophisticated in history, philosophy, and comparative religion than when he had first treated the subject.

Part of what he had written for the *Doctrine of Remission* had been an extended criticism of the reasoning of St. Paul. When Nathaniel Lardner and Caleb Fleming excluded this from the published version, Priestley prepared it for separate publication. He was then dissuaded by Andrew Kippis, "until I should be more known and my character better established" (W 1/1:39). Now, having his own forum, he published a series of papers in the *Theological Repository* "On the Reasoning of the Apostle Paul."[22]

Paul's major offense was the downgrading of the Jewish covenant. There were other problems (e.g., Paul's confining women to places of hearers and learners in the church on "inconclusive" arguments [3:198]) but most of Priestley's criticisms were directed to Paul's suggestions that Jews were no

---

22. Five articles: "Remarks on Romans v. 12–14," "Observations concerning Melchizadeck," "Observations on the Abrahamic Covenant," further "Observations on Romans v. 12 &c," signed "Paulinus," in 2 (1770), 154–58, 283–90, 396–411, and 411–16; and "Paulinus," "Remarks on the Reasonings of St. Paul," 3 (1771): 86–105, 188–212. Rutt conflates these under the general title given in the text above, *Works* 7:365–415.

longer the "peculiar people of God." This is clearest in "Observations on the Abrahamic Covenant," where Priestley declared that the Jewish people had achieved what was promised to Abraham and no Gentile could want to claim any right to those promises. It was this mistaken attitude of Paul that led to his conflating two petty references to Melchizedek (Gen. 14:18–20; Ps. 110:4) into an ill-founded argument (Heb. 7:2–22) that Christ supersedes the old law and order, replacing it with a better. Christians *had*, of course, been blessed by means of the seed of Abraham, as in time, all nations would be blessed.

Then there was the "obscurity" of Romans 5:12–14. Priestley rejected the obvious interpretation: there being no "law" between the sin of Adam and the tablets of Moses, men who died suffered for ancestral sin; suffering that continued beyond Mosaic law and removed only by God's gift of Grace, Jesus Christ. Now aside from its Calvinist implications, which Priestley would not countenance, this suggests that the Old Testament contained no provision for the pardon of sins. It implies that moral law exists only after its formal promulgation. Priestley rejected these ideas. "W. W." (William Willett) sent the *Repository* a vindication of Paul's reasoning that Priestley published, but thought unsatisfying. However, a "now deceased" friend (Samuel Clark) provided an explanation in which Priestley found confirmation for part of his system of historical biblical criticism.

Not only must one take account of changes of language with time and different cultures, one must also understand passages of Scripture in relation to the occasion for which they were written and the issues they addressed. As the apostle had reason for alarm at the tendency to Judaize found by Gentile converts (3:100), Romans 5:12–14 is to be interpreted, from that circumstance, as a criticism of ceremonial law, on which the Jews at the time of Christ had set too high a value.

Priestley still thought Paul argued inconclusively and mistakenly quoted, as prophetic, passages from the Old Testament, but he cautioned: "I am very sensible, what respect is due to writings of such ancient date, to a man of St. Paul's character, sense, learning, and wisdom; and, likewise, how dubious the sense must necessarily be of letters, written upon occasions with which we are not perfectly acquainted; and . . . with what extreme caution, circumspection, and diffidence one should . . . venture to pass any censure upon them" (3:211–12).

A few of Priestley's papers in the *Repository* are not parts of a series. Of these: "Clemens," "Analogy of the Divine Dispensation" is one of the most interesting of his theological writing.[23] A kind of psychological argument for determined theological salvation, it was based on Hartleyan associationism with gestalt overtones. Structured like Hartley's *Observations on Man*, with a description of man's nature and a deduction of the theological consequences, Priestley's "Analogy" begins with the assumption that God created mankind as perfectible intelligent beings. Now the perfection of intelligent beings "consists in comprehension of mind" (4) and this is a consequence of experience, through the association of ideas.

In his exposition of this argument, the "Analogy" has some personal connotations. Children live in the present, have little patience, and are unable to perceive the future value of present pain—in parental discipline, for example. (Priestley had, at the time of writing, three children: Sara, age eight; Joseph Jr., age three; and William, newborn). With age, man is more influenced by ideas, as the stock of these increases, and less influenced by externals. Habit and experience coalesce in association of ideas; sensations are generalized and interpreted—into the laws of optics, for example, and our understanding of the true size of the moon seen at the horizon and the idea of distance or dimension other than length or breadth. (Priestley was working on his *History of Optics* while writing this.) "Even a *sensation* may cease to appear what it originally was . . . be so intimately connected . . . and absolutely lost in associated ideas only, as to be no longer capable of being resolved again into its former state" (11).

By extension of the power of association, man finally perceives acts and their consequences as one undivided thing and the apprehension of the ill consequences of vice and good effects of virtue will then ultimately and naturally make for a virtuous life. This dispensation of God for the improvement of intelligent beings, by the structure of their minds, necessarily leads to happiness and "man being immortal . . . [his] happiness must be infinite upon the whole, though it be limited and finite at any particular time" (4). Thus natural religion leads to an understanding of God; all ideas of pain perfectly uniting with those of the pleasures they have accompanied, and all the events of divine providence constantly appear as "perfectly and infinitely good" (12).

Revealed religion, the "extraordinary dispensations of God to mankind," supports natural religion by analogy. From Old to New Testaments, empha-

---

23. [Joseph Priestley], "Clemens," "An Essay on the Analogy there is between the methods by which the perfection and happiness of men are promoted according to the dispensations of natural and revealed religion," *Theological Repository* 3 (1771); 3–31.

sis is upon the superiority of remote events to present temporary things. The Mosaic books show man existing under a sense of moral government; the patriarchs, adjusting lessons to the capabilities of their civilization, relate duties and obligations to expectations, the prophets constantly emphasize great coming events; and the Christian gospel so focuses on future consequences of present action that it is impossible for a sincere Christian to be addicted to vicious gratification and the pursuit of sin (15–22). Thus the Christian scheme is consonant with human nature. Mankind in history being operated upon by a series of dispensations (the revelation of God's will), we see virtue "to be a thing in which we are more deeply interested, than we could otherwise have known ourselves to be" (31).

Other, shorter, papers are miscellaneous in subject and character. "Observations on Infant Baptism" is an expanded version of the argument contained in the *Institutes* volume of 1773 that the baptism of children is a profession of faith by parents. "Jews and other Asiatic nations" habitually required the head of a household to subject all members of it to religious ceremony, as when the slaves of Abraham were circumcised. This practice carried over to the primitive Christians and continued until the corrupt notion of sacraments prevailed. When the religious practice of primitive Christians is known, it should be followed as it has the implication, at least, of apostolic approval. In "Observations on Christ's Agony in the Garden," Priestley cites medical cases from the *Acta Physico-Medica Norimbergae* to show that descriptions of that event must have been witnessed by the evangelists and that Christ was affected exactly as other men could be, in body and mind.[24]

There is a paper arguing against Arians, in favor of Socinianism, from the Christian system in general and from a corrected interpretation of some passages of Scripture. Priestley thinks that in John 16:28—"I came forth from the Father and am come into the world." Christ speaks of entering his public ministry, "world" meaning "worldly." More notably, Priestley identifies the Arian position with the "*oriental* doctrine" of the preexistence of human souls, the basis of the Gnostic heresy.

> For if the soul of one man might have preexisted, separate from the body, why not the soul . . . of all? . . . the separability of the think-

---

24. [Joseph Priestley], "Liberius," "Observations on Infant Baptism," *Theological Repository* 3 (1771): 231–39; [Joseph Priestley], "Clemens," "Observations on Christ's Agony . . .," *Theological Repository* 3 (1771): 376–82; there was a supplementary note, 3 (1771): 476–77 with further evidence of "bloody sweats." In a much later paper, *Repository* 6 (1788): 302–22, "Pamphilus" returns to the subject.

ing part . . . from . . . [man's] bodily frame . . . is so far from . . . human nature, that it is almost expressly contradicted. . . . His capacity for thinking depends upon . . . his brain. It is injured when this is injured . . . obstructed when this is obstructed; . . . suspended in a state of sound sleep, and . . . may be . . . presumed . . . to cease, when . . . with the rest of the body, this . . . part of the system shall be dissolved.[25]

In "Observations on Christ's Proof of a Resurrection," "Paulinus" thinks that Christ's argument, from Matthew 16:31–32, Mark 12:26–27, or Luke 20:37–38, on the necessity of resurrection is not conclusive, but ad hominem merely to confound his Sadducee questioners. In "Observations on the Apostleship of Matthias," "Paulinus" wonders at the peculiar regard for numbers seven, twelve, forty, and seventy, exhibited in Scriptures, but agrees that there must be twelve disciples, as there were twelve thrones, tribes of Israel, gates to the New Jerusalem, angels, foundations, kinds of precious stones, pearls, and manner of fruit on the Tree of Life. The defection of Judas required replacing, but the choice was Christ's, who selected Paul, not that of the other disciples. Matthias was therefore not a disciple.

"Clemens," in "Observations on the Importance of Faith in Christ," cannot believe the usual interpretation of passages, such as Mark 16:16, which suggest that man can only be saved by faith in Christ. This is contradicted in numerous passages of Scripture and by the nature of God. The Christian religion is a means to the reformation of man and if that end is attained by other means, God will rejoice in it (495). Finally, "Liberius" passes on the suggestion of "an ingenious clergyman" for a different translation of 1 Corinthians 15:27.

In his remarks concluding volume 3, Priestley answered three correspondents: "Q" may be sure that the Synoptic Gospels were written before the destruction of Jerusalem, as none of the writers could have "prophesied" that after the event without overwhelming criticism. The explanation "Pyrro" wants of an "error" in Acts is to be found in Benson, *History of the First Planting of the Christian Religion*. Recognition that Luke gives the

25. [Joseph Priestley], "Liberius," "'The Socinian Hypothesis Vindicated," *Theological Repository* 3 (1771): 344–63; see 355. Note this is not a denial of life-after-death. That, for Priestley, is guaranteed by Scriptures and exemplified by the resurrection of Christ. It is an argument in cited agreement with "Dr. Law and others," for "sleep" after death until resurrection of body and soul at the second coming of Christ.

"proper and natural descent" while Matthew follows the usual style of the Jews and provides a pedigree will supply the comfort "NF" seeks for their differences on Christ's genealogy.[26]

From a modern view, the *Theological Repository* is a curious mixture of uncritical conservatism, of forced anti-Trinitarianism, and of noteworthy insights, but this is to be expected from an enterprise so new in conception. Herbert McLachlan declares: "As a pioneer in Biblical research, the *Theological Repository* is entitled to respect for its scholarly articles on textual, historical, and exegetical subjects, and it stands to its credit that various writers anticipated modern critical verdicts in the fields of translation and hermeneutics."[27] In theological scholarship, as in religious education and disciplinary reorganizing of Dissenting religious life, Priestley made a significant contribution to "rational Christianity."

In doing so, he solved his personal problems as a minister. At Mill Hill Chapel, he adopted a strategy that emphasized his strengths: his interest and experience in teaching, his passion for structure and order, and his fondness for scholarship, while it minimized his weakness in preaching. Whether or not he deliberately designed his role for that purpose, he became, finally, a success as a minister. When he left Leeds in 1773, the young men who had attended his lectures on religion sent him an affectionate letter and an unexpected gift. The congregation went on record as blessing him for his concern for their children and testifying to the mutual love and friendship that had prevailed during his ministry. Nearly fifty years later, when searching for a new minister, they held him up as an example to be followed.[28]

26. [Joseph Priestley], "Paulinus," "Christ's Proof . . .," and ". . . Matthias"; "Clemens," ". . . Faith in Christ"; "Liberius," ". . . I Corinthians xv. 27," conclusion to vol. 3; *Theological Repository* 1 (1773): 300–303, 376–81; 3 (1771), 239–43, 255–56, 477–82.

27. McLachlan, *Unitarian Movement*, 168.

28. See Priestley's answer to the young men at Leeds, and the letter from the congregation, dated Leeds, Mill Hill, 27 December 1772; each printed in the introduction to Priestley's *A Sermon Preached before the Congregation of Protestant Dissenters at Mill-Hill Chapel in Leeds, May 16, 1773 . . . On Occasion of his resigning his Pastoral Office among them* (London, for J. Johnson, 1773) and reprinted in *Discourses on Various Subjects, including several on Particular Occasions* (Birmingham: for J. Johnson, 1787). Also, H. John McLachlan, "Mill Hill, Leeds and a Ministerial Appointment 1817," *Transactions of the Unitarian Historical Society* 10 (1951–54): 26–28.

IX

LEEDS, 1767–1773

Politics

Two of Priestley's major activities during his years at Leeds did not obviously relate to his religious profession. One was his writing on politics, the other his writing and research in natural philosophy (i.e., science). Yet each of these, given their social and intellectual context, must be regarded as part of Priestley's ministry. He was later to describe his work in politics as "inconsiderable" and to insist that he had little interest in the subject.[1] Yet he was to have published, during his years at Leeds, at least six works that are most appropriately described as political and was to add to these, during the rest of his life, at least fourteen more for a total number, in books and pamphlets, equal to that of his scientific publications.

The majority of these, however, related to the repeal of the Test and the Corporations Acts of 1673 and 1661. Priestley, like many of his fellow Dissenting ministers, no doubt regarded this as essentially a religious issue rather than a political one, however it might be regarded by members of the established church. Dissenting ministers had been forced, for more than a hundred years, into political action in defense of their religious activities. By Priestley's day, that kind of action must have seemed a routine part of Nonconformity.

The first ("Cavalier") parliament after the restoration of the monarchy in 1660 (and over the king's remonstrances) had enacted a series of laws,

---

1. See Priestley, *Appeal . . . on the Riots of Birmingham*, 53; also his letter to Richard Price, 21 July 1772 (W 1/1:175), where he notes his not having particularly studied "things of a *political* nature."

known somewhat unfairly as the Clarendon Code: the Corporation Act (1661), Act of Uniformity (1662), Conventicle Act (1664), and Five-Mile Act (1665), all designed to force adherence to the established church. Taken together, these laws essentially excluded persons dissenting from the Thirty-nine Articles of the Church of England: from holding political office, teaching school, attendance at or preaching in a "conventicle," serving as officers in the armed forces, or graduating from Oxford or Cambridge. The "Five-Mile Act" even prohibited a Dissenting minister from coming within five miles of a city, corporate town, or borough.[2] If the minister was to carry out what he felt was his divinely appointed mission, without offending his religious convictions by assenting to the rites and beliefs of the established church, he had to break the law and frequently paid the consequences.

It is true that these laws were not uniformly enforced and there were even attempts to moderate or repeal them, but their intention was repeatedly affirmed. When Charles II issued a royal Act of Indulgence in 1672 he was forced by Parliament to withdraw it and the Corporation Act of 1661 was reinforced with the Test Act (1673), which additionally implied that a Dissenter was, in effect, an outlaw by disabling him from suing or using any bill in course of law, from being guardian, executor or administrator, or receiving a legacy or deed of gift. As political troubles mounted for King James II, he also issued an Act of Indulgence (1687) matching concessions to Roman Catholics with those for Dissenters. Fearing that Dissenters might accept this as revenge for their treatment by the established church, George Savile, Marquis of Halifax, wrote *A Letter to a Dissenter upon the Occasion of His Majesty's Late Gracious Declaration of Indulgence* (London, 1687). "The common danger," Halifax said, "hath so laid open that mistake that all former haughtiness towards you is forever extinguished . . . [and] turned the spirit of persecution into a spirit of peace, charity and condescension." But Halifax was wrong.

Once the danger was over, with William and Mary replacing James II, Parliament approved "An Act for Exempting their Majesteyes Subjects dissenting from the Church of England from the Penalties of Certain Lawes" (1689). Described by Sir James Stephen as "a narrow and jealously-worded concession," this "Toleration Act" nowhere mentions toleration and leaves Dissent a crime, though it exempts from penalties those "criminals" who

2. For a detailed study of these and other laws affecting Dissenters, see Charles F. Mullett, "The Legal Position of English Protestant Dissenters, 1660–1689": "1689–1767"; and "1767–1812"; *Virginia Law Review* 22 (1936): 495–526; 23 (1937): 389–418; and 25 (1939): 671–97.

take the oath of allegiance and supremacy, declare their disbelief of transubstantiation, register their congregations at a bishop's court, and subscribe to thirty-five of the Thirty-nine Articles of the Anglican Church.[3] Roman Catholics and Unitarians were excluded from this toleration and Dissenters included still could not hold public office without taking Anglican communion, still had to pay tithes, and found many of their actions still subject to review in ecclesiastical courts.

They could, however, build and maintain their chapels in the open and they could now openly teach school. In consequence, many high churchmen (clergy and laity alike) began to maintain that Dissenters had been given too many rights and agitated for more restrictions. Some of their wishes were achieved in the Occasional Conformity Act (1711) and Schism Act (1713) of the reign of Queen Anne, forbidding any officeholder from attending a conventicle and requiring schoolmasters to be licensed by a bishop. The death of Anne halted the enforcement of the Schism Act and both acts were repealed in 1718, as soon as the reign of King George I was well established and it became obvious that the high church enemies of Dissent were enemies also of the Hanoverian succession.

From this time, until the French Revolution, concerted persecution of Dissenters ceased. Yet Dissenting ministers, such as Priestley, continued their agitation for repeal of the Corporation and Test Acts. This was partly because Dissenters remained in a position of second-class citizens; unable to hold public office without the impiety of occasional communion in the established church; an illegal action for which indemnity bills were annually passed in Parliament after 1728. There was more, however, to be complained of than deprivation of office.

As second-class citizens, Dissenters were open to intermittent nagging acts of persecution on a local basis, from bishops, parish clergy, country gentry, corporation officers, and so forth. Throughout the eighteenth century, there were repeated examples of magistrates refusing to license a meeting-house, of bishops demanding that teachers apply to a diocese for licensing (as happened to Philip Doddridge in 1733). Dissenters were denied parish relief by local magistrates and burial rights by parish clergy, bequests

---

3. Excluding the four dealing with ritual and discipline; Baptists and Quakers were given some additional concessions. See Richard Burgess Barlow, *Citizenship and Conscience: A Study in the Theory and Practice of Religious Toleration in England During the Eighteenth Century* (Philadelphia: University of Pennsylvania Press, 1962), passim, for criticism of historians for prematurely dating achievement of toleration, and 20 for the quotation from Halifax. The quotation from Stephen's *History of Criminal Law in England* (London, 1883), 2:482, can be found in Mullett, "Legal Position of Dissenters, 1660–1689," 526.

to Dissenting ministers were challenged, rioters disturbing Dissenting worship left unpunished. Lord Hardwicke's Marriage Act of 1753, requiring all marriages except those of Jews and Quakers be performed in establishment churches, opened a new road to torment, for clergymen began refusing to marry Dissenting couples.

When Dissenters, or Dissenting congregations had the energy, time and money to appeal this behavior before the King's Bench, they usually won their cases as the national government did not encourage persecution. Nonetheless, it went on and was almost harder to bear than continued general hostility would have been. Priestley and his fellow ministers had just cause to complain of the Dissenters' political and social status.[4]

The years of Priestley's stay in Leeds added new elements to Nonconformist political complaints. The succession of King George III was followed by administrations that adopted policies many liberal Dissenters felt were economically ruinous and that appeared to threaten their remaining mechanism of political influence: their votes. Leading members of English Presbyterian congregations were likely to be city merchants, whose trade was hurt by government policy toward the American colonies and whose rights of petition and parliamentary representation were challenged by repudiation of the Middlesex elections of John Wilkes. To these issues also Priestley turned his political attention in Leeds and there is no doubt that he was fully supported in his efforts by prominent members of Mill Hill Chapel.

His first political publication, however, was appended to the *Essay on a Course of Liberal Education,* published two years before he went to Leeds. Dr. John Brown, Anglican clergyman and eccentric author of the celebrated *Estimate of the Manners and Principles of the Times* (1757), had published a recommendation that the chief magistrates of the country establish a uniform statute for education. As this would frustrate the intentions of Priestley's course of liberal education, he addressed Brown's arguments in "Remarks on a Code of Education, proposed by Dr. Brown, in a late Treatise, entitled, Thoughts on Civil Liberty, &c."[5]

4. See Mullett, "Legal Position," passim, T. G. Crippen, ed., "Protestant Society for the Protection of Religious Liberty," *Transactions of the Congregational Historical Society* 6 (1913–15): 364–76, and "Report of the Annual Meeting of the Protestant Society for the Protection of Religious Liberty," *Monthly Repository* 14 (1819): 330–36, 388–94, for descriptions of continuing petty persecution, which continued up to 1828 and the repeal of the Test and Corporation Acts.

5. My references are to the first edition of the *Course of Liberal Education* (1765); "Remarks" was subsequently embodied in the second (1771) edition of Priestley's *Essay on Government.*

Brown wished to emulate the educational system of ancient Sparta, impressing the mind from infancy with habits of thought and action to promote public obedience to law. Priestley agrees that education is important to society, but greatly prefers Athens to Sparta. The object of education is not to ensure domestic tranquillity, but to form wise and virtuous men. Education is an art in infancy; establishing a code for it would prevent continuing the trials and experiments needed to perfect it. Establishing one mode of education would produce one kind of men, while the "excellency of human nature is in its variety"; one must not discourage the Newtons, Lockes, Hutchesons, Clarkes, or Hartleys (149, 178).

Brown's scheme would also halt the progress of Britain's constitution toward greater perfection. The science of civil government is also in its infancy; the divine plan shows the historical progress of constitutions through barbarous and imperfect systems. Priestley joins Brown in praise of Britain's constitution—when compared with that of any other country—but he does not agree that it is perfect. For example (in obvious allusion to the Test and Corporations Acts), men of ability and of integrity are incapacitated from serving their country, "while those who pay no regard to conscience may have free access to all places of power and profit" (169).

All friends of civil liberty would be alarmed if the Court assumed direction of the uniform system proposed by Brown, as, "justly," all friends of religious liberty would be alarmed if control went to the established clergy. It would be intolerable interference to "instill [in children] . . . religious sentiments contrary to their parents' judgment and choice" (156). The object of civil society is the "happiness of the members of it," for the sake of which they give up some of their natural rights in order to enjoy the more important ones. It is tyranny to be deprived of too many rights without an equivalent recompense (152).

These references to civil and religious liberty, in the "Remarks" on Brown, prompted some of Priestley's friends, who thought "that some of the views I had given of this important, but difficult subject, were new," to encourage him to write more on the subject.[6] The result was Priestley's most systematic political work, his *Essay on the First Principles of Government*. The *Essay on Government* was first published in 1768, immediately pirated in Dublin; and had its second English edition in 1771, but did not

---

6. Joseph Priestley, *An Essay on the First Principles of Government; and on the Nature of Political, Civil, and Religious Liberty* (Dublin: James Wilson, 1768; from the first ed., London: J. Dodsley, T. Cadell, and J. Johnson, 1768), iv.

reach a third English edition until 1835 and was translated into Dutch (1783). Yet it may well have been Priestley's most effective political writing, though the nature of its importance was not significantly to be felt until the nineteenth century through Jeremy Bentham's political philosophy of utilitarianism.

Even there the importance of its influence has been questioned: although Bentham explicitly acknowledged Priestley as the source of his "greatest happiness principle," he did so in two separate, incomplete, and somewhat aberrant statements. In one of these, he claimed to have passed close to Warrington on a summer trip with his father in 1764: "Warrington was then classic ground. Priestley lived there. What would I not have given to have found courage to visit him? He had already written several philosophical works; and in the tail of one of his pamphlets I had seen an admirable phrase, 'greatest happiness of greatest number' which had such influence on the succeeding part of my life."[7] This anecdote must be challenged, as Priestley had, in 1764, published only his anonymous *Doctrine of Remission*, his *English Grammar* and *Theory of Language*—in none of which does he write on the happiness of the greatest number. Bentham's other acknowledgment is better. In a third-person autobiography, "A Short History of Utilitarianism," he wrote:

> In 1768 he discovered a second set of threads for his system. He returned to Oxford to vote in the parliamentary election as an MA of the university. In a small circulating library of Harper's Coffee House he found a pamphlet by Dr. Priestley which contained the phrase, "the greatest happiness of the greatest number." At once he made it his own. For him the principle of utility was not an identical proposition, as Dr. Priestley offered it, but the grand normative end of government.[8]

This version has greater intrinsic plausibility. Priestley's *Essay On Government* was published in 1768; a copy might well have found its way into

7. John Bowring, ed., *Collected Works of Jeremy Bentham* (London: Simpkin, Marshall and Co., 1838–43), 10:46, quoted by Charles Warren Everett, *The Education of Jeremy Bentham* (New York: Columbia University Press, 1931), 38.

8. Bentham MSS, University College Collection Box 13, p. 360, "A Short History of Utilitarianism" (1829), quoted by Mary P. Mack, *Jeremy Bentham: An Odyssey of Ideas, 1748–1792* (New York: Columbia University Press, 1963), 102–3; reprinted by permission of the publisher. Mack points out that John Bowring published a garbled version of this essay as his own in 1834.

a coffeehouse circulating library, even in Oxford, and Bentham would surely have looked into a pamphlet with that title. Unfortunately, the precise phrase, "the greatest happiness of the greatest number" does not appear in that work—nor in any other written by Priestley. The *Essay* does contain the passage: "all people live in society for their mutual advantage; so that the good and happiness of the members, that is the majority of the members of any state, is the greatest standard by which every thing relating to that state must finally be determined" (17). Clearly Bentham's formula can be derived from that passage, but other writers of the period, Montesquieu, for example, or Cesare Beccaria, from whose *On Crimes and Punishments* (English translation, 1768) Priestley quotes in the *Essay* (63), also wrote passages from which the formula can be derived while Francis Hutcheson's *An Inquiry into the Original of our Ideas of Beauty and Virtue*—which Priestley also knew—contains the explicit expression: "the greatest happiness for the greatest numbers," in its first edition.[9]

The utilitarian idea was, obviously, not original with Priestley and some students of Bentham's work have questioned the Priestley source of the phrase in Bentham's writing. Yet surely Bentham's memory of the person who inspired his thought should be accurate, however fuzzy memory of the date and precise wording of the inspiration might be. That he was aware of the range and importance of Priestley's writings is clear from his frequent references to them in his correspondence while his personal opinion of Priestley ("cold and assuming, pretending to discoveries that were not discoveries at all") was not such as to prompt acknowledgment of an imaginary, flattering, dependence.[10]

Moreover, his description of Priestley's "principle of utility" as an "identical proposition" adds conviction to the identification. For Priestley did regard "the good and happiness of the members" of a society as identically equivalent to that of the "majority of the members" of the state. This is more clearly seen in that passage in the *Institutes of Natural and Revealed*

9. Francis Hutcheson, *An Inquiry into the Original of our Ideas of Beauty and Virtue; In Two Treatises* (London: J. Darby, Wil. and John Smith, W. and J. Innys, J. Osborn and T. Longman, and S. Chandler, 1725), 164: "that action is best, which accomplishes the greatest Happiness for the greatest Numbers." Hutcheson's remark was not made in connection with government and, though the sentiment remains, the wording is much changed as subsequent editions are "corrected" and enlarged. It seems unlikely that Priestley or Bentham would have read the work in its first edition when, by 1753, it had been reprinted at least five times.

10. [Jeremy Bentham], Timothy L. S. Sprigge, ed., *The Correspondence of Jeremy Bentham* (London: University of London, Athlone Press, 1968), 3 vols., passim. Bowring, *Works of Jeremy Bentham*, 10:571.

Religion (1772) where he notes that God's design is to produce happiness, though He may find it necessary to sacrifice the temporary interests of a few to those of a greater number, as a greater sum of happiness can exist in a greater number.[11]

Priestley's utilitarian principle was not entirely devoid of prescriptive intent. He had written, in the anonymous *Doctrine of Remission* (1761): ". . . in a wise human administration . . . the end of government . . . is the happiness of the community" (75). For the most part, however, Priestley's "principle" is not, as Bentham's becomes, a "normative end of government." He should not, therefore, be credited with intentionally abandoning the dogma of natural rights for the relativism of utilitarian philosophical radicalism.[12] No more in politics than in grammar or aesthetics was Priestley ever a cultural relativist. He never doubted but that, in God's creation of society, there was to be found a perfect polity, toward which civilized man progressed.

There is less sense in the *Essay on Government* than in his earlier works, that man would attain that perfection in this world: "our progress towards perfection must be continuously accelerated; . . . nothing but a future existence, in advantageous circumstances, is requisite to advance a mere man above every thing we can now conceive of excellence and perfection" (3–4). There is, however, no doubt in Priestley's mind that mankind is capable, in this world, of continual improvement, has "unbounded" potential (5). Clearly he is a disciple of the doctrine of progress: " . . . men will make their situation in this world abundantly more easy and comfortable, they will probably prolong their existence in it, and will grow daily more happy. . . . Thus, whatever was the beginning of this world, the end will be glorious and paradisiacal, beyond what our imaginations can now conceive" (8).

The instrument for this progress was "society, and consequently government." It was organized society that permitted division of human labor, one man confining his attention to one activity that he could learn well, while "another may give the same undivided attention to another" (6–7). As not all governments favored progress, it was necessary to examine the

---

11. See Chapter VII.

12. Despite, for example, W. R. Sorby, chap. xiv, "Philosophers," in *Cambridge History of English Literature*, ed. Ward and Waller, 10:391, "Priestley . . . set the example, which Bentham followed, of taking utilitarian considerations for the basis of a philosophical radicalism, instead of the dogmas about natural rights common with other revolutionary thinkers of the period."

nature of those that did. To do this, Priestley adopted the fiction of the
social contract, by which people resigned some part of their natural liberty
for the common benefit, but he separated this "natural liberty" into political
liberty, which "consists of the power . . . of arriving at the public offices,
or at least of having votes in the nomination of those who fill them" and
civil liberty, that power over their own actions which members of the state
reserve to themselves, and which their officers must not infringe (13). It
was, perhaps, this division which formed that "more accurate and extensive
system of morals and policy" that his friends encouraged Priestley to describe in the *Essay* (iv). Certainly he emphasized a distinction that had been
slighted by others, though it was to become a commonplace afterward, as
in John Stuart Mill's *On Liberty*, as late as 1859.

Although it was part of his civil liberty that man sacrificed when entering
society and political liberty was what he might acquire in compensation,
the good of mankind does not require a state of perfect political liberty.
The more political liberty people have, the safer their civil liberty is, but
in a large state there must be established forms of government in which
various degrees of political liberty are variously distributed among the
population. Any action of government is ultimately subject to the approval
of the people, but disapproval should not have to take the extreme form of
rebellion. The form combining stability, credibility, and institutionalized
response to public opinion is the best.

For Priestley, this meant the English system of "in some measure, heredi-
tary" monarchy and two houses of parliament: "elective monarchies having
generally been the theatre of cabals, confusion, and misery" (22). The concept of a "joint understanding of all members of a state, properly collected,"
has greatly impaired civil liberty (57). He was, however, not greatly concerned with *forms* of government: any in which the people have little power
over its actions is tyrannical whatever its form (55).

As a Dissenter deprived of major elements of political liberty, Priestley's
emphasis was on civil liberty in its two primary branches: the rights of
education and of religion. In discussing rights of education, he reverted to
his "Remarks" on Brown, repeating arguments and even some phrases
against a code of state education. Some of his objection was fear that a state
system would be controlled by the established church. He can scarcely be
faulted for this; an English state system was delayed until 1870 because the
Church refused to sanction a system it did not control.

But Priestley had other objections. One was simply familial: "I believe
there is no father in the world . . . would think his own liberty above half

indulged to him, when abridged in so tender a point, as that of providing, to his own satisfaction for the good conduct and happiness of his offspring" (93). Equally important, however, was a conviction that any single educational system would produce a uniformity more characteristic of brute creation than of the excellence of human nature (185). Because Priestley's theory of progress was that of a dialectic, an open competition between opposing views, it was necessary that there be a diversity of people.[13] Those arts "stand the fairest chance of being brought to perfection, in which there is opportunity of making the most experiments and trials, and in which there are the greatest number and variety of persons employed in making them" (78). He was sure that a state education would discourage this diversity and, nearly a hundred years later, John Stuart Mill was to agree: "A general State education is a mere contrivance for moulding people to be exactly like one another: and as the mould in which it casts them is that which pleases the predominant power in the government . . . in proportion as it is efficient and successful, it establishes a despotism over the mind, leading by natural tendency to one over the body."[14]

Of course Priestley was against any government authority in matters of religion. At this stage in his political career, he did not favor "immediate dissolution of . . . ecclesiastical establishments," but he suggested that they might be improved by, for example, extending the system of toleration to all members of the community, including papists—for which he was to receive much criticism, even from his friends.[15] He proposed that the articles of the established church be reduced to a declaration of belief in the religion of Jesus Christ put forth in the New Testament—citing Francis Blackburne's *Confessional* as indication of how badly this reform was

13. Whitney, *Primitivism and the Idea of Progress,* 177–79, suggests this emphasis upon diversity makes Priestley's progressivism a "touchstone for other versions" because it is more logically consistent than that of Hartley, Godwin, and others.

14. John Stuart Mill, "On Liberty," in *Essays on Politics and Society,* vol. 18 of *Collected Works of John Stuart Mill,* ed. J. M. Robson (Toronto: University of Toronto Press, 1977), 302. There is no direct evidence that Mill derived this view from Priestley, though he was clearly aware of him. Mill's *Works* contain several references to Priestley, but those relate primarily to his Hartleyan associationism.

15. Correspondence of his friends (e.g., Archdeacon Blackburne, Thomas Hollis, Caleb Fleming, Theophilus Lindsey) to Priestley and one another for the next two years, contain frequent references to their strong disagreement with this idea. Hollis, an obsessed anti-Catholic, sent him a copy of Blackburne's "book on the Present State of Popery," warning against toleration for papists. Priestley replied that even friends could not agree on everything, but that "A lover of liberty like you, must be uniformly and consistently so"; see W 1/ 1:94–100, 104–7.

needed. Church livings should be in proportion to the duties required and clergy confined to ecclesiastical duties, not conducting affairs of state. The progress of civil society is slowed by encroachments on civil and religious liberty and the *Essay on Government* ends with vigorous support for freedom of speech, of the press, and of religion.

The year following publication of the *Essay on Government* was not a happy one for liberal Dissenters, economically or politically, but it provided a plentitude of subjects for Priestley's political pen. The death of Charles Townshend in September 1767 had not lessened the resolve of other government leaders to enforce the duties and the Acts he had pushed against the American colonies. Collapse of the abortive Chatham ministry was followed by administrations set on coercion. Petitions from colonists and merchants seemed to have no influence on a Parliament increasingly seen as unrepresentative of their interests.[16]

Meanwhile John Wilkes returned to England from France, to remind the populace of the political scandals of 1763 and 1764: the illegal general warrant, denial of habeas corpus, questionable waiver of parliamentary privilege, attack on the press, and vindictive treatment of those who had opposed the king's will. Elected to Parliament in March 1786, in April Wilkes was committed to prison and heavily fined on his previous conviction for libel (though his outlawry was lifted on a technical flaw). When a crowd of his supporters was fired upon by an unreproved military, petitions in his favor insolently rejected, and he was again expelled from Parliament, Wilkes became another focus of agitation for reform.

In Parliament, proposals favored liberalism less than they attacked the influence of the crown. The emphasis was on disenfranchisement of revenue officers and placemen, demands for accounting of debts in the Civil List, and questioning of government pensions. These proposals and others, favoring greater political freedom, were supported in Priestley's first political pamphlet of 1769, *The Present State of Liberty in Great Britain and her Colonies.*[17] Published anonymously, "By an Englishman," and written in form of a catechism, "to make plain and intelligible, a just idea of natural

16. The *Leeds Mercury* for 9 January 1770 reported that 11,000 persons had signed a petition of York County presented to the king, but Priestley wrote Lindsey that the Leeds town-clerk, "no friend to it," had not pushed to get signatures; see W 1/1:106, letter of 18 December 1769.

17. [Joseph Priestley], "By an Englishman," *Present State of Liberty . . .* (London: Johnson and Payne, 1769); second edition also 1769.

and civil rights to all the subjects of government" (iii–iv), the *State of Liberty* makes full use of the arguments of his *Essay on Government*.

The interests of Great Britain and of the colonies were the same; it was *"folly"* if not *"iniquity"* to attempt to enslave "fellow subjects," particularly when the liberties of all were in danger. Any power assumed by magistrates beyond that for which the public cannot do better themselves is tyranny and an arbitrary invasion of man's natural rights. In England, people must fear corruption of their elected representatives, the number of placemen and pensioners in the House give thinking men apprehension, septennial parliaments were an arbitrary usurpation of the rights of the people, and too many representatives are now chosen by freemen of inconsiderable towns (18–19). There have been attacks on essential rights and privileges of subjects: evasion of habeas corpus, use of general warrant, restriction of liberty of the press in cases of libel, refusal to seat elected representatives, misuse of military forces (20–21).[18] There must now be compensation for sufferers in the cause of liberty, redress of grievances, exclusion from Parliament of placemen, pensioners and sons of noblemen, the duration of Parliaments shortened, small boroughs abolished, oaths required from candidates against bribery and corruption, and adoption of election by ballot (22–23).

In America, oppressive measures had been taken against a people who had left their home country for love of liberty "in the arbitrary reigns of our former princes" (v). Americans had taxed themselves; they were now to be taxed by others to reduce the burden on government ministry. Taxation of the colonies by Parliament was not the same as taxation of an unequal and imperfectly represented people in England; a tax imposed on any people in this country was a tax also on the imposers. Would not a tax solely on unrepresented towns (e.g., Leeds or Manchester) be resisted by inhabitants of those towns? (25).

English Parliaments had never before attempted to tax provinces sending no members: Ireland taxes itself, as Scotland and Wales had done before sending members. Open and undisguised oppression of direct taxation cannot be enforced without assertion of arbitrary power, while the colonists will continue to accept an equivalent restriction on their commerce. It is

---

18. The last section is in obvious reference to Wilkes, to whom Priestley sent a note (n.d., but probably late 1768 or early 1769): "Dr. Priestley begs Mr. Wilke's acceptance of a copy of his *Essay on Government* as a small acknowledgment for the many personal civilities he has received from him, and more especially for what he owes him as a member of the same community, and a lover of liberty" (Add. MSS 30, 871; published by permission, British Library).

sheer foolishness to attempt to acquire by taxes amounts "infinitely overbalanced" by what we shall lose in trade (30). If, in consideration of the good of the whole, we allow things to remain as they were some years ago, the gratitude and affection of the colonists will be secured, they will concentrate on agriculture and we on manufactures, to our mutual strength and opulence (32).

Although Priestley's *Present State of Liberty* places him, correctly, among advocates of liberal reform in politics, it is not a classic in that movement nor one of his major publications. Grievances at home are merely enumerated, not discussed while the treatment of American grievances is clearly a rehearsal of the views of Benjamin Franklin.[19] It would almost seem that Priestley wrote this pamphlet out of a sense of duty. That clearly is not the case, however, for the other four political publications he produced during 1769, each of which concerns the nature and status of Dissent. Three of these are directly related to one another and the political activism in which Priestley became involved late in the Leeds period. The other, an attack on William Blackstone's *Commentaries on the Laws of England*, was an interruption that became one of his best-known political works, for it was also his first confrontation with a national political figure.

In the course of answering a sermon on church authority, Priestley was referred to the newly published fourth volume of the *Commentaries*, in which Blackstone declared that dissent from the Church of England was a crime, quoted with apparent approval statutes of Edward VI and Queen Elizabeth establishing confiscation of goods and imprisonment for life as punishment for speaking against the Book of Common Prayer, claimed that "peevish and opinionated" sectaries had divided from the Church on matters of indifference, or upon no reason at all, and questioned whether their spirit, doctrine, and practices were any better calculated than those of the papists to make good subjects.[20] Blackstone had recently resigned as Vinerian Professor of English law at Oxford. Solicitor-general to the queen, M.P. for Westbury, Wiltshire, he was the obvious candidate for justice in

19. Priestley could also have learned of American opinions from Hollis via Blackburne and Lindsey, but Blackburne and Hollis were chiefly concerned in opposing extension to the colonies of the Church hierarchy. There seems to be no reference in Priestley's writing or correspondence to the problem of an American episcopacy, on which, see Carl Bridenbaugh, *Mitre and Sceptre: Transatlantic Faiths, Ideas, Personalities, and Politics, 1689–1775* (New York: Oxford University Press, 1962), passim.

20. William Blackstone, *Commentaries on the Laws of England. Book the Fourth* (Oxford: Clarendon Press, 1769), "Chapter the Fourth. Of Offences against God and Religion," especially 42, 51–52.

the court of common pleas, to which he was soon appointed. His *Commentaries* were written with skill and a show of authority. They were well on their way to becoming the classic handbook of law for laymen (recall that Priestley declined publishing his lectures on English law when Blackstone's *Commentaries* appeared; see Chapter 5).

In time it became clear that the *Commentaries* were flawed by rigid conservatism and intolerance, Blackstone's ignorance of civil law, of the role of equity in English law, and of legal philosophy, but that was all in the future. In 1769 Priestley feared with reason that Blackstone's sentiments, unchallenged, might be regarded as the new Ministerial attitude toward Dissent. With advice and encouragement of John Lee, counsel for the freeholders of Middlesex to retain Wilkes as their representative in Parliament, Priestley wrote what Blackstone was later to describe as "a very angry Pamphlet," his *Remarks on Some Paragraphs in the Fourth Volume of Dr. Blackstone's Commentaries on the Law of England relating to Dissenters.*[21]

It *was* an angry pamphlet. In Priestley's opinion, Blackstone's reflections on the Dissenting community were "destitute of candour . . . unsupported by truth, or . . . a decent appearance of argument" (1–2). Why must Dissenters remain silent in face of the malice and nonsense of high churchmen? The general cause of dissent was the Church's claim (e.g., in the twentieth of the Thirty-nine Articles) to a power of decreeing rites, ceremonies, and authority in matters of faith, a power not invested by Christ in any man or body of men (25–26). (Priestley footnoted his own further dissent from the Church's "idolatry," in worshiping more than God, the father [25 n–26 n]).

He corrected Blackstone's grammar (20), and his history, noting that English Dissenters, unlike high churchmen, had always supported the Hanoverian succession! He challenged Blackstone's understanding of law, find-

---

21. Joseph Priestley, *Remarks on . . . Blackstone's Commentaries* (London: for J. Johnson and J. Payne, 1769). There were three English edition (1769, 1773, 1774), a Dublin pirated edition (1770), and seven American editions, most of the latter combined with a more temperate response by the Reverend Philip Furneaux. Blackstone wrote *A Reply to Dr. Priestley's Remarks* (London: C. Bathurst, 1769) and Priestley replied with *An Answer to Dr. Blackstone's Reply,* which was included with the Dublin and most of the American editions of Priestley's *Remarks* and Blackstone's *Reply.* In England, Priestley's *Answer* appeared only as a letter, dated Leeds, 2 October 1769, in the *St. James Chronicle* until it was printed in Priestley's *Works,* 22:328–34. John Lee (1733–93) came from an old Leeds family and had been a member of Priestley's congregation at Mill Hill, spending his vacations in Priestley's home before he married (*W* 1/1:86). King's counsel and solicitor-general in the second Rockingham administration (1782), attorney-general under Portland (1783), he remained a friend to Priestley and to liberal politics and theology all his life.

ing absurd the contention that change in the constitution or liturgy of the Church of England would be a danger to the Act of Union between Scotland and England (28–30). And, finally, he declared "in a free and equal government, a government which leaves men a reasonable share of their natural rights," Dissenters are of all others, the best subjects and "in this light, the Dissenters regard the constitution of Great Britain" (44).

This criticism stung Blackstone into a *Reply*. He protested that his opinions had been misunderstood and misrepresented. He had merely cited the law and regrets any passage that seemed to suggest he esteemed those laws. He undertook to rectify misleading phrases, but insisted that Nonconformity was formally a crime and defended his interpretation of the Act of Union. Priestley's *Answer* "thanks and esteems" Blackstone for his genteel and liberal answer to a hastily written pamphlet. Others also had read the critical passages as slurs on modern Dissent. He was pleased that Blackstone planned to moderate his language in subsequent editions.

Later editions of the *Commentaries* were indeed modified.[22] Wording was added to underline the desirability of toleration and to identify with earlier times the view that Nonconformity was without reasonable cause; the historical remarks were removed, as was the remark that Dissenters made poor subjects. Priestley had cause to be pleased with his first venture into a national political spotlight, but he should have obtained more concessions. For he was right, and Blackstone quite wrong, on the two major issues which remained unchanged—and Blackstone, as the student of law, should have known he was wrong.

The Act of Union had survived a series of laws and decrees that promptly violated its terms (including the imposition in 1712 of a test on the Presbyterian Church) and overcome a move for its repeal in 1713. It was wholly unlikely to be endangered, after more than half a century, by changes in the Church of England. Furthermore, Nonconformity had been judicially declared no crime in 1767. That year the case of *Evans v. the Corporation of London* was tried finally before the House of Lords, to determine whether a Dissenter could be fined for declining to serve in an office from which he was excluded by the Test and Corporations Acts. Lord Chief Justice Mansfield, speaking for the unanimous ruling in Evans's favor, declared that the

---

22. William Blackstone, *Commentaries . . . Book the Fourth* (Oxford: Clarendon Press, 1770), "fourth edition" (but surely the second of Book 4 though fourth of earlier books), 51–52. Some six years later, when Priestley appeared as a witness in a case tried before Blackstone, the judge was "exceedingly civil, and took several occasions of paying me compliments"; see Priestley to Lindsey, 25 March 1775 (W 1/1:267).

Act of Toleration, in removing the penalties for Dissent, had made the Dissenters' way of worship not only lawful, but established.

Having successfully negotiated his detour into explicit political controversy, Priestley returned to his efforts to define the religious and social roles of Dissent and their relation to the established church. He had begun quite simply, with *Considerations on Church-authority*, and was diverted to Blackstone. Then, shocked with the ignorance of Church members on the most elementary convictions of Dissent, he went on to *A View of the Principles and Conduct of the Protestant Dissenters*. Andrew Kippis and Richard Price then persuaded him that Dissenters were quite as much in need of instruction and he wrote, anonymously lest his observations and advice be eclipsed by controversy over their source, *A Free Address to Protestant Dissenters, as Such*.[23]

The immediate purpose and the audience of each were different: Balguy's Sermon on church authority was welcomed "because of the advantage to religious liberty of keeping the subject continually in view" (*Cons.* v). *View of the Principles* had "the merit of books on travel, informing people of the manners and customs of those to whom they were strangers" (6–7); while *Free Address* concerned itself with the distressing situation of the Dissenting interest (iii). There would inevitably be some overlap in three political pamphlets, of roughly one hundred pages each, written in less than a year, but the congruence here is more than casual. Taken together, the three, approaching the same subject from different directions, constitute a nice summary of Priestley's view of where liberal Dissent fitted in the design of progress and why.

The clergy and members of the Church of England (and, therefore, all government officials) knew surprisingly little about the lives, beliefs, and practices of Dissenters; many even confused English with Scottish Presbyte-

---

23. Joseph Priestley, *Considerations on Church-authority; occasioned by Dr. Balguy's Sermon, on that Subject; preached at Lambeth Chapel and published by Order of the Archbishop* (London: J. Johnson and J. Payne, 1769). Most of this printing was destroyed in a fire early in 1770 and *Considerations* was then incorporated, with the earlier *Remarks on . . . Brown*, into the second edition of the *Essay on Government*. Priestley had originally intended to add to the title of *Principles and Conduct*: "containing some Strictures on Dr. Blackstone's Commentaries and his Reply," but wrote to Lindsey, "Dr. Blackstone's is so little considered in the body of the work, that I do not know whether I should mention him on the title page or not" (Letter of October–November 1769, Priestley MSS, Dr. Williams's Library; Rutt's version [*W* 1/1:101–4] is incomplete). In the end, the pamphlet appeared as Joseph Priestley, *A View of the Principles and Conduct of the Protestant Dissenters, with respect to the Civil and Ecclesiastical Constitution of England* (London: J. Johnson and J. Payne, 1769): 2d ed. 1769. [Joseph Priestley], "A Dissenter," *A Free Address to Protestant Dissenters, as Such*

rians. Unfortunately, as Dissenters agreed only in dissenting from the doctrines and discipline of the established church, they had no single person, or body of persons who could speak for them. Now, clearly, there was need for understanding. Priestley attempted to provide some of what was needed, though he doubted that many Dissenters would be entirely happy with him as their spokesman.

Yet there were principles, both negative and positive, that underlay that general opposition to church establishment. Dissenters believed that there were no spiritual grounds to the claims of the Church. They disdained human authority in matters of religion, which should be based entirely on what is in the New Testament (*View* 8). Even advocates of church authority now avoided spiritual claims, arguing on grounds of its utility to religion and the state (*Cons.* 61). But these arguments, also, were demonstrably wrong.

That every organization requires officers and rules is true, but that does not mean that these need be provided by government. Discipline in the early church was based upon its spiritual authority. Had that church been under rules of government, it could never have opposed the establishment of its own day. Our Lord and his apostles did not apply to civil governments for permission to change religion or to propagate it (*Cons.* 10, 31–38). Nor was there need for state authority to spread Christianity. In fact, introduction of that authority by Constantine led to establishment of the Roman hierarchy—which scarcely recommended it (*Cons.* 50–52).

Assignment of ministers to churches was unnecessary. The early church chose its own ministers, as did modern Dissenting congregations. The occasionally resulting disputes within them were no worse than the perpetual squabbles between clergy and parishioners over tithes (*Cons.* 58). Hierarchical selection of the clergy politicized it and led to pluralism, nonresidence, inadequate salaries for lesser clergy, and such waste in the revenues of the universities (appendages, in a manner, to church establishment) that the Scottish universities, with trifling funds, were superior in instruction and usefulness (*View* 51–52).

Nor were magistrates, untrained in theology and primarily concerned with other matters, capable of determining rites and doctrine: "low as is my opinion of the persons who composed the synods and general councils

---

(London: for J. Johnson and J. Payne, 1769), 2d ed. enl. (London: J. Johnson, 1771); there were third (1774), fourth (with Priestley's name, 1788), and fifth (1815) editions of the *Free Address*. I used the first edition of *Considerations* and second of *View* and *Free Address*.

of former times, I cannot help thinking them more competent judges of articles of Christian faith than any king of England, assisted or not assisted by an English parliament" (*Cons.* 24). Certainly it was a curious argument of the Bishop of Gloucester (Warburton) that the civil magistrate, being disinterested, was more likely to decide according to truth in matters of religion than were churchmen.

The need of the state for a religious establishment was equally illusory. There was no reason to suppose that civil society could not subsist without the sanctions of a state church. Holland, Russia, Pennsylvania and some other American colonies failed to guard a prevailing mode of religion and were not troubled by civil strife (*Cons.* 48). Dissenters did not disclaim human authority in earthly matters (*View* 29). Most Dissenters, and therefore their ministers, were members of the middling classes with a strong desire for peace and tranquillity. Priestley had no objection to the presence, in his congregation, of a magistrate there to keep public peace and safety, so long as there was no attempt to set doctrines or determine rites for that religious meeting (*Cons.* 66).

The political history of Nonconformity had been misread and misrepresented. True, the Puritans, after a long period of intense persecution, had responded in kind when given the chance. Events had proved them the weaker party and their religious heirs, Dissenters, had learned the lesson. They were consistent supporters of civil, political, and religious freedom. In fact, it was only with the support of the Presbyterians in 1660 and the Dissenters in 1688, that the Church itself had been preserved, the latter instance being the "second time that the Church of England was rescued out of the most imminent danger, by men, for whose satisfaction they would not abate a ceremony" (*View* 77–78).

Since then it had been the Dissenters who had been the continual supporters of the Hanoverian succession and though recent events (especially the "Luttrell affair": the seating of Luttrell as member for Middlesex when Wilkes was elected) had lessened Dissenters' enthusiasm for government ministries, this had not been unique to Dissenters, nor based upon principles unique to Dissent (*View* 42). There is no evidence that Dissenters, in general, have any fondness for equality or republican maxims and no thinking person favors anarchy. Though they reject efforts to control their conduct respecting a future world, they submit freely to laws for this one. " . . . I value," wrote Priestley, "my life, my liberty, my property, and my ease, as much as any other man; and therefore I . . . have as sincere an affection for a system of laws and government that can secure them to me, as any other person whatever" (*View* 31).

These were essentially negative justifications for Nonconformity to Church authority. There were other, positive, reasons as well. Only through Dissent was there a chance for genuine toleration and a continuation of religious reform (*Free* 18). The Church of England had been established in the early stages of reformation and, though it retained unsavory vestiges of Popery, had remained unchanged after two hundred years despite revolutionary changes of thought throughout Europe (*View* 13–14). But establishments never reform themselves, reformation always being imposed *ab extra*. Differing Dissenting sects did treat one another illiberally, and Priestley wished they would exhibit less animosity over their differences, but they also acted as checks upon one another and inspired emulation through their spirit of free inquiry (*View* 80–81, *Free* 121). And "freedom of inquiry . . . will produce its natural offspring, truth; and truth has charms, that require only to be seen and known, in order to recommend itself to the acceptance of all mankind" (*Free* 30).

As a consequence of that freedom, it may, perhaps, be possible, in some remote period, that one set of religious sentiments and one method of conducting public worship will be approved and adopted by everyone (*View* 80). In the meantime, let Dissenting attitudes toward civil and religious liberty be diffused throughout the nation. Let there be a spread of toleration. It was absurd for any Protestant to maintain the right to resist Popish government and, at the same time, insist on the right to impose upon Protestants of other denominations (*Cons.* 8). All rational Dissenters objected to the sentence of damnation pronounced upon unbelievers in the Athanasian doctrine of the Trinity. Priestley even suggests that some men may be saved though they are not Christians! (*View* 16–18, *Free* 55–60).

Those who imbibe their opinions from Voltaire or Rousseau, being indifferent to all religions, think any strong defense of religion to be "unpolite, bigotted, and illiberal" (*Free* 61–69), while churchmen complain of indecency when Dissenters defend their opinions. But surely, in equity, civility should be required from both sides. "Why should insolence on one side be not answered by contempt on the other?" (*Cons.* vi–viii). Dissenters owe to themselves and to God the duty of asserting their principles and attacking the erroneous ideas of others, but they must accept the sincerity of those with whom they differ. Especially they must not insult or ridicule those retaining principles they once held themselves (*Free* 60, 55, 99). Penal laws against Dissent are not now enforced and it would be just and wise to repeal them, but should persecution return, "we had rather be among those against whom it is directed, than among those who direct it" (*View* 91).

Given the importance, to religion and to the nation, of Dissent, there was cause for concern over the cooling zeal of Dissenters (*Free* 16). Priestley doubted that any one had left Dissent for the established church from "real *conviction of mind*, and after mature consideration" (*Free* 109). Still, there had been desertions for reasons of social policy, marriage, or self-interest and, within Dissenting congregations, there was less earnestness and support than in the days of persecution. Dissenting ministers, perhaps inferior to the clergy in classical learning, were not so in philosophical knowledge, nor in politeness and were usually superior in their knowledge of theology. Now they and their congregations were sinking fast to the level of the established church in their reading and knowledge of scripture (*View* 87).

In total numbers, Dissenters were actually increasing, due to recruits from the Methodists, " . . . raised up by Divine Providence, at a most seasonable juncture." Though they communicate with the Church of England, they are not attached to the hierarchy, the parish churches, or the Book of Common Prayer. They will, in time, come to think freely and supply the places left by rational but lukewarm Dissenters absorbed into Church or irreligion (*Free* 118–19).

Earnest rational Christians must now rally to their responsibilities. It *was* hard to be taxed to maintain the establishment and then raise money for education, salaries, and housing of Dissenting ministers and maintenance of meeting-houses. Nor was it likely that this would ever be a smaller burden, for prices would rise. But the income of a Dissenting minister was now barely enough to maintain himself and family. There was no way he might provide for his family after his death. The number of young persons being educated for the Dissenting ministry was, therefore, decreasing every year (*Free* 86–89). To make the ministry acceptable, wealthy gentlemen must recognize its importance. The cost was relatively small, especially compared to the sacrifices of primitive Christians or of the Puritans who left everything and fled to the "inhospitable coasts of North-America" for refuge from persecution.

To deserve support, the ministers must set an example of confidence in Divine Providence, "under the circumstances of a scanty and precarious provision." They should shun impropriety in behavior and, without austerity or conceit, be examples of Christian living to their congregations.

> Let . . . the principal object of your attention . . . [be] the proper
> duty of your profession, and let no taste you may have for any of

the *polite arts,* as music, painting, or poetry, nor a capacity for the improvements in *science,* engage you to make them more than an *amusement.* . . . Let not even the study of *speculative theology* prevent your applying yourselves chiefly to the advancement of virtue among your hearers. (*Free* 99–100)

The extraordinary spate of Priestley's political publications during 1769 was followed by an equally unusual silence respecting the inevitable answers to them. He acknowledged the accommodating response of Blackstone, but the only detailed reply to the other answerers was practically extorted from him by the not-very-anonymous author, William Enfield, of *Remarks on Several Late Publications* relative to Dissenters of 1770.[24] It is hard to know what Enfield intended to accomplish by his *Remarks.* He had recently been appointed to fill the tutorial position at Warrington that Priestley had once held and perhaps, Priestley thought, had "the no very blameable desire of gaining reputation as a writer" (*Works* 22:533). He appears also to have been a Unitarian of the Channing-Martineau type, without their romantic sensibility but with the same expectation that by smooth accommodations and emphasis upon similarities rather than differences, "no insuperable bar should exist to an entrance into the established church," or, at least, into a more general acceptance by the higher ranks of polite society.[25] He didn't, that is, dispute the truth of what Priestley wrote so much as he wished it had not been written, or not with so much "vehemence of temper." He also complained of the "laborious, and in great measure, fruitless course of study" of students for the Dissenting ministry and wished that there might be more attention by Dissenters to innocent diversions of music, dancing, polite literature, and theatrical entertainment.

24. Priestley wrote of events leading to his answer in "A Letter to a Friend in Manchester," dated 20 September 1770, "to be shewn to a few friends" (reprinted in *Works* 22:533–35). He and Enfield had corresponded and even corrected proofs together for the last of these publications, without Enfield's raising any objections. Then, in late March, while Priestley was in London, a copy of *Remarks* was shown him, with an offer to withdraw it because of his losses in the fire at Johnson and Payne's. Priestley refused this "charity" and printing of *Remarks* was begun. Enfield begged for a public answer, hoping he would not be treated like Mr. Venn, but "any notice was better than none." After *Remarks* had been distributed and Enfield saw a draft of Priestley's reply, he wanted to suppress the entire controversy, but then it was too late.

25. Turner, *Warrington Academy,* 39. Enfield's major objections are to the *Free Address to Dissenters, as Such,* which has been described as a widely approved pamphlet that "did much to restore the tone of those to whom it was directed"; see Earl M. Wilbur, *A History of Unitarianism: in Transylvania, England, and America* (Cambridge: Harvard University Press, 1952), 297.

Priestley's response, in *Letters to the Author of "Remarks on Several Late Publications relative to Dissenters,"* was predictably scathing.[26] This was the work of one of the *"new* species of Dissenters . . . of young gentlemen and ladies, who have as little of the spirit as they have of the *external appearance,* of the old *Puritans"* (4–5). As the author (Priestley rigorously concealed Enfield's identity) agreed in all major particulars, his objections seem to stem from fear that reviving the interest of Dissenters in their principles would rouse the sleeping lion. Accepting these grounds for lack of persecution gave up the principle of religious liberty by acknowledging a right to require submission.

Dissenting ministers and their families are actually starving, while the author defends the right of opulent Dissenters to "innocent" gratifications and amusements (46–49). The studies he recommended are more suitable to fine gentlemen than to learned divines; they are, of themselves, not evil, but, as St. Paul has said: "all things are lawful unto me, but all things are not expedient" (55). *Remarks* seems to suggest that "absurd and monstrous opinions in religion must be suffered to die away *without notice,* whereas a taste for any species of pleasure or amusement is an object of so much importance, that every thing must be risqued for it" (63).

Enfield, rather naturally, was left smarting by this treatment, but went on to become, if not a friend, at least friendly with Priestley. His biographer reports that the failure of imminent attempts to ease differences between clergy and Dissenting ministers made him a more decided separatist and "with a rare magnanimity, he acknowledged his former error, recommended those books which he had before disapproved, and took every opportunity of shewing his respect and approbation of that distinguished person, to whose conduct he had most objected and who certainly had treated him with no particular respect."[27]

Priestley's indirect involvement in those abortive attempts at accommodation may have been one of the reasons for his silence on political matters for the two and a half years remaining of his stay in Leeds. There was, of course, his writing on religious matters (ten works appearing during the period) and his scientific activities, but occupational pressures seldom pre-

26. Joseph Priestley, *Letters to the Author of "Remarks . . ."* (London: Joseph Johnson, 1770), one ed. only. A short, pained, reply from Enfield and a shorter (4 pp.) *Answer to A Second Letter to Dr. Priestley,* dated 6 September 1770, with the note: "To be given to the Purchasers of the former Letters," ended the exchange on much the same note.

27. Turner, *Warrington Academy,* 39. Enfield's *Institutes of Natural Philosophy* (London: J. Johnson, 1785), was dedicated to Priestley.

cluded Priestley's taking on additional chores. The general political situation, however, appeared to have improved during that time, though a commercial crisis in 1772 caused a financial crisis and there were food riots.

The unstable Grafton ministry had ended in January 1770 and was succeeded by the long-term ministry of Lord North. That promised future difficulties for the American colonies, but immediate tensions were relieved by repeal of the Townshend Acts (except for that on tea). Time was thus gained for negotiation over the claim by Parliament to its right to tax the Americans. Wilkes was released from prison in April 1770; he would shortly be elected alderman and Lord Mayor of London and then, once more, to Parliament—this time to be accepted. Many of the political problems that most concerned Priestley seemed on their way to solution and even the major one, that of state authority in religion, was being brought to the attention of Parliament.

Publication of Archdeacon Blackburne's *The Confessional* (1766; 2d ed., 1767; 3d ed., 1770) brought into the open the dissatisfaction of many clergymen over their obligation to subscribe to the Thirty-nine Articles. Priestley's writings, from 1767 through 1769, heightened their discontent as he repeated denunciations of man-made creeds, doubted that any thinking person could honestly subscribe to more than a third of the articles, and called upon honest clergy to repudiate them.[28] Early in 1770, a group of clergy organized to rid themselves of the burden of subscription.

By July 1771, the group, led by Blackburne and Priestley's friend, Theophilus Lindsey, meeting at Feather's Tavern, London, formed an association to petition for relief from subscription. They advertised in the newspapers for additional signers. In December 1771 some undergraduates at Cambridge petitioned the Vice-Chancellor for relief from subscription on taking their degrees. In February 1772 a petition, signed by two hundred and fifty clergy and lay graduates in law and medicine at the universities, was presented to Parliament asking that a declaration of belief in the Scriptures be substituted for subscription to the articles, for ordination and for admission into the universities.

The proposal to receive and consider the petition was vigorously debated and soundly defeated on 6 February 1772. Priestley had no role in the

---

28. See, for example, *View of the Principles of Dissenters*, 19–20: "The most learned and respected members of the Church of England have been foremost in their labors to explode them [the doctrines of the Thirty-nine Articles]. . . . They agree with us [Dissenters] in thinking them a disgrace to the Established Church, and heartily wish they were fairly rid of them."

Feather's Tavern Association. Indeed, he advised Lindsey in February 1770 that an alliance between reforming clergy and Dissenters would do the former no good, making them "more obnoxious to the superior clergy." Nonetheless, he was fully informed of their activities and his correspondence was enlarged by at least one of the other leaders of the association, John Jebb of Cambridge.[29]

During the debate over admitting the Feather's Tavern petition, Lord North observed that as Dissenting ministers and schoolmasters received no emoluments or opportunities for preferment from the church, the government would not oppose a petition for relief from their subscription to thirty-four of the Articles. At once, the General Body of Protestant Dissenting Ministers in and about London appointed a committee to solicit such a petition and draw up an appropriate bill, which was presented to the House of Commons on 3 April 1772. The House accepted the bill without division and approved it on second and third readings by large majorities, but the House of Lords threw it out by a majority (including nearly all the bishops) of nearly four to one. Early in 1773, another petition and a second bill again passed the House of Commons and was again defeated in the House of Lords.

Priestley was only peripherally involved in these activities as well. His name was on the initial petition of 1772 and he was directly concerned in getting signatures for the second, having, as he wrote to Richard Price in November 1772, to "divide, subdivide, and distinguish" before the ministers in the Leeds neighborhood would acquiesce.[30] Certainly he kept informed on the progress of the applications; he was a spectator at the House

29. Priestley to Lindsey, 21 February 1770; he also notes to Lindsey, 23 August 1771, that Lindsey chaired the first general meeting of the Association, "more to your honour than being at the head of any convocation or general Council" (*W* 1/1:114 and 145, respectively). An entry from about this date, in the diary of Sylas Neville (1741–1840), notes that Jebb's father, Dr. Jebb, dean of Cashel, called Priestley: "the divine Priestley"; see Basil Cozens-Hardy, ed., *The Diary of Sylas Neville, 1767–1788* (London: Oxford University Press, 1950), 111.

30. Priestley to Richard Price, 11 November 1772, *SciAuto*, no. 53. "Rigid calvinists," much influenced by the Reverend Mr. Hitchen (brother-in-law of Dr. Priestley's cousin, Mr. Joseph Priestley), who had written opposing the application to Parliament, "came to oppose and wrangle," but were brought in line by "judicious and fair management." See T. G. Crippen, ed. "A View of English Nonconformity in 1773," *Transactions of the Congregational Historical Society* 5 (1911–12): 205–22, 261–77, 372–85, for the names (including Priestley's) on the first petition; W. H. Summers, ed., "Gibbons' Diary," 2:22–38, for a contemporary record of some of the activities of the General Body of Dissenting Ministers respecting the Dissenters' bills. S. Maccoby, *English Radicalism, 1762–1785* (London: George Allen and Unwin, 1955), 177–80, describes in some detail the petitions, bills, and reactions relating to these events.

of Commons at first reading of the bill and was immediately informed by John Lee of its failure in the House of Lords.

Although opposition by the bishops almost guaranteed the failure of the second, or any subsequent, attempts, still it seems surprising that Priestley should have been backward in supporting these activities. Yet the answer is easy to find. In the first place, as he told Lindsey in March 1772, he was basically opposed to any application to Parliament on religious matters; that would imply that it had power over such things. Second, having brought himself to accept an application ("it is only desiring a person to recede from a claim which never has been, and never will be, acknowledged") he rightly believed that his favoring the bill would count against it: "So sensible was I of the predicament in which I stand with Dissenters in general . . . that wishing well to our late bill, I studiously avoided appearing in it, and wrote not a syllable about it for the use of the public."[31]

He had, however, already written enough on American affairs, and on political and civil liberty, to call himself favorably to the attention of William Petty, Lord Shelburne and to set into motion a train of negotiations that would take him from Leeds to Calne. It is unlikely, nonetheless, that Shelburne's patronage of notable figures would have stretched to include Priestley without the reputation in the sciences he had also gained during his six years in Leeds.

31. Priestley to Lindsey, 2 March 1772 and 9 March 1772 (*W* 1/1:160, 161). [Joseph Priestley], "Author of the 'Free Address to Protestant Dissenters as Such,'" *A Letter of Advice to those Dissenters who conduct the Application to Parliament for Relief from Certain Penal Laws, with Various Observations relating to similar Subjects* (London: Joseph Johnson, 1773), 36, written after the second bill was defeated. Priestley notes that the Bishop of Llandaff read extracts from his writings in the House of Lords, to show what heresies would be promoted by passage of the bill.

X

LEEDS, 1767–1773

Electricity, Perspective, Optics

However strongly Priestley resolved to exercise his "capacity for the improvements in science" as an amusement only, he was equally determined to pursue that amusement. Scarcely had he committed himself to the move from Warrington when he began to ensure continuation of his scientific studies. He wrote to John Canton in April 1767 about obtaining apparatus to replace that he had had available through the academy. Curiously the apparatus he intended to purchase was mostly optical—a microscope for opaque objects, a telescope—though he promised Canton a winter's campaign of electricity when he got to Leeds.[1] His services at Mill Hill commenced in September and by the end of the month he was already writing to his electrical correspondents.

The first letters, however, were about chemical experiments as well as electricity. Some experiments with "mephitic air"—bleaching vegetable color or making the "most delightful *Pyrmont Water*"—were, no doubt, partially a consequence of the ready supply of carbon dioxide from the neighboring brewery, as he was later to insist. Those on "inflammable air" haven't that pretext and all of the experiments on "airs" are better seen as a continuation of the chemical interests displayed during Priestley's Warrington years.[2]

---

1. Priestley to Canton, 21 April 1767; *SciAuto*, no. 14.
2. Priestley to Canton, 27 September 1767; *SciAuto*, no. 15. Priestley claims to have written "a post or two ago" to Dr. Watson, who might be thought, as a M.D., to be more interested than Canton in chemical topics.

Pneumatic studies, though increasingly evident in his correspondence, did not yet constitute a major concern. He was still committed, partially by the success of his *History of Electricity,* to a continuation of electrical investigations. Priestley's first letter from Leeds relating to science introduced the "lateral effect" of electrical discharge and "Priestley's Rings," about which he later wrote papers for the Royal Society. And while performing those experiments, before bringing them to a state for publication, he attempted a financial exploitation of the success of the *History.*

When the Priestleys began making electrical apparatus for sale is not clear. Timothy Priestley, when describing the kite he made for Joseph in 1766, also mentioned a proposal by his brother of "a partnership, for making electrifying machines." The preface to the *History,* written early in 1767, recommended obtaining an electrical machine. In November 1767, Priestley sent Canton a battery (of Leyden jars) similar to that he was using: "As to the price, I cannot tell you till I see you." In the same mail he sent a copy (in manuscript or sheets) of his *Familiar Introduction to the Study of Electricity* with a message for Richard Price that he had given up the scheme of selling electrical machines in London.

Yet the verso of the half-title of *Familiar Introduction,* published early in 1768, contains an advertisement for the machines, ready "about the middle of March" (see Fig. 7). Made under the direction of Dr. Priestley, orders could be placed at Leeds, or with the Reverend Mr. Priestley, at Manchester, "under whose immediate inspection a considerable part of the machines and apparatus will be made." Remittances were to be sent to Mr. Johnson, bookseller in London, with whom orders might also be placed. As an added commercial touch, each purchaser of a machine would get a copy of the *Familiar Introduction,* in which the reader would find the statement: "I would advise all persons who propose to understand the subject of Electricity . . . to provide themselves with an electrical machine, or at least desire . . . their friends to show them the experiments. Without this, I should despair of making any person whatever a master of the subject."[3]

Of course, the *Familiar Introduction* was, itself, an exploiting of the *History of Electricity.* It also used what his friend Lindsey called Priestley's "talent for making dark things perspicuous" (W 1/1:121). Written, Priestley

3. Priestley, *Familiar Introduction to Electricity,* viii. For Timothy's statement, see Chapter 6; see also Priestley to Canton, 17 November 1767, *SciAuto,* no. 17. As late as 1770, Priestley hoped to sell machines in London, through the shop of John Dolland; see letter to Canton, 28 February 1770, *SciAuto,* no. 26. Abraham Rees, ed., *Cyclopaedia: or Universal Dictionary* (London: Longman, Hurst, Rees, Orme and Brown, et al., 1820), 2:plates; plate

Fig. 7. Priestley's Electrical Machine, from *The History and Present State of Electricity*, 1767.

8 is of Priestley's electrical machine. David Brewster, ed., *The Edinburgh Encyclopaedia* (Philadelphia: Joseph and Edward Parker, 1832), 8:335–36, has a description with an illustration, fig. 4, plate 247.

says, at the persuasion of John Aiken of Warrington and some friends in Leeds (but surely also with the encouragement of its London booksellers), it was intended to make the subject of electricity intelligible to those beginners who found the *History of Electricity* beyond their comprehension. It cannot have taken much time in writing, for it was short (51 pages, in the 4th ed.), with four plates from the *History*. It essentially repeated the operational entertaining-experiment part of that work. No sentence was copied, varied expression being calculated to assist people better to understand the principles (vi).

There was a caution to young electricians about use of "large" batteries of Leyden jars, because of the danger of their shocks, but "large" was not defined and, as in the *History*, quantitative information about the charging of jars was flawed by separation of the phrase "the thinner the glass is, the stronger the charge which it is capable of bearing" from the implication that the charge was also proportional to the area of coated surface (29, 45). Priestley's discovery that electrical discharges follow all available paths was made after publication of *Familiar Introduction*, which still said that electrical fire goes by the shortest path or the best conductor.[4] And perhaps it is only lack of historical context that made repetition of the phrase "supposed to" when briefly discussing the electrical matter or fluid, seem an echo of his developing doubts of the existence of such a fluid, sui generis (9, 50). The *Familiar Introduction* clearly does not go beyond the *History*, yet it seems to have filled a need, for there were four editions of it (1768, 1769, 1777, 1786) in the next twenty years.

No doubt an award to Priestley of the Copley Medal of the Royal Society would have enhanced the financial success of his commercial venture (as membership in the Society increased sales of his *History*) and Benjamin Franklin would surely have approved of this as a by-product when he recommended the award. His efforts to obtain that medal for Priestley, however, derived from a conviction that the work in electricity deserved that honor. His recommendation, made first to the president and council of the Society in November 1767, was rejected, at least in part, because some councilmen were unaware of the original experiments in the *History of Electricity*.

Franklin then returned to the Society, as a whole, in March 1768 with "A brief Account of that Part of Doctor Priestley's Work on Electricity, which relates the new Experiments made by himself." In thirteen sections,

4. Priestley to Franklin, 1 November 1768; *SciAuto*, no. 24. *Familiar Introduction*, 33.

the "brief Account" lists most of Priestley's purely empirical discoveries: for example, the confirmation of electric wind, conductivity of charcoal and hot glass, and physiological effects of shocks from large batteries. With singular infelicity, it fails to list those discoveries that, in retrospect at least, had the most long-range significance: the inverse-square law observation, studies on comparative resistivities, and developing notions of how to compare strengths of electrical discharges.[5]

Probably few of his hearers would have valued these additions, however, for most Englishmen of the time seemed uninterested in quantitative studies of electricity. In any event, the Society awarded the Medal to John Ellis in 1767 and to Peter Woulfe in 1768. And perhaps that was just as well. Priestley was already committed, by the success of his *History of Electricity* and the momentum of a series of experiments on which he was embarked, to a continuation of some work in electricity. Had he obtained the Copley Medal, he might have felt obligated to concentrate upon investigations in an area that was soon to develop in directions beyond his particular talents.

In the preface to the first edition of the *History*, Priestley had promised to print "Additions, as new discoveries are made" and requested any person making such discoveries to communicate them for inclusion in the "Additions" or in subsequent editions. This was written when Priestley still felt that history was his forte in science, but as late as 1770, he could write that the *History* had made some people electricians, "whose labors I consider, in some respects, as my own." And, in 1772, after the appearance of the French translation of the *History*, he was corresponding with Alessandro Volta and Thorbern Bergman (at least) respecting their claims to electrical discovery, to be included in future editions.[6]

As it happened, though Priestley hoped to write a continuation of the *History* some years hence [preface, 3d ed., 1775], neither the "Additions" nor the later editions that were published included any "new discoveries" but Priestley's own. Only the second edition, of 1769, really made substantial changes in the text, though both the second and third reprinted some of Priestley's experimental papers from the *Philosophical Transactions* of the

5. Franklin to Ld. Morton, [November 1767], to John Canton, 27 November 1767; "Brief Account," read 10 March 1768; *SciAuto*, nos. 18, 19, 21. Franklin observed to Canton that technicalities had been raised in council and he suspected opposition existed to awarding the Medal to Priestley.

6. Priestley to Lindsey, 4 November 1770, 9 March 1772; Priestley MSS, Dr. Williams's Library (Rutt versions, W 1/1:121–23, 161–63 incomplete). Priestley to Volta, and Bergman, 14 March 1772, *SciAuto*, nos. 45, 46.

Royal Society. The textual changes in the second edition were partly the result of books people had sent Priestley after reading his printed "Catalogue of Books written on the Subject of Electricity" and noted those indicated as "not seen." Priestley could then read these primary sources and make corrections and additions.

Changes also occurred, however, because Priestley was aware, from the publication of the first edition, of a major deficiency in it. The majority of his references to German and Scandinavian researches had been secondhand versions, in English, French, or Latin. By November 1767, he was already relearning "High Dutch" (German) to remedy that situation and though he later reported that "several of the Dutch books proved very tiresome and empty," the most significant improvements of the second edition were derived from his having read the work of Daniel Gralath, in the *Versuche und Abhandlungen der Naturforschenden Gesellschaft in Danzig* (1747, 1754, 1756).[7] There were other minor changes: the secretary of the Royal Society sent information about Musschenbroek, Canton wrote a fuller description of electrifying the air in a room, and here and there some wording has been modified, usually to restate a phrase Priestley had found by experience to be perhaps too positive or too imprecise. With this, the second edition, "corrected and enlarged," was completed; there would be fewer changes in subsequent editions, which never covered the period past 1766, and for practical purposes Priestley's historical chore was done—so far as electricity was concerned.

There were still some experimental problems to be worked out, however, and to these Priestley had also turned his attention from the first month of his settling in Leeds. Within the next three years he was to present five papers to the Royal Society: "An Account of Rings consisting of all the Prismatic Colours, made by Electrical Explosions on the Surface of Pieces of Metal," read 10 March 1768; "Experiments on the lateral Force of Electrical Explosions," read 23 February 1769; "Various Experiments on the Force of Electrical Explosions," read 2 March 1769; "An Investigation of the lateral Explosion and of the Electricity communicated to the electrical Circuit in a Discharge," read 29 March 1770; and "Experiments and Observations on Charcoal," read 5 April 1770.[8]

7. Priestley to John Canton, 17 November 1767, 24 May 1768, *SciAuto*, nos. 17, 22. Note that Priestley had studied "High Dutch" while still a youth in Heckmondwike; see Chapter I.

8. Joseph Priestley, "Account of Rings . . . ," *Philosophical Transactions* 58 (for 1768): 68–74; "On Lateral Force," and "On Force of Explosions," *Philosophical Transactions* 59 (for

The first four of these papers relate, in one way or another, to phenomena associated with what is now called coronal discharge. The process is a complex one and Priestley cannot really be said to have contributed much to the understanding of it. (Indeed there are still arguments about what exactly is going on.) Nonetheless, taken together, these papers represent Priestley's finest work as an electrical experimenter. His best-known discovery, of the inverse-square law, was an anomaly in his characteristic nonquantitative approach to science and interpreted observations initially made by Franklin. These "electrical explosion" papers began with empirical observations, developed new techniques for measuring impedances, and ended with descriptions of what Sir Oliver Lodge was later, and independently, to study as "side-flash" and of oscillatory spark discharge. Any one of these papers could have started fruitful lines of inquiry had anyone paid attention to it.

Priestley's investigations began with an attempt to reproduce, in a laboratory, what he called "lateral force," the force that, in lightning affected people and objects near to but not in the path of the stroke. Taking spark discharges of a battery of Leyden jars through a pointed rod to a plate of metal, he was diverted by observing a pattern of concentric colored rings that appeared on the plate, surrounding the spot hit by the spark. The "Ring" paper is a relatively trivial one, though it describes the only discovery eponymally linking Priestley to electricity. He describes experiments in which he varied the metal of the plates, the distance and angle between rod and plate, the sharpness of the point on the rods.

There is no indication of interest in conditions regulating the discharge; Priestley's attention is concentrated on the differences in size, intensity, and color of the rings. He speculated that they are produced by *laminae*, formed from surface-heating of the metals, recognizing the appearance as a variation of thin-film colors reported in Newton's *Opticks*. He suggests that these rings, like Newton's thin-films, may be of "admirable use to explain the colours, and perhaps, in due time, the constituent parts and internal structure of natural bodies" (69), but his interest and that of the Royal Society, in "Priestley's Rings," seems primarily a simple fascination in their appearance.

Returning to "lateral force" phenomena for the second paper, he found that he could laterally disperse objects from near the path of an electrical discharge through air or even through an imperfectly conducting wire. The

---

1769): 57–62, 63–70; "Lateral Explosions," and "On Charcoal," *Philosophical Transactions* 60 (for 1770): 192–210, 211–27.

greater the force of discharge, the larger the lateral force seemed to be. Now one of the explanations for lightning damage away from the stroke path is the return stroke; that is, the sudden release of induced charges through objects within the electric field, which collapses with the lightning stroke. Richard Price approached that as an explanation of Priestley's observations with his suggestion that bodies were driven off by mutual repulsion from suddenly acquired electricity. Priestley rejected that proposal because he found no evidence of attraction between bodies and discharging rod before it delivered its spark, nor did the equilibrium of electric fluid seem disturbed in an insulated brass rod placed near the discharge of the battery. He finally reached an unsatisfying conclusion that the lateral force was produced by expulsion of air from the path of discharge, though that force was manifest through paper, foil, and even glass and was scarcely influenced by taking the discharge through vacuum or compressed air.[9]

Had Priestley ended his experiments there, his would have been just another insufficient attempt to unravel the complex problem of lightning strokes, but his observations prompted him to experiment on the force and momentum of the discharge and his third paper introduces an interesting speculation and defines two methods for measuring resistances. As electrical discharges along the surface of a leaf left a well-defined path, Priestley took discharges over leaves cut into right and acute angles, to observe whether electric fluid, forced to abrupt changes in direction, might not be diverted by its momentum from a strict following of the sharp angles. He also attempted to detect a decrease in the force of a discharge when taken through wires made to turn through a number of sharp angles. The paper reports no sign of diversion of path or decrease in force and drops the subject.

In his correspondence, however, Priestley remarks that these experiments, with that reported in the *History of Electricity,* where cooling hot glass had not, as Franklin predicted, drawn electric fluid from a copper plate: " . . . make me inclined to think, there is no electric fluid at all, and that electrification is only some [new?] modification of the matter of which any body consisted before that operation." Franklin responded that Priest-

---

9. Price's suggestion is referred to in Priestley to Canton, 12 November 1767; *SciAuto,* no. 16. For some general treatment of coronal discharge and lightning, see B. F. J. Schonland, *The Flight of Thunderbolts* (Oxford: Clarendon Press, 1950) and for some earlier attempts to understand the phenomena, Oliver J. Lodge, *Lightning Conductors and Lightning Guards* (London: Whittaker and Co., 1892). Lodge, incidentally, wrote a biographical sketch on Priestley, unaware of his work on electricity.

ley wrongly believed that "Electric Matter goes in a Ball with a projectile Force," whereas it is really a constant stream.[10] That response may explain why Priestley did not publish the speculation. He went on, instead, to describe methods of comparing force of discharge.

The first method was a variation of a technique he had used to test conductivities of different metals and reported in the *History of Electricity*. Now, however, he inserted a short length of iron wire (a fuse) into the circuit and examined lengths of circuit at which the discharge of his battery just melted the fuse. He was surprised that the force melting the wires varies with circuit length, forgetting his statement of the same "law" in his *History* (65; see Chapter VI).

In a second method of measuring conductivities, he placed the ends of a circuit of wire close enough that the discharge might choose to jump the air-gap in preference to following the wire. He reports: "In this method the different degrees of conducting power in different metals may be tried, using metallic circuits of the same length and thickness and observing the difference of the passage through the air in each" (69). He also observed, however, that "the whole fire of an explosion does not pass in the shortest and best circuit; but that, if inferior circuits be open, part will pass in them at the same time." Having raised a number of questions on electrical resistances and circuits that might demand exact quantitative answers, Priestley characteristically drops this line of investigation to return to that on electrical discharges.[11]

The fourth paper in this series is probably the most interesting and certainly the most prescient of any that he was to write in his electrical career. While it was true that objects near the path of an electric discharge did not appear to acquire a charge, it had also been observed that an insulated conductor near or in contact with, but not part of, the discharging circuit of a Leyden jar could sometimes deliver a shock (or spark) to another

10. Priestley to Canton, 12 November 1767; Franklin to Canton, 27 November 1767, *SciAuto*, nos. 16, 19. No doubt Priestley had thought of the discharge moving as a projectile, as did many others. Bewley's review of the *History of Electricity*, for example, declared: "There is reason to suppose it [electric fluid] moves through bodies, in the form of a ball or cylinder," *Monthly Review* 37 (1767):451. But Franklin's response is no answer to the problem, for his stream of electric fluid must commence to move, change direction, or stop and any such change might be expected to show in Priestley's experiments. The answer, at this time, was a substance without inertia (an imponderable fluid) and Priestley's suspicions here mark his first doubts as to the existence of such substances.

11. One might observe that he was dealing with transient electrical currents and the questions implied by his work could not easily be solved with those conditions.

insulated conductor. There was, that is, occasionally an "explosion" lateral to that which discharged the jar. The "lateral explosion" paper reports on the experiments Priestley performed to determine the circumstances in which the "side-flash" (as Lodge called it) would occur and the nature of the electricity communicated by it.

Finding first that the greatest lateral explosions occurred when the discharge circuit was imperfect, Priestley arranged his apparatus with one end of a long iron rod in contact with the side of the jar and the other about a quarter of an inch from an insulated cardboard tube, seven feet long and four inches in diameter, covered with tinfoil ("the larger the insulated body . . . the greater quantity of the electric fluid it was capable of receiving, or parting with, and . . . the more sensible the effect would be" [194]). Discharging the jar through an interrupted circuit, the expected spark between iron rod and tube appeared—but no electrical charge could be found on the tube, though the experiment was repeated "above fifty times."

Greatly puzzled, Priestley rearranged his apparatus and found a small amount of charge now communicated to the tube, though far less than the size of the lateral spark would suggest. Naturally he concluded that his first observations were wrong and began to elaborate experiments to find conditions under which communicated electricity would be positive or negative. This seemed to depend upon the nature of the interrupted circuit and of the part from which the lateral spark was taken, a spark from one part giving a negative and from another a positive charge to the insulated tube. If then the iron rod were connected to a circuit "balance point," no lateral spark should occur—but one did, again leaving no charge on the insulated tube! Priestley was forced to conclude that there were circumstances in which "an electric spark must enter and pass out again, within so short a space of time as not to be distinguished, and leave no sensible effect whatever" (204).

He had, in effect, tuned his circuit so that an oscillatory Leyden-jar discharge was obtained. He did not know what he had done nor connect the phenomenon to the general nature of Leyden-jar discharge. He did know, however, that there were median conditions of arrangement of the circuit, before a lateral spark would be taken "which shall enter and leave the insulated body without making any sensible alteration in the electricity natural to it" (207). Discussing these conditions in a general, non-quantitative way, he left it to later observers, Lord Kelvin and Joseph

Henry, working within a different concept of the nature of electricity, to predict and again observe oscillatory discharge.[12]

The paper "On Charcoal" marks Priestley's transition from electricity to chemistry. Writing to Lindsey in February 1770, he had described this paper as a report on "many new experiments, electrical and chemical," while explaining that he was "now taking up some of Dr. Hales's inquiries concerning air."[13] And, indeed, the significance of the paper is its chemical emphasis. Priestley had announced his discovery of the conductivity of charcoal in the *History of Electricity*. Now he proposed to examine the differences in its conductivity as a function of the methods and substances of its manufacture.

He found all wood charcoals equally conducting if sufficient heat was used in making them. There was no significant relationship between dimensional changes and conductivities. He then made and tested other "charcoals," from animal matter and oils. He appears to have obtained "lustrous carbon" from vegetable oils, for he reported heating turpentine or olive oil in a glass tube covered with sand and finding the tube lined with a "whitish glossy matter" that resisted acids and scraping but disappeared when the glass was made red-hot in air. The glossy matter conducted electricity well and Priestley concluded that it was a "kind of charcoal only white instead of black" (223–26).

Having read "Macquer and other chemists" on charcoal, he had a clearer idea of phlogiston—"the inflammable principle . . . that is fixed and united to the earth in plants"—than he had when writing the *History of Electricity*. But he still believed it likely that the conductivity in both charcoal and metals is caused by phlogiston. In charcoal, the phlogiston must be firmly united to its base by intense heat. This cannot be done in air, for there seems to be something in the atmosphere that can unite with phlogiston, on the principle of chemical affinities, the moment it is separated from its base.

This gave Priestley the idea that lead and other base metals might, like charcoal, be made better conductors by intense heating in the absence of air (where, he conjectured, they might not calcine or vitrify). Lead, thus treated, did not calcine, but there was no improvement in its conductivity.

12. See Oliver Lodge, *Modern Views of Electricity* (London: Macmillan and Co., 1892), 46–47 and lecture 3, passim, on oscillations of Leyden-jar discharge and the tuning or damping of circuits.

13. Priestley to Lindsey, 21 February 1770; W 1/1:112–14.

Still, Priestley believed, these metals may have their quality altered, and be improved in other respects by this process, ". . . though they should not be changed into gold by it. The specific gravity is not changed . . . so that, alas! it is still lead" (227).

With the minor exception of transmitting, via Franklin, William Henly's "Account of a new Electrometer," published in the *Philosophical Transactions* for 1770, Priestley's career as an electrician closed with this series of papers. Henceforth, though he might use electricity as an *instrument* of investigation, it was not to be the *object*. He did not even carry out the experiments on possible changes in reflection, refraction, or inflection of light by electricity that he had once proposed. But before he could give full attention to those experiments on air he had already begun, there were some science-writing chores—one minor, one major—to be completed.

The minor chore was the writing of a teaching volume on perspective drawing, companion to his primer on electricity. At Warrington, he had been unable to find anyone to draw the plates for his *History of Electricity* and had, finally, done them himself, learning the elementary rules of perspective in the process. He then wrote, in the section: "Branches of knowledge peculiarly useful to an electrician," of the *History*: "Lastly if an electrician intend that the public should be benefited by his labours, he should, by all means, qualify himself to draw according to the rules of PERSPECTIVE; without which he will often be unable to give an adequate idea of his experiments to others."

Now when Priestley learned something, he immediately wanted to teach it. And his personal experience, his antecedent "advertisement," and an increased public interest in art, signaled by the founding of the Royal Academy of Art in 1768, prompted Priestley's publisher, Joseph Johnson, to encourage his writing *A Familiar Introduction to the Theory and Practice of Perspective*, dedicated to the academy's first president, Sir Joshua Reynolds. There were a number of English treatises on perspective available before Priestley's appeared. As he typically did in these circumstances, he consulted many of them before writing his own.

From one of these, in fact, William Emerson's *Perspective; or, the Art of Drawing the Representation of all Objects upon a Plane* (1768), he took extracts and examples. He claimed, however, that improvement was possible in all the books he had seen and, as most of them seem to have developed out of Brooke Taylor's *Treatise on Linear Perspective* (1715), that is probable. Taylor had condemned all books before his own as too concerned with technique, at the expense of theory, but even the distinguished Swiss mathematician Johann Bernoulli found Taylor's work "abstruse to all and unintelligible to artists for whom it was more especially written."[14] Priestley intended his *Introduction to Perspective* to provide "any person[s] who ever think of drawing" with information sufficient to enable them, in less than a week (and a few hours would be enough for those who knew geometry), to "make themselves masters of every thing essential to this art" (ix).

It is a very practical book, with rules for preparing a drawing board with sets of lines for the five different varieties of position that Priestley felt were needed. Other books introduce more theory and technical terms than necessary; the little theory in the *Perspective* is owed to Mr. Joseph Priestley of Halifax (no immediate relation) who also assisted in revising and correcting the work. There are one hundred and thirty-two pages of text, twenty-three fold-out plates, some with small pasted lappets.

It was "nearly ready for the press" in June 1769, when Priestley wrote Anna Laetitia Aikin; "Come see us before it is quite printed, and I will engage to teach you the whole art and mystery of it, in a few hours. If you come a month after, I may know no more about the matter than anybody else."[15] But it was not yet printed in December and still open to corrections in February 1770.

When finally issued, it was only moderately successful. It went to a second edition in 1780, but during the decade between the first and second editions, there were at least four other English treatises on perspective that also went to second editions. Another, dated 1776, was produced in Leeds by Henry Clarke, a teacher said to be a friend of Priestley, but it had no reference to Priestley's *Perspective*.[16] In 1800, the preface to James Malton's *The Young Painter's Maulstick; Being a Practical Treatise on Perspective* could still claim that there was no practical study of the doctrine of perspective, "in an easy, familiar, and engaging manner." As late as 1808, an anonymous drawing-master is said to have recommended "a book by one

14. Quoted in Lawrence Wright, *Perspective in Perspective* (London: Routledge and Kegan Paul, 1983), 165. See also Gavin Stamp, *The Great Perspectivists* (London: Trefoil Books, 1982). Neither Wright nor Stamp, incidentally, mention Priestley's *A Familiar Introduction to the Theory and Practice of Perspective* (London: J. Johnson and J. Payne, 1770).

15. Priestley to Anna Laetitia Aikin, 13 June 1769; copy in the Yates Collection of Priestleyana, no. 15, Library, Royal Society of London. There is reference here to Priestley's tendency to forget something he has learned, once he's made use of it.

16. Henry Clarke, *Practical Perspective. Being a Course of Lessons exhibiting Easy and Concise Rules for drawing justly all Sorts of Objects. Adapted to the Use of Schools* (London: for the author, sold by J. Nourse, 1776). See *Gentleman's Magazine* 88 (1818): 465–66, for obituary of Clarke, who became a professor at the Royal Military Academy, Sandhurst.

Priestley, that's as good as any other."[17] There seems to have been nothing about the *Perspective* to make it better than any other book on the same subject.

There was, however, one singularity. At the last minute and at the request of John Canton and Richard Price, the last page of the preface was cancelled, to introduce the statement: "Since this Work was printed, I have seen a substance excellently adapted to the purpose of wiping from paper the marks of a black-lead-pencil. It must, therefore, be of singular use to those who practice drawing. It is sold by Mr. Nairne, Mathematical Instrument-maker, opposite the Royal Exchange. He sells a cubical piece of about half an inch for three shillings; and he says it will last several years."[18] This appears to be the first published recommendation for the use of rubber erasers; the second edition of 1780 does not include this statement and returns to the older usage of bread-crumbs as erasers.

It is reasonable to assume that most of Priestley's writing ventures, aside from theology, were undertaken with the encouragement of friends and publishers, expecting they might add to his scanty income. This was true of revised editions of the *History of Electricity, Familiar Introduction to Electricity,* and *Familiar Introduction to Perspective,* written at Leeds. Yet a major science writing project was begun without encouragement or sure prospect of financial return. His "History of Experimental Philosophy," which, in the end, covered only the science of optics, cost more in time and money than he could easily afford.

He had suggested, in the preface to the *History of Electricity,* dated March 1767, that a history of all experimental philosophy ought to be written, but such a immense work should be undertaken in parts, by several persons. In less than a year, euphoric from the reception given the *Electricity,* he changed his mind. He described his decision: " . . . I then considered the history of *all* the branches of experimental philosophy as too great an undertaking for any one person; but . . . a nearer view has familiarized it to me, and I now look upon it not only without dread, but with a great deal of pleasure; considering it not only as a very practical business, but even as an agreeable *amusement.*"[19]

17. James Yates, in a note appended to his copy of Priestley's letter to Anna Laetitia Aikin (see note 15) repeats the anonymous recommendation made to a fellow student, Joseph Hunter.

18. Priestley, *Introduction to Perspective* (1770), xv; Priestley to Canton, 28 February 1770; *SciAuto,* no. 26.

19. Joseph Priestley, *The History and Present State of Discoveries relating to Vision, Light and Colours* (London: J. Johnson, 1772), iii; commonly known as the *History of Optics.*

He took time to survey the "extent of the work" and his resources for it and, in March 1770, printed his first proposals with a "Catalogue of the Books, of which Dr. Priestley is already possessed, or to which he has Access, for compiling The History of Experimental Philosophy," and a list of "Books Wanted." This first proposal and catalogue was followed by several revisions, particularly as he found that books available from the libraries at Manchester or Warrington were not really practically accessible. Eventually he was to purchase a substantial number of books and borrow others from the library of the Royal Society. From the first, it was apparent that he could not sell the project to booksellers: " . . . it cannot be supposed that booksellers should be able to give any person an equivalent, even for the expences in which it will necessarily involve him, without considering the value of his *time* and *labour*."[20]

Even when he offered the single *Optics* volume to them in January 1771, he could not get a favorable response. He tried to find a patron for the series, applying to the Duke of Northumberland on the advice of friends and with support from Lindsey, sometime tutor to Northumberland's son. Northumberland was friendly, though scarcely enthusiastic, and apparently assisted with gifts of books sufficient to justify the dedication of the "first" volume to him, but there was no financial support. And costs mounted quickly. Priestley estimated them at £100 in July 1770 and then supposed the total would be at least £100 to £200 more. By the time the *Optics* was being printed, the lists of "memoirs of all the philosophical societies, *of note* in Europe" that he "either had or expected very soon to possess," contained some seventeen names and their cost alone was more than £200.[21] Eventually, against his wishes, he published by subscription and, of the nine booksellers who shared publication of the third edition of the *History of Electricity* in 1775, only Joseph Johnson assisted in that of the *Optics*.

Priestley's initial plan was to collect the materials for the entire general history (estimated to fill six or eight quarto volumes and be completed in something over six years). He would read and take notes from each work, then distribute the materials into appropriate sections. In this way "any philosophical treatise will go through the writer's hands only once, to com-

---

Priestley to Canton, 17 November 1767; *SciAuto*, no. 17, asked that Price be told he had not given up his design of writing the general history of experimental philosophy.

20. Joseph Priestley, *Proposals for Printing by Subscription. The Histories and Present State of Discoveries relating to VISION, LIGHT, AND COLOURS* (Leeds: n.p., 1 February 1771; Leeds, 1 December 1770), *SciAuto*, no. 28a.

21. Priestley to Franklin, 19 April 1771; *SciAuto*, no. 32.

monplace all the materials it contains" and thus escape the necessity of examining a great number of volumes more than once (*Optics* v). Although he was working at it six hours a day by the end of July 1770, he still had not determined on which subject to focus for the initial volume.

At first he thought of doing the history of magnetism, but by November had chosen light and colors because, as he wrote in his proposals for 1 December 1770, "the books he had collected furnished him with more materials for it than for any other." By December, he looked upon his "History of Discoveries relating to Vision, Light and Colours to be as good as finished, as little remains to be done besides transcribing which however is necessarily slow and tedious."[22] In April 1771, the printing of one thousand copies was begun; in October and November, Richard Price and John Canton were reading sheets for corrections and the *History of Optics* was finally published in March 1772.

Although Priestley thought the subject a finer one for history than electricity and as much in need of historical treatment, the *History and Present State of Discoveries Relating to Vision, Light and Colours* was not a success. It had only one edition and was translated only into German (1774–75), though it was to be the only history of optics in English for 150 years and the only one in any language for more than fifty. It was favorably reviewed at the time, but faintly praised by Thomas Young in 1807:

> The late Dr. Priestley rendered an essential service to the science of optics, considered as a subject for the amusement of the general reader, by an elegant and well written account of the principal experiments and theories, which had been published before the year 1770. But this work is very deficient in mathematical accuracy, and the author was not sufficiently a master of the science to distinguish the good from the indifferent.[23]

22. Priestley to Lindsey, 23 December 1770; *SciAuto*, no. 31.

23. Thomas Young, *Lectures on Natural Philosophy*, 1:480–81; The *Optics* is also listed in the bibliography appended to Young's *Lectures*, with the asterisk indicating a work of superior merit and originality, 2:280, 323. Bewley, *Monthly Review* 47 (1772): 304–19, wrote: "the Author has, with the greatest industry, digested . . . every essential particular relating to light and vision, that he could collect from the numerous publications, foreign and domestic, respecting that science: interspersing occasionally some original observations and remarks made by himself, or such as have been communicated to him by his many valuable philosophical friends." As late as 1827, the *Optics* was recommended as "a work which is at once interesting to the profound reader, and yet adapted to the humblest capacity," E. S. Fischer, *Elements of Natural Philosophy*, trans. John Ferrar (Boston: Hillard, Gray, Little and Wilkens, 1827).

There were 477 subscribers, including libraries, for 531 copies (Franklin for example, subscribed for twenty copies and John Lee for five) at a price of £1.1.0, but Thomas Cooper later reported that "by far the majority of these defaulted payment."[24] Even with all the subscriptions paid, they would not have covered the cost of the books purchased for the project and Priestley resolved to drop it. "If I do work for nothing, it shall be upon theology."[25]

Immediate financial return, however, should not have been a significant consideration. In the preface to the *Optics,* he had pointed out that " . . . the expense of executing any one part completely [of the entire history of experimental philosophy] is not much less than the whole will require," and he had nearly completed collecting the library for that whole. It appears, rather, that he had lost interest in the project before completing this first volume of it. He publicly left open the option to continue it, while previously, and privately, declaring that he would not do so. Not even the prospect of a "history of discoveries relating to AIR," hinted at in the *Optics,* was to divert his attentions in science from original experiments to the history of them, after the trouble and disappointment of this experience.

Yet there is no reason to doubt Priestley's continued belief, asserted in the *History of Optics,* that such histories were valuable: to facilitate the advancement of useful science (i), but also as an illustration of the Works of God, which, like their author, are infinite. The progress of real knowledge may, then, be expected to continue not merely in a uniform manner, but constantly accelerated (30). The "historical method" was, in Priestley's rhetorical theory, the best for engaging the attention and communicating knowledge with the greatest ease, certainty, and pleasure (vii). The care for scholarly standards in that method, which distinguished the *Electricity,* was also to be seen in the *Optics*: "I have always been careful to refer, at the bottom of the page, to the very authors that I have made use of in compiling the work." He does not quote any author he has not actually consulted and when he could consult an original account, he did so, those at second-hand being, in general, "exceedingly lame and superficial" (ix).

Remarks such as " . . . it is not my business to note the mistakes of great men, but to record their useful labours" (181), are reminders, however, that Priestley's purpose is not historical, but didactic. History, indeed, suffers rather badly in offhand summaries: up to the period of the Renaissance,

24. Thomas Cooper, appendix no. 1, 286, *Memoirs of Joseph Priestley.*
25. Priestley to Canton, 18 November 1771; *SciAuto,* no. 39.

" . . . we have hardly seen one just idea concerning the *cause* of any appearance, nothing of the true *philosophy* of light or vision" (30); Descartes "was more indebted for the reputation he so long enjoyed to the fertility and boldness of his imagination . . . than to the accuracy of his judgment and the real discoveries that he made" (97); "I cannot say that I have much to advance in favour of any of Gassendi's peculiar opinions concerning vision" (116).

Such comments reflect Priestley's unrequited affection for experimental discovery prior to theory. Newton's theory of refraction, he wrote, "is grounded upon experience only" (334). Yet Priestley was aware that "observations of those . . . not acquainted with the causes of appearances" were not dependable (49). He knew both the short- and long-range value of speculations and theories: " . . . it is by no means necessary to have just views, and a true hypothesis, a priori, in order to make real discoveries" (181). Every desideratum is an imperfect discovery, as every doubt implies some degree of knowledge (773).

A summary of optics, he declared, was shorter at the time he wrote than it would have been a hundred years earlier, and longer than it would be in another century, for many things that enquiries would ultimately reduce into general observations were then recited as unconnected and independent (768–69). It was those many seemingly unconnected and independent particulars that were surely the cause of Priestley's abandoning his science-history project. Unlike the *History of Electricity*, the *History of Optics* was written at a time when the materials were "too many" and too mathematical to make a good history of Priestley's type. Within a month of his concentration on light and colors, he wrote to Franklin:

> I have just dispatched the discoveries of Newton and his Contemporaries, and from his time to the present have such a number of Memoirs, dissertations, tracts, and books on the subject of Light and colours to read, compare and digest, as, I think, would make any person not practiced in the business of arrangement, absolutely despair. Till I had actually taken a list of them, I did not think there had been a tenth, or a twentieth part so much upon the subject. And other subjects, I see, will be much times more embarrassing than this.[26]

26. Priestley to Franklin, 21 November 1770; *SciAuto*, no. 30.

Priestley lacked the knowledge in optics to discriminate successfully between what should be emphasized and what neglected as he attempted to cope with his mass of materials.

Moreover, his work was to be on "experimental philosophy" and he explicitly declared his intention "to make everything . . . of much value in it, perfectly intelligible to those who have little or no knowledge of Mathematics" (viii). Even in 1770, however, optics was preeminently a mathematical science and Priestley admitted that philosophical discoveries in optics and those properly mathematical were so naturally intermixed that it would deviate from nature to separate them (230). Sometimes he tried to give mathematical relationships, on focal lengths, for example, or magnification, in words; frequently he adopted some version of the tactic: "For the theories and problems deduced . . . from these new principles of optics . . . I must refer my reader to the memoirs themselves" (467).

Neither of these subterfuges would suffice, however, when mathematics was an essential element in the conceptualization of a theory Priestley attempted to describe. He did not understand the works of Descartes, Fermat, Maupertuis, or Leibniz on the principles of least time and least action and dismissed them with "this method of arguing from final causes, could not satisfy philosophers" (105–6). And he had no conception of, and therefore could not really discuss, the envelope of wavelets that distinguished the "wave theory" of Christiaan Huygens and its revival with Leonhard Euler.

His descriptions of experimental work presented a somewhat similar problem. For the *History of Electricity*, Priestley had repeated most of the experiments he had to describe and knew, at first hand, the manipulations and the difficulties involved. But optical experiments were a different matter, even when he had acquired the requisite apparatus. The early demonstrations of reflective and refractive angles were easy enough to repeat. Newton's experiments were so carefully described in the *Opticks* that it took small ingenuity to reproduce them. But later, more sophisticated, work required a nicety of quantitative judgment that Priestley had neither the time nor the inclination to acquire.

Most of the experiments that he described in detail were, therefore, those that were fundamentally simple and qualitative in conception. Some related to ideas on chemistry and the nature of matter with which he was already familiar when he commenced writing. Bouguer's experiments on the comparison of light intensities, for example, which Priestley thought the most considerable work on light since Newton's, were treated exhaustively (though Priestley neglected similar work by Lambert). His one really favorable reference to Euler described experiments on the variation of refraction with temperature.

Repeated discussion of phosphorescent materials provided an opportunity to speculate on the physical nature of light. Do phosphors, as Dr. Slare supposed, contain the pablum of flame? Heat promoted the expulsion of the light "imbibed" by these substances and Canton believed that the attraction between light and particles of natural bodies was overcome by the strong vibrations produced in heating them. This supported the concept of materiality of light, as did the change in color and inward texture of some bodies when exposed to light (373–79). One such "body" was the silver dissolved in aqua fortis (nitric acid)—and a problem that was to complicate Priestley's chemistry surfaced here for the first time. He had no conception of chemical compounding. For him the silver dissolved in aqua fortis remained there in the form of the silver metal.

Priestley showed a consistent interest in the relationships among substances, light, and colors. He experimented on the effects of the light of electric sparks, transmitted through flint glass of varying thicknesses, onto "artificial" phosphorus. He noted the differences of these from the effects of sunlight. He described, in some detail, Melvill's examination of the quantities of different colors produced when various salts are dissolved in burning alcohol. Melvill queried whether luminous bodies did not differ from one another according to the colors they emit most plentifully (756–60). In this category, surely, was the query, a semi-anticipation of Olbers's paradox: "What becomes of the light stopped in bodies?" "If light was subject to no laws but those of reflexion and refraction, no place into which light was admitted would ever be dark, and the light of the whole universe would be continually increased" (779).

When he was writing the *History of Electricity* Priestley had the help and encouragement of his London advisers. This was missing for the *Optics*, on which none of them was expert. He enlisted, instead, the aid of John Michell, who had come as rector to Thornhill, near Leeds, from Cambridge University the year Priestley came to Mill Hill from Warrington. Michell (1724–93) was not a "discoverer" in optics, as Franklin, Watson, and Canton had been in electricity, but he had done important work in astronomy and was to do more. Moreover, he possessed the knowledge and interest in quantification that Priestley lacked.

It was Michell's experiment with a length of wire, pivoted at its middle, that rotated when light was focused on a vane at one end, that Priestley used to counter an argument that light could not be particulate. From the amount of light hitting the vane and the velocity of rotation of the wire, the momentum of light particles and therefore their total mass could be computed. Assuming the matter of the sun had the density of water, the sun would have lost no more than ten feet in radius in six thousand years of emissions (387–90). It appears that Michell also performed many of the other sophisticated quantitative optical experiments reported in the *History of Optics*.[27]

Michell also introduced (or rather reintroduced) Priestley to the concept of matter as particles surrounded by concentric spheres of attractive and repulsive forces. Priestley had first seen this idea in the *Compendious System of Natural Philosophy* (1735–43), by John Rowning, which he read as a student at Daventry Academy. It is far from clear that Priestley remembered Rowning in this connection, though he used plates from the *Compendious System* for illustrations in the *Optics* and practically paraphrased Rowning in his remarks on Newton's aether as a physical cause of attraction and repulsion: " . . . [this] hypothesis seems to labour under as many difficulties as the hypothesis of the mechanical production of the motions of light without attraction or repulsion" (778). He would, however, have been prepared by Rowning for consideration of the idea. That preparation had then been reinforced by the use of spheres of attraction and repulsion in Hartley's *Observations on Man* and in Stephen Hales's *Vegetable Staticks* (4th ed., 1769), whose "inquiries concerning air" Priestley had taken up while "commonplacing" books for the general history of experimental philosophy.

Michell, however, had taken the idea beyond anything that Priestley could have read in Rowning, Hartley, or Hales. He had conceived that the ultimate particles of matter might be no more than geometrical points, surrounded by the spheres of repelling and attracting forces. He was supported, in this conjecture, by the *Philosophiae Naturalis Theoria* (1763 ed.) of the Abbé Roger Joseph Boscovich, who had independently developed much the same theory, but far more elaborately.[28]

27. See, for example, the letter from Michell to William Herschel, 12 April 1781, in *Science and Music in Eighteenth-Century Bath*, ed. A. J. Turner (Bath: University of Bath, Exhibition Pamphlet for Holburne of Menstrie Museum, 1977), no. 184, pp. 98–102. Michell referred Herschel to Priestley's *History of Optics*, and Michell's contributions to it. William Nicholson, *An Introduction to Natural Philosophy* (London: J. Johnson, 1790), 256, also referred to Michell's experiments, described in Priestley's *Optics*.

28. For Rowning and Hartley, see Chapter 2; for Hales, D. G. C. Allan and R. E. Schofield, *Stephen Hales; Scientist and Philanthropist* (London: Scolar Press, 1980), 41–47. There is a modern edition of an English translation of Boscovich's *Theoria*: R. J. Boscovich, *A Theory of Natural Philosophy* (Cambridge: MIT Press, 1966), and it is discussed in detail in Schofield, *Mechanism and Materialism*, 236–42.

Michell introduced the theory and Boscovich to Priestley's attention in connection with the materiality of light and as an alternative explanation to Newton's for the colors of thin plates and films. Indeed, the only genuinely argumentative discussion in the *Optics* is Michell's adaptation of Boscovich's theory to that explanation. Priestley thought the Boscovich "hypothesis" of the penetrability of matter and its immateriality so new and important that he outlined it in some detail and reverted to it, or at least to its spheres of attraction and repulsion, throughout the *Optics* (390–94, 308–10, 510–11, 537, 771, 803–4). He also repeats, as though on Michell's authority, the general view about matter he had first learned from Doddridge's *Pneumatology*: " . . . we know nothing more of the nature of substance, than that it is something which supports properties" (393). This concept of matter was to be of central importance in Priestley's metaphysical-theological speculations for the rest of his life.

Had Priestley learned nothing more, in his preparation for the *History of Optics*, than the matter theory of Boscovich and Michell, that alone should have reconciled him to his labors. But Priestley did not see it that way—or perhaps he was further impelled by that theory to get on with other work. For the major problem with the *History of Optics* is that the subject never compelled his serious attention. By the time he began to write the *Optics*, he had already discovered his interest in experiments on airs and he clearly completed the *History* as a chore. For most of its 812 quarto pages have the earmarks of "commonplaced" notes, strung together chronologically, with topical variations (on the physiology of vision, for example, or optical instrumentation) within periods.

There are six such periods, from antiquity to the time since Newton, a general summary of the contemporary English (primarily Newtonian) theory of light, and a short section of Desiderata, mostly derived from the Queries in Newton's *Opticks*. Many of the sections that most graced the *Electricity*—the comparison of contending contemporary theories, constructive descriptions of apparatus and experiments, "practical maxims" for young experimenters, and the ingenuous descriptions of Priestley's own original experiments—are absent from the *Optics*.

Moreover, the work shows marks of haste. That but two months elapsed between Priestley's decision to concentrate on light and colors and his pronouncing the work essentially completed did not promise well. That ill-promise is confirmed in a page listing twenty-two "Errata of the press and other small inaccuracies," which, in fact, did not catch them all, and twenty-four pages of "additions," mostly supplied from comments of friends who

had seen the proofsheets. Priestley forestalled some criticism with his statement: "It cannot be expected . . . that a work of this extensive kind . . .
should be free from imperfections and errors, and the uniformity and other
advantages, which must result from its being executed by a single person,
will, it is hoped, be a sufficient plea for the greater indulgence which he
will necessarily stand in need of" (iv).

But the errors were not simply those of a single person grappling with
a subject about which he was not fully a master. Priestley's correspondence
with Price and Canton, during the months of October and November 1771,
while they were correcting sheets, tells a rather different story. Some of the
"errors" found were differences of interpretation on subjects where Price
or Canton disagreed with Priestley or Michell, but many more (and there
were *many* more) seem to have been simply mistakes.

Priestley protested that this wasn't the result of carelessness, that Michell
and others had helped in a careful initial correcting of the press. All of the
mistakes found would be rectified. Finally, he is pleased to have Canton's
approval of the work as a whole and he hopes that Price "to strangers . . .
will speak as favourably of my work as you can."[29] It was a sad ending to
a project begun with such optimism; it was, moreover, an ending that Priestley could ill afford.

In October, he told Price that when the volume was printed, "I shall
not be worth a groat, except my books, and copies of the work." By
November, he knew that subscriptions would not even pay for the books.
During the same period he estimated his losses from the *Theological Repository* at about £30 and gave up that project. A Dissenting minister on a
salary of £100, with a growing family (three children), could not afford
projects that cost him half his salary, with no prospects of improvement
when, as he complained in his *Free Address to Protestant Dissenters, as Such,*
congregations no longer felt responsible for the future of the minister's
children. No wonder, then, that less than a month after the *Optics* disaster
became clear, he jumped to accept what seemed an opportunity for a respectable position, with a handsome provision for himself and family.

29. Priestley to Canton, 9 and 22 October, 18 November 1771; *SciAuto,* nos. 35, 38, and
39; Priestley to Price, 19 October, 23 November 1771; *SciAuto,* nos. 37, 40.

XI

LEEDS, 1767–1773

Cook, Pyrmont Water, Chemistry, Shelburne

The story of the abortive invitation to Priestley to accompany Captain Cook on his second voyage to the South Seas is confused and fragmentary. This is partly because the most relevant letters to Priestley were (if he saved them) destroyed in the Birmingham riots of 1791. Nonetheless, it is possible to piece together a coherent account of this episode, one that varies significantly from the version implied in Priestley's *Memoirs* and frequently retold.

The Board of Admiralty agreed, by 13 September 1771, to send another expedition to the South Seas and probably determined that James Cook be given command, though the appointment was not made formal until late November. The expedition was to find rest and refitting stations for ships rounding Cape Horn. It would also allow Cook to continue the search for the "great Southern Continent" that he had begun on his previous voyage in the *Endeavour*. On voyages of this kind, the Admiralty frequently permitted the Royal Society, or private individuals at their own expense, to mount accompanying expeditions for collecting botanical, zoological, mineralogical, or ethnological information.

Naturally Joseph Banks, who had won deserved praise for his scientific activities accompanying Cook on the *Endeavour*, was asked, by Lord Sandwich, First Lord Commissioner of the Admiralty, if he wished to go again. Banks (1743–1820), a very rich young man, with a towering ego further raised by the admiration he received on his return with Cook, had been soliciting that invitation since the end of the first voyage. He seized his opportunity and, indeed, attempted to seize the entire project.

Having experienced the facilities on a typical naval expedition, he knew them inadequate best to serve a scientific one and began his plans to correct that situation. On the *Endeavour*, his party had consisted of nine persons: himself; his chief assistant, botanist Daniel Solander; a secretary; two draftsmen; two field assistants; and two servants. This time his plans included fourteen persons besides Solander and himself: an artist, three draftsmen, two secretaries, six servants, and two horn players (and may also have included smuggling on board, at the Madeira stop, of a young woman disguised as a gentleman-botanist).[1]

Then, on 25 October, the Astronomer-Royal, Nevil Maskelyne, proposed to the Board of Longitude that it send instruments on the voyage for the improvement of navigation, along with two proper persons to use them and to instruct the officers in their use. The Royal Society was to supply instruments and approve plans for their use. On 28 November, the Board agreed to send the observers and to defray the expenses; the Astronomer-Royal and the professors on the Board were asked to look for persons qualified and willing to serve and to report to the Board's meeting on 14 December.[2] And someone on the Board seems, injudiciously, to have asked Banks for suggestions as well.

Banks looked upon this as an opportunity to increase his party by two more scientists. Before a week passed, William Eden wrote to Joseph Priestley and shortly after, Daniel Solander wrote to Dr. James Lind, each asking his correspondent, in Banks's name, to accept nomination for one of the astronomer posts. Why Priestley had been selected as an "astronomer" is far from clear. He had written, in the preface to the *History of Electricity*, a plea for scientific societies of the world to join in sending ships for "the complete discovery of the face of the earth," but he had never published on astronomical questions and the *History of Optics* had yet to appear. He certainly was not a naturalist, to appeal to Banks's exclusive scientific concerns.

1. Details of the planning of Cook's second voyage, and other voyage information not specifically credited, have been collected from the invaluable work of John C. Beaglehole, ed., *The Endeavour Journal of Joseph Banks, 1768–1771* ([Sydney]: Trustees of the Mitchell Library, State Library of New South Wales, Angus and Robertson, [1962]), vol. 1, esp. 71–79; original MS:Z Safes 1/12, 1/23, published form quoted and cited by permission. See also, for a less critical appraisal of Banks, Harold B. Carter, *Sir Joseph Banks* (London: British Museum [Natural History], 1988), 96–101.

2. The professors of astronomy and of mathematics at the Universities of Oxford and Cambridge were ex officio members of the Board of Longitude. See John C. Beaglehole, ed., *The Voyage of the Resolution and Adventure: 1771–1775* (Cambridge: Cambridge University

One may also wonder at the selection of James Lind, for there were two of these, each a former ship's surgeon, each a graduate of Edinburgh, and each equally unknown to Banks. Dr. James Lind (1716–94) had written a major, unrecognized, *Treatise on Scurvy;* Dr. James Lind (1736–1812), who received Solander's letter, had published two astronomical papers in the *Philosophical Transactions* (1769). Did Banks conflate the two Linds when he suggested the invitation?[3]

Exactly what Eden's letter to Priestley said cannot be known, as the letter has not survived. The general tenor of its contents can be gathered, however, from Priestley's answer and, more especially, making due allowances for differences in writers' personalities, from Solander's letter to James Lind (1736–1812): An exciting opportunity had been opened by the prospect of a new voyage of discoveries into the South Seas. Banks is going, and has resolved to spare no expense in making everything agreeable to those who go with him. "The Board of Longitude have resolved to send out two astronomers and wish very much that one of them at least, should be a philosopher at the same time. They are willing to be very liberal in their reward."

> Mr. Banks . . . is by the Admiralty Board, the Board of Longitude, and all of them that have any thing to do with the Equipment of this Expedition, continually consulted and very much attended to. He and myself have been Desir'd by the Board of Longitude to think of, and propose to them proper persons fit for these undertakings. . . . Will you . . . give us leave to propose You, to the Board of Longitude as willing to go out as an Astronomer? . . . Figure to yourself what real pleasure you would find, in being so useful to Mankind. . . . A field of very great extent is now open to You. . . . How we all have . . . not before this, apply'd to You, . . . we will talke over, when we walk the quarterdeck together. . . . we shall do wonders if you only will come and assist us.[4]

Solander's letter did not quite say that Banks had the nomination of the astronomers for the voyage, but Lind might be excused for thinking that

---

Press, for the Hakluyt Society, 1961); all citations and quotations by permission of the Hakluyt Society: "Appendix III. The Board of Longitude and the Voyage," 719–26.

3. The timing of the letters to Priestley and Lind is uncertain, as will be seen below.

4. Daniel Solander to [Dr. James Lind], n.d., but during the first week of December 1771, lacking last page and signature; printed in Beaglehole, *Voyage,* 901–3.

his acceptance of this invitation was all that would be required for his selection. And that, certainly, was how Priestley understood Eden's letter to him, when he responded, on 4 December 1771, that he was honored by the proposal and that, "provided a suitable provision can be provided for my family"—which he left to the management of Eden and John Lee—he was inclined to accept. On 5 December, in some excitement, he wrote to Richard Price to tell him of the invitation. He even took the proposition to the heads of his congregation. Then the project vanished, so far as he was concerned, into thin air.

Price wrote immediately that he had spoken to Banks at the Royal Society and "found that the proposal which had been made to you was the product of some mistake. He told me he had writ to you the day before to set the matter right." Eden and Banks each wrote (clearly not "the day before"). Their letters have not survived, but Priestley's answers, on 10 December 1771, have frequently the tone of paraphrases. To Eden, he wrote that he was sorry that "seeming to jump at a proposal that was never made to me" had brought painful anxiety to my friends and family and an occasion for rejoicing to my enemies. To Banks, he wrote:

> You now tell me that, as the different Professors of Oxford and Cambridge will have the naming of the person, and they are all clergymen, they may possibly have some scruples on the head of religion; and that on this account you do not think you could get me nominated. . . . If . . . this be the case, I shall hold the Board of Longitude in extreme contempt, and make no scruple of speaking of them accordingly, taking for granted that you have just ground for your suspicions.[5]

From this base, Priestley wrote, in his *Memoirs:* "Mr. Banks informed me that I had been objected to by some clergymen on the Board of Longitude, who had the direction of the business, on account of my religious principles; and presently after I heard that Dr. Forster, a person far better qualified for the purpose, had got the appointment" (W 1/1:79–80). From

---

5. Priestley to Eden, 4 December 1771, Priestley to Price, 5 December 1771; *SciAuto*, nos. 41, 42. Price to Priestley, n.d. [soon after 5 December], transcribed from Price's shorthand draft on Priestley letter of 5 December by Dr. Beryl Thomas, Aberysthwyth; printed in *Correspondence of Richard Price*, ed. D. O. Thomas and Bernard Peach (Durham: Duke University Press, 1983), 1:124–25, quoted by permission of the Press. Priestley to Eden, Priestley to Banks, 10 December 1771; *SciAuto*, nos. 43, 44.

that passage and Charles R. Weld's publication of Priestley's letter to Banks in his *History of the Royal Society,* has come the story, repeated in biographical dictionaries, encyclopedias, and even in J. C. Beaglehole's introduction to Cook's journal of the *Voyage of the Resolution and Adventure.*[6]

But what evidence exists, beyond Priestley's paraphrase of Banks's letter, of objections, by any of the ex officio clergymen members of the Board of Longitude, to Priestley? They did not have right of choice or veto of nominations by the Board. And, from the timing of negotiations: Board approval of the project, 28 November 1771, Eden letter to Priestley, circa 1 December, Banks and Eden letters to Priestley, circa 5 December, there was scarcely time enough to discover opposition from professors at Oxford or Cambridge. Further, James Lind, whose religion was that of the established church of Scotland, did not receive an appointment either.[7]

When the Board met on 14 December 1771, only two names were placed in nomination, William Bayley and William Wales, both by Nevil Maskelyne. Each was approved by the Board and the appointments were confirmed. Now Bayley had been assistant to Maskelyne at the Greenwich Observatory and gone to observe the 1769 transit of Venus at North Cape; he ultimately became headmaster at the Royal Naval Academy at Portsmouth. Wales had gone to Hudson Bay to observe the same transit and published papers in the *Philosophical Transactions* on astronomical observations; he became mathematical master to naval candidates at Christ's Hospital School and secretary to the Board of Longitude. Each of them was a far better candidate for the rather demanding observational post to which he was appointed than either Priestley or Lind would have been.[8]

Rather than impugn the judgment and tolerance of the Board of Longitude (and Banks never quite said that they did object to Priestley, only that he thought they might), it is reasonable to suggest that Banks, with what Beaglehole has described as his "unbounded conceit," had issued invitations for which he had no authority. Though there was scarcely time for him to learn of objections from professors at Oxford and Cambridge, there was

6. Charles R. Weld, *History of the Royal Society* (London: J. W. Parker, 1848), 2:56–57; see also *Encyclopaedia Britannica,* 11th ed., 22:322b and Beaglehole, *Voyage,* xxix.

7. Banks and his friends obtained a resolution, of 7 April 1772, approved in the House of Commons, that a sum, not exceeding four thousand pounds, be granted by his Majesty, to be applied to encouragement of discoveries toward the South Pole. This money was to be used to pay Lind, who accepted Banks's invitation though the Board of Longitude did not appoint him; eventually the money went to Johann Reinhold Forster and his son Georg.

8. The detailed instructions of the Board of Longitude to Wales are printed in Beaglehole, *Voyage,* 724–26.

ample opportunity for him to discover that the Royal Society would insist upon astronomers for the use of their instruments. Moreover, Maskelyne, never pleased with encroachment on his preserves, would have denied Banks, on those grounds alone, any selection he wished to make. Throughout his life averse to losing face, Banks would never have admitted his mistake to Priestley.[9]

There certainly exist sufficient other examples to illustrate Banks's presumptuousness respecting this voyage. To house his party, its equipment, supplies, and specimens and to provide it with adequate workspace, he demanded that the deck of the *Resolution* be raised about a foot, a spardeck be laid on her from quarter-deck to forecastle, and a roundhouse be constructed for Cook, as Banks was to have exclusive use of the captain's cabin. Over the strong objections of the Navy Board, but with the consent of Lord Sandwich, all this was done. It increased draft of the ship from fourteen feet to seventeen feet. When she was tried at the Nore, early in May 1772, she was so dangerously "crank," that the pilot refused to take her. Cook declared that the ship must be returned to her original state; the Navy Board agreed, and the work was done.

Banks predictably objected. Indeed, he had a temper tantrum: "He swore & stamp'd upon the Warfe, like a Mad Man," and ordered everything of his removed from the ship. He wrote a "very foolish document" to Sandwich, in which he claimed that the Navy Board had been ordered to purchase ships so that he might be enabled to serve the public. Then, discovering that the navy had a different view of the primary purpose of the voyage, Banks refused to go along. In the end, he, Solander, and Lind went to Iceland, while Cook, the ships' companies, Wales, Bayley, and, to fill the positions left by withdrawal of Banks's party, Johann Reinhold Forster and his son Georg, left on the *Resolution* and *Adventure* for the South Seas on 13 July 1772, to return on 30 July 1775.[10]

This ended Banks's involvement with Cook's second voyage, but, curiously, not Priestley's. In March 1772, Priestley was in London to read his magisterial paper on different kinds of air before the Royal Society. That

9. "Banks finally had the effrontery to say: 'I had had influence enough to prevail on the board of Longitude to send with us Mess. Bailey & Wales as astronomers,'" Beaglehole, *Voyage*, xxix n.

10. This entire episode is described in Beaglehole, *Journal*, 73–76. A copy of Banks's letter to Sandwich and the Navy Board's answer to it may be found in H.C. Cameron, "The Failure of the Philosophers to Sail with Cook in the Resolution," *The Geographical Journal* 116 (1950): 49–54.

was also the month in which the *History of Optics* was published and, perhaps in that connection, he dined with the Duke of Northumberland. His lordship presented to the party a bottle of water distilled from sea water by a "new" method for which Dr. Charles Irving had petitioned Parliament for an award. The party agreed that the water was perfectly potable, but unpleasant because of its flatness.

Now Priestley had been carbonating water (making "artificial Pyrmont Water") at least since September 1767 and immediately declared that he could restore "briskness" to the water by impregnating Irving's distilled water with fixed air (carbon dioxide). This might also provide the navy with a way of preventing scurvy! Encouraged by the company, the next day he put together, in his London lodgings, apparatus for generating fixed air from chalk and acid and forcing it through New River water. He carried a bottle of this carbonated water with him on a visit to Sir George Savile, mentioning its possible use to the navy.

Savile immediately sent a note to Lord Sandwich, who saw the two men, read a short "proposal" Priestley had written, and promised to present the idea to the Board of Admiralty. On 1 April 1772, the Admiralty secretary referred Priestley's proposal "for rendering salt water fresh" to the president of the College of Physicians and asked Priestley to demonstrate the method to the college. He did so at a meeting attended also by Franklin and, ironically, by Joseph Banks and Daniel Solander. By 23 April, the College of Physicians having reported that Priestley's method "communicated no noxious qualities to the water," Captains Cook and Furneaux (of the *Adventure*) were instructed to experiment with water impregnated with fixed air by Priestley's method, as an antiscorbutic.[11]

Cook acknowledged his instructions and Priestley sent an explanation of the process, with drawings of the apparatus, which was passed to Cook and Furneaux by 13 May 1772. Though denied an opportunity to go with Cook on the voyage, Priestley had attempted a contribution to its success. And Cook's second voyage was, of course, a great success, not least in its medical achievement. Only one man died of disease on the *Resolution* in three years and eighteen days and no one died of scurvy.

That achievement, however, had nothing to do with Priestley's artificial Pyrmont water, which, in fact, seems never to have been tried, Cook having "never failed to take in water wherever it was to be procured, even when

11. For details of this "Pyrmont Water" episode, see Joseph Priestley, *Experiments and Observations on Different Kinds of Air*, 2d ed. (London: [1775] J. Johnson, 1776), 2:269–75;

we did not seem to want it."[12] And had Priestley's suggestion been tried in cases of scurvy, it would have done no good. For scurvy is the result of vitamin C deficiency; carbonated water is not antiscorbutic.

Yet Priestley had cause to believe it might be. William Brownrigg had identified the medicinal properties of a class of mineral waters with the "mephitic air" it contained, an air identified, by the time Brownrigg's work was published, as fixed air (CO_2). Joseph Black had demonstrated that mild alkalis become caustic when they "lose the fixed air they contain." Sir John Pringle had observed that putrefaction was checked by fermentation. David Macbride's *Experimental Essays* (1764; 2d ed., 1767) had explained that the reason was the fixed air given off during putrefaction. Putrefaction, he deduced, was in fact the loss of fixed air. Scurvy, thought to be a putrefactive disease, might therefore be cured by a restoration of the fixed air lost by the body.

Priestley knew Macbride's work in 1767, for he refers to it in the *History of Electricity* (598–99). Had he not known it, or appreciated its significance, Bewley's review of the *Electricity* would have told him: "We shall take the occasion of observing, for the honour of philosophy, that there is reason to hope that the catalogus medicamentorum will ere long be enriched with another promising article, the pure result of philosophical researches: we mean fixed air, a substance successfully investigated by the late excellent Dr. Hales, and which appears in a fair way of being soon happily applied to the relief of putrid disorders, and particularly of the sea-scurvy, in consequence of the ingenious experiments of Dr. Macbride, and his very natural practical deductions from them."

When Priestley announced to Canton, in September 1767, "I make most delightful Pyrmont Water, and can impregnate any water or wine &c. with that Spirit in two minutes time," he was on to a good thing. He did not, however, publicly follow it up for five years.[13] The reception given his

---

Beaglehole, *Voyage*, appendix 8, "Calendar of Documents," 922–29. Franklin to Priestley, 4 May 1772; *SciAuto*, no. 47, reports the presence of Banks and Solander at the demonstration.

12. James Cook, "The Method taken for preserving the Health of the Crew of his Majesty's Ship the Resolution during her late voyage round the World," *Philosophical Transactions* 56 (for 1776): 402–6. Some twenty years prior to Cook's sailing, a true remedy for scurvy had been described in James Lind's (not the Lind who accompanied Banks) *A Treatise of Scurvy* (Edinburgh, 1753), which recommended fresh fruit, especially citrus, as some earlier works had done; see Louis H. Roddis, *James Lind, Founder of Nautical Medicine* (New York: Henry Schuman, 1950), 56.

13. Priestley to Canton, 27 September 1767, *SciAuto*, no. 15; [William Bewley] review of *History of Electricity*, 453. This part of the review appeared in the December number of the

carbonation technique by the Admiralty, when he did reveal it, was so gratifying that he decided to publish a description of the method. That would make his discovery more generally useful "to my countrymen, and to mankind at large."

His *Directions for Impregnating Water with Fixed Air* caused a sensation.[14] A brief reference to the process, in his paper to the Royal Society, was one of the bits of medical news from London reported to William Cullen by Benjamin Bell. It was the only part of that paper that J. H. Magelhaens described, sending also a précis of the pamphlet, in a letter to Paris on 5 July. He then sent several copies of the *Directions*, along with an abridgment, to Trudaine de Montigny on 7 July. Trudaine immediately sent the abridgment and a copy of *Directions* to Lavoisier, recommending that it be translated. A French translation soon appeared, first in *Rozier's Journal* and then separately. And when Priestley was given the Copley Medal of the Royal Society in 1773, his "discovery" of artificial Pyrmont water was a significant cause for the award.[15]

Yet Priestley himself said it was hardly a discovery, and his pamphlet describing it was a very simple publication. Dated Leeds, 4 June 1772, with a dedication to Lord Sandwich and a scant "historical" preface, *Directions* pointed out that water will imbibe fixed air brought into contact with it, but the process was more effective if the water and air were agitated in the same vessel. Fill a vessel with water, invert it in another also full of water, introducing into it fixed air, made from chalk and highly diluted oil of vitriol (sulphuric acid). Priestley generated the air in a phial, connected to a bladder from which a flexible leather pipe extended through the water into the inverted container, which was shaken briskly as the water was displaced by air, until nearly all the air was absorbed, when the process was repeated.

Detailed instructions and observations were given. A plate of apparatus was included, along with "hints as have occurred to myself, or my friends,

---

Monthly Review and cannot have started Priestley's experiments with carbonating water, unless he was given substantial advance notice of Bewley's comments.

14. Joseph Priestley, *Directions for Impregnating Water with Fixed Air, In order to communicate to it the peculiar Spirit and Virtue of Pyrmont Water, And other Mineral Waters of a similar Nature* (London: J. Johnson, 1772), 8vo., (iv) 22 pp.

15. Benjamin Bell to William Cullen, 30 March 1772, in *Account of the Life, Lectures, and Writings of William Cullen, M.D.*, by John Thomson (London: Wm. Blackwood and Sons, 1859), 648–50. Magelhaens to [unknown], 5 July 1772; to Trudaine, 7 July 1772, Trudaine to Lavoisier, 14 July 1772; nos. 186, 187, 188 in *Oeuvres de Lavoisier: Correspondence*, ed. Rene Fric (Paris: Alben Michel, 1964), fasc. II.

with respect to the medicinal uses of water impregnated with fixed air; and also of fixed air in other applications" (17). This is the work for which Priestley became known to the world as a chemist; a description of how to make "soda water," wrongly assumed to have medicinal value.

The incongruity of Priestley's becoming known as a chemist for a paper on carbonation of water was remedied, in 1773, with the publication of his paper, "Observations on different Kinds of Air." This paper immediately brought Priestley's work to the attention of European chemists. He undoubtedly (and deliberately) speeded dissemination of his paper by writing of it to Thorbern Bergman, in Sweden in October 1772. By mid-June 1773, a full translation of the "Observations" had appeared in *Rozier's Journal* and the following year there was a translation into Italian and selections had appeared in German.[16]

The "Observations" made Priestley a leader of pneumatic chemistry, not only because of the discoveries it announced—and there were many of these: the beginning observations leading to a knowledge of photosynthesis, the isolation and identification of nitrous air (nitric oxide) and vapor of spirit of salt (later called acid air or marine acid air; anhydrous hydrochloric acid), and the development of the nitrous air test (eudiometry)—but also because of the simple apparatus and manipulative techniques he had developed and described.

It is a brilliant paper, insightful and challenging; it is also confusing and frustrating, both in form and content. The printed version is endorsed, at its beginning, with "Read March 5, 12, 19, 26, 1772," but even a casual reading of the text will show that material has been added from researches during the summer of 1772 and examination of the manuscript shows editing, emendations, and additions throughout, to bring the earlier part into line with the "additions" to his paper that Priestley read to the Royal Society 26 November 1772. It is not possible, therefore, easily to tell from the text when Priestley performed many of the experiments described in it, though it has the deceptive appearance of a chronological treatment, within sections, of his work on the various subjects discussed. Only by reading his

16. Joseph Priestley, "Observations on different Kinds of Air," *Philosophical Transactions* 62 (for 1772): 147–264; see also the MS in "Letters and Papers," decade 5, no. 284, Archives of the Royal Society. Henry Guerlac, "Joseph Priestley's first papers on Gases and their Reception in France," *Journal of the History of Medicine* 12 (1957): 1–12, clearly demonstrated that this was not published before February and probably not until March 1773. He also described a small-issue "preprint" of the paper (London: W. Bowyer and J. Nichols, 1772) which must, however, have appeared after 26 November 1772, when Priestley read an addition to his paper to the Royal Society.

correspondence with his friends does it become completely clear that a substantial part of Priestley's work was done after the March readings.[17]

There are other curious aspects to this paper. In its entire 118 pages, the word "chemical" (or any of its variations) appeared but once, when reporting the assistance given by William Hey, surgeon at Leeds General Infirmary, in examining his Pyrmont water, "by all the chemical methods that are in use" (153), for traces of sulphuric acid. Inevitably the paper was filled with accounts of chemical processes, but it is noteworthy how frequently Priestley seemed unaware of what was happening. Now some of this is surely to be expected. Much of what he was doing was entirely new; no one else knew what was going on either. Priestley was frequently wandering in an untracked forest of reactions without any landmarks to guide him. It is far too easy, years after the event, when chemical processes have been rationalized into neat theories, to criticize misdirections and shortcomings.

In fact, he was an unusually perceptive observer, especially considering that he didn't know what to look for. It is never wise to assume that Priestley wrongly described a reaction. Many of the "mistakes" he appeared to make (mistakes, that is, if the processes were performed with modern equipment, modern reagents, and with modern theories to define the essential and the accidental aspects of a reaction) are to be explained by noting precisely what Priestley did and with exactly what substances. Nonetheless, there are frequent examples of his ignorance of known chemical reactions, of his use of terminology that suggests a basic lack of understanding of what composition and decomposition were all about, and even of a stubborn persistence in the use of processes that he himself, and in the same paper, had indicated were undesirable.

This is scarcely surprising, considering the way in which his studies in "chemistry" were begun. There is no intention here to continue the myth, initially perpetrated by Priestley in his *Memoirs* and repeated uncritically by others, that his experiments relating to "the doctrine of air" were a "consequence of inhabiting a house adjoining to a public brewery" (W 1/ 1:75). As has previously been noted, his experiments on air began while he was in Warrington and his first chemical experiments in Leeds included studies of inflammable air, which cannot be obtained from a brewery, as

17. His letter to Franklin, 13 June 1772, *SciAuto*, no. 48, describes for the first time, work on nitrous air and acid air. The letter of 21 October 1772, Priestley to Bergman, is *SciAuto*, no. 52.

well as of fixed air, which can. The experiments on fixed air, moreover, continued well past August 1768, when Priestley moved away from the brewery to his new house. The real consideration is that in pneumatic chemistry, he was self-directed and nearly self-taught.

On electricity, his studies commenced with the guidance of electrical experimenters; on optics, his work was aided, at least, by John Michell. In chemistry, his investigations were begun, and continued for some six years, apparently in the absence of advice from any experienced chemist. Had his correspondence with William Bewley survived there might be some indication of chemical instruction beyond that short course with Matthew Turner at Warrington:[18] "His letters to me would make several volumes, and mine to him still more." Those letters are lost, and none of his other pre-1772 correspondents had any particular interest in chemistry. His letters to Franklin, Price, and Canton show they admired his success but failed to suggest methods or interpretations.

Nor was there much guidance to be found from Leeds associates. Ten years after Priestley left, there was organized a "Leeds Philosophical and Literary Society," which included William Hey, John Michell, John Smeaton, and William Dawson, with whom Priestley supposedly leagued while in Leeds. There might have been a nucleus in Leeds for an earlier philosophical association, but the "Leeds Phil. & Lit" lasted only three years and there is no suggestion of any earlier cohesion.[19]

Hey's interests were in surgery and medicine. Priestley described conversing with Hey on philosophical subjects (*W* 1/1:77), but their associations seem to relate mainly to medical issues. The Leeds General Infirmary, probably at Hey's suggestion, purchased a Priestley electrical machine for electrifying patients. Hey tested (successfully he thought) fixed air for its medicinal virtues. He also did the chemical tests for Priestley's "Observa-

18. Priestley, *W* 1/1:79. Bewley was an apothecary and major science reviewer for Griffith's *Monthly Review*. See Roger Lonsdale, "William Bewley and *The Monthly Review:* A Problem of Attribution," *Papers of the Bibliographical Society of America* 55 (1961): 309–13; and *Dr. Charles Burney: A Literary Biography* (Oxford: Clarendon Press, 1965), 41–42, 106. There is an uninformative eulogy of Bewley, said to be by Charles Burney Jr., "Character of the Philosopher of Massingham," *London Magazine* 1, n.s. (1783): 258–59. Bewley's review of the *Electricity* discusses at length the chemical potential of electric fluid, even speculating on the possibility of transmutation, "Review," 458–59.

19. Alfred Mattison, "Mill Hill Chapel and Dr. Priestley," MS, Leeds Reference Library. E. G. R. Taylor, *Mathematical Practitioners of Hanoverian England* (Cambridge: Institute of Navigation, at the University Press, 1966), lists Henry Clarke, math. teacher in Leeds and Manchester, and Thomas Lister, Halifax clockmaker, supposed to make orreries after Priestley's design, as Priestley's friends, but his writings refer to neither, nor to William Dawson.

tions" while Priestley assisted him in studies on blood coagulation. Hey was, however, an enthusiastic Methodist and, though he requested Priestley's pneumatic trough as a memento, it is doubtful that the two were friends. Priestley was not invited to join the medical society Hey organized in Leeds in 1768.[20]

John Smeaton, engineer and instrument maker, had a home in nearby Austhorpe where Priestley frequently visited him. He was given an air pump of Smeaton's design in September 1772 and joined Smeaton, in April 1773, in recommending Jeremiah Dixon for membership in the Royal Society. Smeaton, however, was not a chemist.

Nor was John Michell a chemist, though he gave Priestley most of the metals and semimetals used in the experiments of the "Observations." If Michell contributed to Priestley's chemical ideas, it was in the Boscovichean matter theory, which may, in the long run, have confused more than it helped. Priestley often visited Michell at Thornhill and Michell claimed once to have entertained the visiting Dr. Franklin and Sir John Pringle at dinner with Priestley and Smeaton. That must have been in the spring of 1773, when Priestley was on the point of leaving Leeds. One dinner cannot create continued philosophical discourse, and there is no sign of concerted efforts for such discourse in the Leeds community.[21]

Whatever Priestley learned of chemistry, beyond his own experience, he must have learned from his reading. Several names of chemical writers are specifically mentioned in Priestley's letters, papers, or books prior to the summer of 1772. Black, Boerhaave, Brownrigg, Cavendish, Hales, Macquer, and Neumann are each there, while there are at least fifty chemical titles, exclusive of articles in journals, in that list of books from which Priestley was "commonplacing" for the history of experimental philosophy. His interest, from the beginning, however, was in airs and, for that, most of Priestley's reading into chemical literature would have seemed irrelevant.

Hales, Black, and Cavendish were the only writers significantly touching on his interests and he clearly found Hales the most important of the three. Black's work, he never quite understood. Though Priestley occasionally

20. William Hey, *Observations on the Blood* (London: J. Wallis, 1779), 27; see also Rimmer, "William Hey of Leeds . . .;" Steven J. Anning, *The General Infirmary at Leeds*, vol. 1, *The First Hundred Years, 1767–1869* (Edinburgh: E. and S. Livingstone, 1963).

21. For Priestley, Michell, and Smeaton, see W 1/1:78. John Michell to [Sir Charles Blagden?], 27 July 1785; Misc. MSS Collection, American Philosophical Society, Philadelphia. Priestley reports obtaining metals from Michell in a letter to Price, 11 November 1772, *SciAuto*, no. 53.

determined the specific gravities of his "airs," he scarcely ever used gravimetrics as a means of analysis as Black did and Lavoisier was to do. Cavendish's interest in precise measurement also did not appeal, but Priestley spoke with Cavendish during March 1772, when he read the first part of his paper, and his burst of creative activity during the summer of that year may have resulted from that conference. Hales's work, however, had been influencing Priestley since 1770 and would continue to do so, positively and negatively, until the end of his life.

The long chapter, "A Specimen of an attempt to analyze the Air by a great variety of chymio-statistical Experiments," in Stephen Hales's *Vegetable Staticks* (1727; 4th ed., 1769), had set forth the view that air (a single, simple, substance) was modified by a variety of adventitious particles from other matter, which changed its properties, particularly its essential quality, its elasticity. Air could, by the penetration of matter within the spheres of repulsion of its particles, become fixed in ("wrought into the composition of") other substances, changing their properties. It could, however, also be released from such substances and resume its elasticity, by processes of fermentation (mixing with acids, alkalis, water, and so forth) or distillation (heating).[22] Black and Cavendish demonstrated that there were permanent "airs" besides "common air, but Priestley adapted Hales's ideas to these new circumstances. The new airs could also be fixed, in an inelastic state, in bodies and remain there until they were released. That adaptation was manifest in Priestley's research strategy, which essentially involved subjecting one substance after another to Hales's "fermentation" or "distillation" processes to see what other new airs might be released from them—and Priestley found many of these.

That adaptation also, however, kept him from appreciating what contemporary chemists were learning about composition and decomposition, from realizing that many of his processes were transforming substances, not just releasing them. Early in the "Observations," when complaining about the too-broad generality of the names assigned to airs, he wrote that inflammable air deserved the name "fixed air" equally with that usually given the name, as each was originally part of some solid substance and exists there in an inelastic state (148). Later, he described the operation of organic combustion as "flame disposes air to deposit the fixed air it contains" (163). Commonly he wrote of a species of inflammable air (H_2) as having been "in metals."

22. Allan and Schofield, *Stephen Hales*, chap. 4, "The Vegetable Staticks," 30–47.

Consistent with his adaptation of Hales's ideas, he tended to identify his airs by the processes involved in their production, or reactions, rather than by the substances from which they derive. He introduced his section on "inflammable air" with the observation that he generally made it as Cavendish did, from iron, zinc, or tin with acids, but sometimes extracted it by heating animal or vegetable substances (170). Earlier he had noted a difference in the smell of the airs from these two processes (158), but neither then nor through most of a long career in pneumatic studies, did he clearly distinguish between the hydrogen produced in the one and the carbon monoxide (usually mixed with methane and other inflammable hydrocarbons) obtained in the other.

He grouped together experiments in which something carboniferous had been burned in confined space, with those in which it was sulphur that had burned. He was not sure but that the air remaining in these cases was the same as that remaining when lead or tin were calcined, or iron and sulphur reacted, in confined space. In all these cases, the volume of remaining air was reduced by about the same amount, though the modes of reduction strangely differed. Knowing that respiration and putrefaction produce fixed air (though the latter, sometimes, strangely produced inflammable air as well), he treated the air in containers in which either process occurred as the same, though their behavior was different in many respects, and the latter "air" was foul.

Finally, because he tended to regard his processes in a Halesian mechanical way, rather than a chemical compositional way, he persisted in the hope that physical manipulation would change one air, or combination of airs, into another. This is particularly frustrating to a modern reader, for Priestley clearly demonstrated in the "Observations," that Hales was wrong in supposing air diminished in bulk by having the "elasticity of the mass impaired." Diminished air was rather lighter, than heavier, than common air as it would have to be had its density increased (164–65). He also disproved a report that cold or compression would restore air in which a candle has burnt out; nor will heat or rarefaction injure common air.

Yet, in one experiment after another, described in "Observations," Priestley subjected his "airs" to washing in water, to standing for long periods over water, or mixing with and agitation over water. The results were invariably confusing, to Priestley and his readers, for what seem to be the same airs treated in what seem to be the same ways sometimes remain the same and sometimes vary over a range of properties.

Priestley had the clue to these contradictory results and never grasped it. Early in his experiments he made use of an animal bladder, in transferring airs, but he soon noted that bladders do not sufficiently prevent the airs in them from mixing with external air; nor do corks adequately confine and separate airs (158, 176). He wrote: "I have never thought the communication between the external and internal air sufficiently cut off, unless glass, or a body of water, or in some cases, quicksilver, have intervened between them" (252). Yet he also wrote that airs confined in containers over water could not be maintained separate, the water continually receiving from one container and giving to another, ". . . as water receives some kind of impregnation from . . . every kind of air to which it is contiguous" (186).

He repeatedly observed that when some of his airs were kept over water, the smell of them came through (172, 191). He wrote: "As these agitations were made in jars with wide mouths, and in a trough which had a large surface exposed to the common air, I take it for granted that the noxious effluvia . . . were first imbibed by the water and thereby transmitted to the common atmosphere" (201). Then, three pages later, he stated that violent and continued agitation over water never failed to render noxious air fit for respiration! No wonder he found contradictions when confining inflammable air (H_2) over water for periods up to three years. No wonder the properties of his airs changed in time or with agitation over water.

Priestley can scarcely be faulted for not understanding the causes of these difficulties. Believing, as he and most of his contemporaries did, that air particles are normally confined to positions about which they vibrate elastically, he had no reason to suppose a process of particle transport. He had, however, observed some of the phenomena today explained by kinetic transport and diffusion theory (and perhaps also some ionic migration) and he failed to profit by his observations. In a long career as a pneumatic chemist, Priestley was continually to describe experiments whose results were vitiated by that failure.

The "failure," however, his contemporaries shared and the problem is one most elementary gas experiments today endure. Experiments with gases are still performed by displacement of water (or mercury) in containers inverted in a pneumatic trough, the apparatus and method Priestley used to his considerable advantage—and confusion. Although he later described his apparatus as ". . . in fact, nothing more than the apparatus of Dr. Hales, Dr. Brownrigg, and Mr. Cavendish, diversified and made a little more simple," he has typically overstated his debt.[23]

23. Joseph Priestley, *Experiments and Observations on Different Kinds of Air* (London: [1774] J. Johnson, 1775), 6.

He brought together bits of apparatus from several sources, made of them a coherent armory for attacking pneumatic chemistry, described the whole, pictured it in a plate, and discussed, throughout the "Observations," how its parts were to be manipulated. That is one of the reasons that Priestley's paper, for all its preconceptions and the confusions into which he was led by them, made such a strong and favorable impression. Readers of it were convinced that they could collect the apparatus and repeat the experiments themselves.

Of course, the major reason that preconceptions and confusions failed to disturb Priestley's readers is that they went unnoticed. What they saw was a long paper—longer than any emphasizing gases written to that time—filled with many experiments, discussed with apparent ingenuousness and candor. If the paper started more questions than it answered, the questions were interesting ones. Priestley wrote that his observations were imperfect, the courses of his experiments incomplete, and asked that other persons pursue them (147). And above all, there were positive achievements.

The most surprising of these achievements, its evolution somewhat concealed in the separated parts in which it is discussed in the paper, was the discovery of the "restoration" of air by vegetation. In the *History of Electricity* Priestley had considered the change which air "underwent by passing through fire, or through the lungs, &c. and whether it was not possible to restore it to its original state by some operation or moisture" (598–99). This is the problem that he designates, in the "Observations," as a great object of philosophical inquiry, the answer to which he had persistently sought for six years. Considering the "consumption" of air by fires of all kinds, volcanos, animals breathing, and so forth, Priestley was convinced that there must be some provision, by God in nature, for remedying the injury which the atmosphere receives (162, 183).

He systematically examined all the natural processes that air can undergo: heating, chilling, expansion, contraction, contact with earth, mixture with other kinds of air, agitation in water. Eventually he found that water does not absorb airs equally, that it "decomposes" common air, taking one part and leaving the rest (247). This differential solubility of different gases may indeed be one way of maintaining a balance in the atmosphere, but the principal natural method, Priestley discovered, was by plants in a vegetating state.

"In what manner the process in nature operates, to produce so remarkable an effect," Priestley wrote, "I do not pretend to have discovered" (166), but vegetation did restore bad air. As he did not know the "manner," his demonstration of the process was confused and frequently contradictory. He described a number of instances in which plants, growing with exceptional vigor in air vitiated by combustion or respiration, returned the air to its former state. He had not yet recognized the agency of light in the process; he didn't understand that plants used the "fixed air" in the atmosphere in a complex process of photosynthesis. He had, however, started a program of investigation that would involve numbers of chemists and plant physiologists for a number of years.

The two other substantial achievements were dependent, at least in part, on Halesian strategy and Cavendish suggestions. Hales had described a peculiar form of air, obtained from spirit of niter (nitric acid) and a particular pyrite. Cavendish thought the kind of pyrite was probably irrelevant. Priestley tried some experiments with nitric acid and metals and, being considerably skilled in gaseous manipulation by the summer of 1772, isolated what he called nitrous air (nitric oxide, NO).

He found he could obtain it "from" most metals. It did not differ much in specific gravity from common air, did not precipitate lime water, was not inflammable. It preserved animal matter from putrefaction; most significant, when mixed with common air, there was effervescence, red fumes were produced, and common air was reduced in bulk in proportion to its fitness for respiration or combustion. (Nitric oxide combines with the O_2 in common air, producing soluble nitrogen dioxide, NO_2). This "nitrous air test" provided a way of examining airs for their "degree of goodness" (there being no reaction between "bad air" and nitrous air) that humanely replaced the mice (216), which had, anyway, given uncertain results (183).

Some experiments described in Cavendish's *Philosophical Transactions* paper of 1766 led to Priestley's other material discovery, that of marine acid air (anhydrous hydrochloric acid). Heating spirit of salt ($HCl.H_2O$) released a noncondensable vapor which was miscible in water, but could be collected over mercury. Heavier than common air, it was not inflammable, nor would it support combustion. It was very reactive, extracting phlogiston from most substances, forming with it, "such an union as constitutes inflammable air; which seems to shew that inflammable air . . .consists of the union of some acid vapour with phlogiston" (239).

This is one of the few references to phlogiston in the entire paper. Considering the malefic effect the concept of phlogiston is supposed to have had on Priestley's chemistry, the scarcity of such references comes as a surprise. All of them appear to have been added to the paper after its initial, March, reading. This suggests that the use of the concept was explained to

Priestley by Cavendish and none causes so significant a variation in his thinking as to constitute an added handicap to his interpretation of experimental results.

Changes in substances and airs, when the two were brought together, was a consequence of something removed from the substance and added to the air; sometimes that thing was phlogiston. Acid air removed phlogiston and became inflammable, as was expected of a substance charged with the principle of inflammability. Calcination of metals (removal of phlogiston) in common air diminished the bulk of the air; the remainder was insoluble and, charged with phlogiston, ought to burn. It did not, but that ". . . may depend upon some particular mode of combination, or degree of affinity, with which we are not acquainted" (233). Inversely, inflammable air exposed to substances with affinity for phlogiston (e.g., oil of vitriol, spirit of niter) ought to be reduced in bulk, but there was no sensible effect (178). When moist iron filings and sulphur were confined in fixed air, iron calcined, air bulk was somewhat diminished, and the remainder was insoluble. Had phlogiston removed from the iron combined with fixed air to make common air (162)?

The remainder of the "Observations" can be summarized briefly. Some of it was repetition, with additions, of much of the material published in the Pyrmont water pamphlet. The fixed air obtained at a brewery was far from perfectly pure, but a brewery provided a continuous supply that was pure enough for many purposes. (Priestley stopped using it after ruining a vat of beer mixing the fixed air over it with ether [156]). Fixed air neither burned nor supported combustion; it was heavier than common air, easily miscible with water, and its solution was enhanced by cold (though not in ice) and pressure. It precipitated lime water, and did not appear to conduct heat or electricity more readily than common air. Mr. Bergman declared it made an acid when dissolved in water and Mr. Lane had shown it readily dissolved iron. Mr. Hey, and others, conducted experiments to determine the efficacy of fixed air in curing putrid disorders, and these were described in an appendix (257–64).

When brimstone was burned in air confined over lime water the latter remained transparent. There may, however, have been a "precipitation of the fixed part of the air" which unites with the lime and vitriolic acid to form a selenetic salt (163). No air Priestley made conducted electricity, but the different colors of sparks taken through different airs suggested that they were not all equally good nonconductors; the purple or red spark in inflammable air indicating better conductivity than white indicated for fixed air (175). Nitrous air mixed with inflammable air burned with a green flame (216); acid air extinguished flame, which burnt with a light blue flame just before going out (238). And finally, a set of comments looked to future studies: Candles burn and animals live in the air extracted from saltpeter (185).

All kinds of factitious air [i.e., those produced artificially]:

> . . . on which I have yet made the experiment are highly noxious to animals, except that which is extracted from saltpetre, or alum; but in this even a candle burned just in common air. In one quantity which I got from saltpetre, a candle not only burned, but the flame was increased. . . . It effervesced with nitrous air as much as the best common air ever does, and even a candle burned in it very well. . . . This series of facts, relating to airs extracted from nitre, appeared . . . very extraordinary and important, and, in able hands, may lead to considerable discoveries.[24]

Publication of Priestley's "Observations" marked the conclusion of his career in science as anything but an original experimenter on airs. The final editing of this chemical paper, for printing in the *Philosophical Transactions* in December 1772, also marked the end of Priestley's years at Leeds. His disappointment over the Cook's-voyage proposal apparently brought home to his friends the importance of extricating him from an increasingly unrewarding situation at Mill Hill Chapel.

He was always to describe his position there in commendatory terms. He himself was happy, his salary was larger than that of most Dissenting ministers, his time was at his own disposal. But his schemes led to a variety of expenses, putting it out of his power to do much for his family now and he could not provide for their future.[25] He had thoughts of going to America, where he could more easily make provision for his children. Benjamin Franklin must have instituted inquiries in that direction. In March 1773, John Winthrop wrote from Harvard: "I am extremely concerned to

24. Priestley, "Observations," 245–46. This was Priestley's first notice of oxygen— $2KNO_3$ when heated producing $2KNO_2 + O_2$—but he failed to follow up, despite this expression of interest in the air. "Discovery" of oxygen was postponed until 1775.

25. When compared to the incomes of his, frequently lesser, contemporaries, Priestley's salary of 100 guineas and a house was small. Lindsey's rural living at Catterick brought him £400. William Hey was surely making £500; by the period 1796–1819 he averaged £1400 a year; while John Hunter, the London surgeon was making £5000 in the 1770s.

hear that Dr. Priestley is so poorly provided for . . . and wish it were in my power to do him any kind of service. It would give me great pleasure to see him well settled in America. . . . A man of his abilities would do honor to any of the colleges. At present there is no vacancy among them, but if there were, I believe, Sir, you judge perfectly right, that his religious principles would hardly be thought orthodox enough."[26] Happily, by the time that response arrived, the issue had been settled, through the curious conjunction of Richard Price and Lord Shelburne.

Price and William Petty, Lord Shelburne, had become friends as early as 1769. Price's economic views and cautiously liberal theology and politics appealed to Shelburne while the latter's support of the American colonists and political-economic reform attracted Price. Early in 1771, Shelburne, taking a Continental tour to recover from his wife's death, asked Price to uncover some "gentleman of character and extensive knowledge and learning," from the private lives where they are generally hidden, to act as a companion and assist him in directing the education of his children. One cannot but wonder whether that request was not an invitation for self-nomination; if so, the intention failed. "Great men," such as described by Shelburne, were scarce and Price could think only of the Reverend George Walker as possibly the kind of man Shelburne was looking for. Walker was, however, recently married and disinclined to accept a position that would interfere with his domestic comfort.[27]

In Price's letter of [5] December 1771, informing Priestley of the error of Banks's proposal, he additionally inquired whether Priestley knew of any "young and single man among the Dissenting Ministers . . . who is possessed of extensive knowledge in History, language, philosophy and qualified for bringing up and training . . . two children." Priestley did not respond. During the early months of 1772, Price, or Shelburne, or both, transformed that position into a way of helping Priestley and an approach was made to him in July.

Priestley's correspondence during the next several months, to Price, Lindsey, Franklin, and Shelburne, is a curious mixture of pneumatic chemistry, politics, theology, and discussion of Shelburne's proposal. The latter

26. Winthrop to Franklin, 4 March 1773; *SciAuto*, no. 55.
27. Richard Price to Shelburne, 22 May 1771; Bowood MSS, now housed in the Bodleian Library, Oxford. This is clearly but one of the letters of a long correspondence and I owe the transcript that I have seen to the courtesy of Drs. D. O. and Beryl Thomas, Aberystwyth; also George Walker, *Essays on Various Subjects, to which is prefixed, a Life of the Author* (London: J. Johnson, 1809), 1:lxxi–lxxii.

seems to have been extraordinarily sensitive to Priestley's amour-propre, not overpowering him with financial arguments, but assuring him that the position entailed necessary and useful work for which he was particularly qualified. Priestley spent months of indecision: worry about maintaining his independence, about the different social level in which he would be expected to move, whether he must give up preaching.

His friends everywhere watched and worried with him. It was in this connection that Franklin wrote to Priestley his famous letter on "prudential algebra," advising him to set down pros and cons in columns on a single sheet, weighing items against one another, canceling "equals," and finally settling on the side where items remain. In late September, Priestley resolved to refuse the offer. Price wrote Franklin: "Indeed, I don't know whether to be glad or Sorry on account of his rejection of Lord Shelburne's proposal. I love him and am heartily concerned for him and wish he were better provided for." In October, Josiah Wedgwood wrote Thomas Bentley about "Priestley's noble appointment," hoping that "he is to go on writing and publishing with the same freedom he now does, otherwise I had much rather he still remained in Yorkshire."[28]

By November Priestley wavered toward approval and, at length, accepted the proposal. On 20 December 1772, he addressed a letter of resignation to the Mill Hill congregation. They formally accepted on 27 December and had chosen his successor by mid-February 1773. Priestley preached his farewell sermon on 16 May, and his letter to the young men at Mill Hill, thanking them for their letter and unexpected gift, is addressed from Calne, Wiltshire, 14 June 1773. For the next seven years, Priestley was to act as librarian-companion to Shelburne, superintend the education of his sons, with a tutor (Thomas Jervis) under him, and collect information on subjects of parliamentary discussion. His long years of preparation were over, his apprenticeship in the provinces completed. From 1773, Priestley had entered a new world of service, responsibility, and national exposure.

28. Franklin to Priestley, 19 September 1772; Albert H. Smyth, ed., *The Writings of Benjamin Franklin* (New York: Macmillan, 1907), 8:20–21; Price to Franklin, 30 September 1772, Franklin Papers, Vol. 3, 2, no. 122, American Philosophical Society; Wedgwood to Bentley, 4 October 1772, Wedgwood Letter-books, 6:162 [E.18411–25], Josiah Wedgwood and Sons, Burslem, Staffordshire.

Epilogue

Joseph Priestley lived another thirty years after his move from Leeds, dying at Northumberland, Pennsylvania, in February 1804. Those years can rather neatly be divided into near-decades of different residences and different major concerns:

From 1773 to 1780, he lived at Calne, Wiltshire as Lord Shelburne's librarian-companion. He was less companionable than Shelburne had anticipated, while his religious and political activities were limited by his position. His major accomplishments during the Shelburne "decade" were in science and metaphysics. He completed the work of five of the six volumes of his pneumatic studies, *Experiments and Observations* (1774, 1775, 1777, 1779, and 1781), in which he was to announce discovery of five more gases: ammonia, sulphur dioxide, nitrous oxide, and nitrogen dioxide, as well as oxygen—the discovery for which he is most famous.

During the same period he wrote a series of books that, together, outlined his scientific and religious philosophy. He attacked Scottish Common Sense Philosophy, brought David Hartley's associationism to the attention of the literati, and described a curiously spiritualized materialistic monism that opened him to furious charges of atheism from his contemporaries. It has, more recently, been the source of bemused speculation for modern philosophers and historians.

Shelburne remarried in 1779 and the Priestley connection became a social and political embarrassment. Given a choice, Priestley elected to break off the formal association. He accepted a previously contracted pension, and moved to Birmingham in 1780. This period, until the last year, 1791, was in Priestley's view the happiest of his life. He became a member of the Lunar Society of Midland scientist-entrepreneurs, where his pragmatic concerns in science could be exploited. He wrote several papers for the Royal Society, and the last volume of his *Experiments and Observations* (1786). His scientific endeavors were chiefly directed to arguments with the French chemist, Antoine Lavoisier, and Lavoisier's disciples. Lavoisier based a "revolution in chemistry," on Priestley's oxygen and the related discovery that water was a compound. It remains an unresolved problem why Priestley resisted Lavoisier's suggestions, reiterating defenses for various forms of the doctrine of phlogiston. It is now generally agreed that previous charges of ignorance and ineptness are inadequate explanations. Philosophic and theological concerns seem to be involved, but specific details are subject to argument.[1]

His happiness, however, primarily stemmed from his employment as minister to one of the largest and most opulent Dissenting congregations in England: Birmingham's "New Meeting." Once again he was doing the work he thought the most important of any in the world. The major activities of his Birmingham years were religious, theological, and political—the latter related to advocating repeal of the Test Acts and disestablishment of the Church. He wrote here his longest theological works, more than seven volumes, attempting to prove that the earliest Christians were Unitarian, and became involved in another set of acrimonious disputes.

The beginning of the French Revolution, in 1789, introduced a new element into the situation. Like many of his liberal contemporaries, Priestley greeted the unfolding events with enthusiasm; they were the signs of worldwide reforms of government and religion, perhaps even of the beginning of the Millennium. Establishment figures were increasingly less happy and soon instituted formal opposition to liberal reform movements and figures. Priestley became a symbol of the dangers of reform to Church and King. Somehow (details are lacking) a mob was organized in Birmingham that destroyed New Meeting House, Priestley's home, and those of many other Dissenters. Priestley was driven from Birmingham by these riots of 1791. He moved temporarily to Hackney where he preached and taught natural philosophy at Hackney New College.

Political opposition escalated into persecution and Priestley had reason to fear for his liberty and safety. Finally, in 1794, he reluctantly left England for the United States. There he was, at first, greeted warmly as an old friend of the American Revolution and for his scientific and theological reputation. Though he was invited to settle in New England, New York,

---

1. On this, John G. McEvoy ("Joseph Priestley, 'aerial philosopher,'" *Ambix* 25 [1978]: 1–55, 93–116, 153–75; 26 [1979]: 16–38) and I have been in unresolved contention. Perhaps a second volume would bring us closer to agreement, although historians and philosophers rarely think alike!

and Philadelphia, he chose to go up the Susquehanna River to Northumberland, Pennsylvania.

He hoped to establish a community of English "rational Christians," in north-central Pennsylvania, with himself as minister and head of a college. These plans failed, as the abortive treason trials of Thomas Hardy, John Horne Tooke, and John Thelwall reduced the threat of persecution in England and stemmed the expected flood of liberal, Dissenting emigrants. Still, Priestley chose to remain in Northumberland, where he built a house, now a museum in his honor. With his family resettled, Priestley entered the last phase of his life, the anticlimax.

He attempted to replicate his former living patterns as much as possible. He reestablished his research laboratory, writing to England for equipment he could not supply from Philadelphia. His experimenting went on and papers continued to be read, now to the American Philosophical Society on annual trips to Philadelphia. The experiments and the papers were, however, dominated by attempts to disprove Lavoisier's chemistry and increasingly his work was ignored, though he did have the satisfaction of forcing the discovery of carbon monoxide.

He discovered that Americans were not more generally tolerant of religious heresy than the English and he could not establish a local Unitarian congregation, though he held worship services in his home until just before his death. He preached in Philadelphia on his visits there and continued to write and publish on religious and theological subjects. During the Adams administration, there was a brief flare-up of political abuse because of his "French sympathies," but the election of Jefferson to the presidency gave Priestley, as he was to declare, for the first time in his life, a government to live under that was not hostile to his ideas and ideals. Indeed, Jefferson befriended the aging scholar, asking for advice on curricula for projected educational reforms in Virginia and praising his religious writings.

His favorite son, Henry, died in Northumberland, followed shortly afterward by the death of his wife. Increasingly tired and ill, he moved the family of his son, Joseph Jr., into his house and lived with them until his death there just under the age of seventy-one.

Select Bibliography

Works by Priestley

Additions to the [Free] Address to Protestant Dissenters, on the Subject of the Lord's Supper, with some corrections of it; and a Letter to the Author of the Protestant Dissenter's Answer to it. London: J. Johnson 1770.

An Address to Protestant Dissenters on Giving the Lord's Supper to Children. London: J. Johnson, 1773.

An Appeal to the Public on the Subject of the Riots in Birmingham. 2d ed. Birmingham: by J. Thomson, for J. Johnson, 1792.

[A Lover of the Gospel]. *An Appeal to the Serious and Candid Professors of Christianity. On the following Subjects, viz. i. The Use of Reason in Matters of Religion. ii. The Power of Man to do the Will of God. iii. Original Sin. iv. Election and Reprobation. v. The Divinity of Christ. and vi. Atonement for Sin by the Death of Christ. To Which are added, a Concise History of the Rise of those Doctrines.* 3d ed. London: J. Johnson, T. Cadell, J. Gore, J. Grigg, J. Harrop, T. Banks and W. Eyres, W. Edwards, and B. Binns, 1771.

A Catechism for Children and Young Persons. Leeds: for J. Johnson, 1767.

A Chart of Biography . . . Published according to Act of Parliament, 2 February 1765. London: J. Johnson, 1765.

Considerations on Church-authority; occasioned by Dr. Balguy's Sermon, on that Subject; preached at Lambeth Chapel and published by Order of the Archbishop. London: J. Johnson and J. Payne, 1769.

Considerations on Differences of Opinion among Christians; with a Letter to the Rev. Mr. Venn, in Answer to his Free and Full Examination of the Address to Protestant Dissenters, on the Subject of the Lord's Supper. London: J. Johnson and J. Payne, 1769.

Correspondence (Collections) and Journal

Memoirs and Correspondence, 1733–1787; vol. 1, part 1 in *Theol. & Misc. Works,* ed. J. T. Rutt. This is the major collection of printed correspondence, but is very poorly edited. Cited herein as W 1/1.

A Scientific Autobiography of Joseph Priestley: Selected Scientific Correspondence. Edited by Robert E. Schofield. Cambridge: MIT Press, 1966; herein cited *SciAuto.*

"Joseph Priestley's Journal while at Daventry Academy, 1754," *Enlightenment and Dissent* 13 (1994): 49–113.

Correspondence (Separately Published)

To John Seddon, 9 April, 10 April, 1 May, 6 May, 19 May 1762, in [Robert B. Aspland,] "Brief Memoir of Rev. John Seddon, of Warrington, with Selections from His Letters and Papers," *Christian Reformer* 10, n.s. (1854): 625–29.

A Course of Lectures on Oratory and Criticism. London: J. Johnson, 1777.

A Course of Lectures on Oratory and Criticism. Edited by Vincent M. Bevilacqua and Richard Murphy. Landmark in Rhetoric and Public Address. Carbondale: Southern Illinois University Press, 1965.

A Course of Lectures on the Theory of Language and Universal Grammar. Warrington: W. Eyres, 1762.

A Description of a Chart of Biography: with a Catalogue of all the Names inserted in it, and the Dates annexed to them. 2d ed. Warrington: for the author, 1765.

A Description of a New Chart of History, containing a View of the principal Revolutions of Empire that have taken Place in the World. 5th ed. London: J. Johnson, 1781.

Directions for Impregnating Water with Fixed Air, In order to communicate to it the peculiar Spirit and Virtue of Pyrmont Water, And other Mineral Waters of a similar Nature. London: J. Johnson, 1772.

Discourses on Various Subjects, including several on Particular Occasions. Birmingham: for J. Johnson, 1787.

Disquisitions relating to Matter and Spirit. London: J. Johnson, 1777.

The Doctrine of Divine Influence on the Human Mind, Considered in a Sermon [on Matt. 18:3–20]. Bath: R. Cruttwell for J. Johnson, 1779.

The Doctrine of Philosophical Necessity Illustrated. Being an Appendix to the Disquisitions relating to Matter and Spirit. 2d ed. Birmingham: for J. Johnson, 1782.

An Essay on a Course of Liberal Education for Civil and Active Life. With Plans of Lectures on: I. The Study of History and general Policy. II. The History of England. III. The Constitution and Laws of England. To which are added, Remarks on a Code of Education, proposed by Dr. Brown, in a late Treatise, intitled, Thoughts on Civil Liberty, &c. [London:] C. Henderson, T. Becket, and deHondt, J. Johnson and Davenport, 1765.

An Essay on the First Principles of Government; and on the Nature of Political, Civil, and Religious Liberty. Dublin: James Williams, 1768.

An Essay on the First Principles of Government, . . . 2d ed. London: J. Johnson, 1771.

Examination of Dr. Reid's Inquiry into the Human Mind . . . Dr. Beattie's Essay . . . and Dr. Oswald's Appeal. 2d ed. London: J. Johnson, 1775.

Experiments and Observations on Different Kinds of Air. 2d ed. London: J. Johnson, 1775.

Experiments and Observations on Different Kinds of Air. Vol. 2. 2d ed. London: J. Johnson, 1776.

[A Lover of the Gospel]. *A Familiar Illustration of Certain Passages of Scripture relating to the Power of Man to do the Will of God, Original Sin, Election and Reprobation, the Divinity of Christ, and Atonement for Sin by the Death of Christ.* London: J. Johnson, 1772.

Familiar Introduction to the Study of Electricity. London: J. Dodsley, T. Cadell, and J. Johnson, 1768.

A Familiar Introduction to the Theory and Practice of Perspective. London: J. Johnson and J. Payne, 1770.

Familiar Letters addressed to the Inhabitants of Birmingham. Birmingham: for J. Johnson, 1790.

[A Dissenter]. *A Free Address to Protestant Dissenters, as Such.* 2d ed., enl. London: J. Johnson and J. Payne, 1769.

A Free Address to Protestant Dissenters, on the Subject of Church Discipline: with a Preliminary Discourse, concerning the Spirit of Christianity, and the Corruptions of it by False Notions of Religion. London: J. Johnson, 1770.

A Free Address to Protestant Dissenters on the Subject of the Lord's Supper. 2d ed. London: J. Johnson, 1769.

Institutes of Natural and Revealed Religion. 3 vols. London: J. Johnson, 1772, 1773, 1774.

Institutes of Natural and Revealed Religion. 2 vols., 2d ed. Birmingham: for Joseph Johnson, London, 1782.

Lectures on History and General Policy. Birmingham: for J. Johnson, 1788.

[Author of the "Free Address to Protestant Dissenters as Such"]. *A Letter of Advice to those Dissenters who conduct the Application to Parliament for Relief from Certain Penal Laws, with Various Observations relating to similar Subjects.* London: Joseph Johnson, 1773.

Letters to the Author [William Enfield] *of "Remarks on Several late Publications relative to the Dissenters, in a letter to Dr. Priestley."* London: J. Johnson, 1770.

Letters to a Philosophical Unbeliever. Vol. 1. 2d ed. Birmingham: for J. Johnson, 1787.

Memoirs of Dr. Joseph Priestley to the Year 1795 . . . with a continuation by his son. Northumberland: J. Binns, 1806.

Miscellaneous Observations Relating to Education. More Especially, as it respects the Conduct of the Mind. To which is added, An Essay on a Course of Liberal Education . . . Bath: for J. Johnson, 1778.

A New Chart of History . . . Engraved and published according to Act of Parliament, 11 April 1769. London: J. Johnson, 1769.

No Man Liveth to Himself, a Sermon preached before the Assembly of Protestant Dissenting-Ministers, of the Counties of Lancaster and Chester, met at Manchester May 16, 1764, to carry into Execution a Scheme for the Relief of their Widows and Children; and Published at their Request. Warrington: W. Eyres, 1764.

Philosophical Empiricism. London: J. Johnson, 1775.

[An Englishman]. *The Present State of Liberty in Great Britain and her Colonies.* London: Johnson and Payne, 1969.

The Proper Objects of Education in the Present State of the World. London: J. Johnson, 1791.

Remarks on Some Paragraphs in the Fourth Volume of Dr. Blackstone's Commentaries on the Laws of England relating to Dissenters. London: J. Johnson and J. Payne, 1769.

Reply to the Animadversions on the History of the Corruptions of Christianity. Birmingham: for J. Johnson, 1783.

The Rudiments of English Grammar; adapted to the Use of Schools. With Observations on Style. London: R. Griffiths, 1761.

The Rudiments of English Grammar, Adapted to the Use of Schools; with Notes and Observations, for the Use of those Who have made some Proficiency in the Language. London: for J. and F. Rivington, T. Lowndes, S. Crowder, T. Becket and Co., and J. Johnson, 1771.

A Scripture Catechism, consisting of a Series of Questions, with References to the Scriptures instead of Answers. London: J. Johnson, 1772.

[Anon.]. *The Scripture Doctrine of Remission. Which sheweth that the Death of Christ is no proper Sacrifice nor Satisfaction for Sin: but that pardon is dispensed solely on account of Repentance, or a personal reformation of the Sinner.* London: C. Henderson, R. Griffiths, T. Beckett, and P. A. De-Hondt, 1761.

A Serious Address to Masters of Families, with Forms of Family-Prayer. 2d ed. London: J. Johnson, [1771].

A Sermon Preached before the Congregation of Protestant Dissenters at Mill-Hill Chapel in Leeds, May 16, 1773 . . . On Occasion of his resigning his Pastoral Office among them. London: J. Johnson, 1773.

A Syllabus of a Course of Lectures on the Study of History. Warrington: William Eyres, 1765.

The Theological and Miscellaneous Works of Joseph Priestley, LL.D. F.R.S. &c. Edited with notes, by John Towill Rutt. 25 vols. New York: Kraus Reprint, [London, 1817–31] 1972. Includes the following works:

> "An Answer to Dr. Blackstone's Reply." Leeds, 2 October 1769, 22:328–34.

> "A Letter to a Friend in Manchester." 20 September 1770, 22:533–35.

> "Letters and Queries addressed to the anonymous Answerer of an Appeal to the Serious and Candid Professors of Christianity; to the Rev. Mr. Thomas Morgan, and to Mr. Cornelius Cayley." 21:3–28.

> "Memoirs and Correspondence, 1733–1787." vol. 1, part 1; herein cited W 1/1.

> "The Rudiments of English Grammar; A Course of Lectures on the Theory of Language, and Universal Grammar; and On Oratory and

Criticism." vol. 23. Note that Rutt's version of 1824 does not agree with Priestley's of 1762.

Two Discourses; I. On Habitual Devotion, II. On the Duty of not living to Ourselves; Both Preached to Assemblies of Protestant Dissenting Ministers, and published at their Request. Birmingham: Piercy and Jones, for J. Johnson, 1782.

A View of the Principles and Conduct of the Protestant Dissenters, with respect to the Civil and Ecclesiastical Constitution of England. 2d ed. London: J. Johnson and J. Payne, 1769.

Edited by Priestley

Laws for the Regulation of the Circulating-Library in Leeds; and A Catalogue of the Books belonging to It: To which are prefixed the Names of the Subscribers. Leeds: Griffith Wright, 1768.

Elwall, Edward. *The Triumph of Truth, being An Account of the Trial of Mr. Elwall, for Heresy and Blasphemy.* 2d ed. Leeds: J. Binns, 1771.

The Theological Repository. Vol. 1 (2d ed., 1773); 2 (1770); 3 (1771); 4 (1784); 5 (1786); 6 (1788). London: Joseph Johnson. Individual items by Priestley separately listed.

Papers: Scientific (chronological)

"An Account of Rings consisting of all the Prismatic Colours, made by Electrical Explosions on the Surface of Pieces of Metal." *Philosophical Transactions* 58 (for 1768): 68–74.

"Experiments on the lateral Force of Electrical Explosions." *Philosophical Transactions* 59 (for 1769): 57–62.

"Various Experiments on the Force of Electrical Explosions." *Philosophical Transactions* 59 (for 1769): 63–70.

"An Investigation of the Lateral Explosion, and of the Electricity communicated to the electrical circuit in a Discharge." *Philosophical Transactions* 60 (for 1770): 192–210.

"Experiments and Observations on Charcoal." *Philosophical Transactions* 60 (for 1770): 211–27.

"Observations on different Kinds of Air." *Philosophical Transactions* 62 (for 1772): 147–264.

Papers: Theological

Review of "A New English Translation of the Psalms from the original Hebrew . . . by Thomas Edwards." *Monthly Review; or, Literary Journal* 12 (1755): 485–89.

["Clemens"]. "An Essay on the One Great End of the Life and Death of Christ, Intended more especially to refute the commonly received Doctrine of Atonement." *Theological Repository* 1 (2d ed., 1773): 17–45, 121–36, 195–218, 247–67, 327–53, 400–430.

["Clemens"]. "An Essay on the Analogy there is between the Methods by which the Perfection and Happiness of Men are promoted according to the Dispensations of Natural and Revealed Religion." *Theological Repository* 3 (1771): 3–31.

["Clemens"]. "Observations on Christ's Agony in the Garden." *Theological Repository* 3 (1771): 376–82; with supplementary note, 476–77.

["Clemens"]. "Observations on the Importance of Faith in Christ," *Theological Repository* 3 (1771): 239–43.

Editor. "Conclusion." *Theological Repository* 3 (1771): 477–82.

["Liberius"]. "Observations on Infant Baptism." *Theological Repository* 3 (1771): 231–39.

["Liberius"]. "The Socinian Hypothesis Vindicated." *Theological Repository* 3 (1771): 344–63.

["Liberius"]. "A Criticism on I Corrinthians xv.27." *Theological Repository* 3 (1771): 255–56.

["Paulinus"]. "Remarks on Romans v. 12–14." *Theological Repository* 2 (1770): 154–58.

["Paulinus"]. "Observations concerning Melchizadeck." *Theological Repository* 2 (1770): 283–90.

["Paulinus"]. "Observations on the Abrahamic Covenant." *Theological Repository* 2 (1770): 396–411.

["Paulinus"]. "Observations on Romans v. 12 &c." *Theological Repository* 2 (1770): 411–16.

["Paulinus"]. "Remarks on the Reasonings of St. Paul." *Theological Repository* 3 (1771): 86–105, 188–212.

["Paulinus"]. "Observations on Christ's Proof of a Resurrection, from the Books of Moses." *Theological Repository* 1 (1773): 300–303.

["Paulinus"]. "Observations on the Apostleship of Matthias." *Theological Repository* 1 (1773): 376–81.

Manuscripts

Catalogue of Books in the Library of the Warrington Academy; Library, Harris Manchester College, Oxford.

Diary of the Reverend Henry Crooke; Clarke MSS, Leeds District Archives

Letters (chronological)

Matthew Turner to John Seddon, 16 March 1762; Library, Harris Manchester College, Oxford.

Joseph Priestley to John Seddon, 9 April 1762, Boyd Lee Spahr Library, Dickinson College, Carlisle, Pennsylvania; 19 May 1762, Wedgwood Papers, The John Rylands University Library of Manchester.

Joseph Priestley to [Samuel Pegge], 26 February 1764; Thomas Bentley to Pegge, 27 February 1764; Priestley to Pegge, 21 April 1764; MS Add. C244, fols. 13 and 15, Bodleian Library, Oxford.

Joseph Priestley to Thomas Birch, 11 July 1764; Birch Papers, Add. MSS British Library.

Joseph Priestley to John Wilkes, n.d., but probably late 1768 or early 1769; Add. MSS 30, 871; British Library.

Joseph Priestley to Anna Laetitia Aikin, 13 June 1769, copy, no. 15, Yates Priestleyana Collection, Royal Society of London.

Richard Price to Benjamin Franklin, 30 September 1772; Franklin Papers, vol. 3.2, no. 122, American Philosophical Society, Philadelphia, Pennsylvania.

Josiah Wedgwood to Thomas Bentley, 4 October 1772; Wedgwood Letterbooks VI, p. 162 [E. 18411–25], Josiah Wedgwood and Sons, Burslem, Staffordshire.

Joseph Priestley to Newcome Cappe, 13 April 1777; Burndy Library, MS 1257B, Smithsonian Institution Libraries, Washington, D.C. 20560 [Part printed W 1/1:298–99.]

John Michell to [Sir Charles Blagden?], 27 July 1785; Misc. MSS Collection, American Philosophical Society, Philadelphia, Pennsylvania.

"Letters and Papers," Decade V, no. 284: "Observations on different Kinds of Air," Archives of the Royal Society, London.

Mattison, Alfred, "Mill Hill Chapel and Dr. Priestley," MS, Leeds Reference Library, Leeds, Yorkshire.

Priestley Papers, Dr. Williams's Library, London.

Minute-Book, 1741–55; Independent Church, Heckmondwike; Upper United Reformed Church, Heckmondwike, Yorkshire.

Minute-Books, MS Da 31.5, *Senatus Academicus*, 1:156–57; Edinburgh University Library.

Minutes of the Circulating Library, Warrington, 1760–67; Warrington Municipal Library, Warrington, Cheshire County Council.

Minutes of the Proceedings of the Trustees of Warrington Academy; Warrington Municipal Library, Warrington, Cheshire County Council.

MS Journal Book Copy, 1763–66, 25:707; Royal Society, London.

Yates Priestleyana Collection; Royal Society, London.

Other Primary Sources

AEpinus, Franz Ulrich Theodosius. *Aepinus's Essay on the Theory of Electricity and Magnetism.* Edited by R. W. Home. Princeton: Princeton University Press, 1979.

Annet, Peter. *Expeditious Penmanship: or, Shorthand Improved, &c.* London: for the Author, sold by R. Baldwin, n.d. [British Library suggests 1750].

Baily, Francis. *An Epitome of Universal History, Ancient and Modern, from the Earliest Authentic Records to the Commencement of the Present Year.* London: J. Johnson & Co., and J. Richardson, 1813.

[Barbauld, Anna Laetitia]. *A Selection from the Poems and Prose Writings of Mrs. Anna Laetitia Barbauld.* Boston: James R. Osgood, 1874.

Beale, Catherine Hutton, ed. *Reminiscences of a Gentlewoman of the last Century: Letters of Catherine Hutton.* Birmingham: Cornish Brothers, 1891.

Belsham, Thomas. "A List of Students educated at the Academy at Daventry." *Monthly Repository* 17 (1822): 163–64, 195–98.

[Bewley, William]. Review of "History and Present State of Discoveries Relating to Vision, Light and Colours." *Monthly Review; or, Literary Journal* 47 (1772): 304–19.

————. Review of "History and Present State of Electricity." *Monthly Review; or, Literary Journal* 37 (1767): 93–105, 241–54, 449–65.

[Blackburne, Francis]. *The Confessional: or, a Full and Free Inquiry into the Right, Utility, Edification, and Success of Establishing Systematical Confessions of Faith and Doctrine in Protestant Churches.* 2d ed. London: S. Bladon, 1767.

Blackstone, William. *Commentaries on the Laws of England. Book Fourth.* Oxford: Clarendon Press, 1769.

————. *Commentaries on the Laws of England. Book Fourth.* 4th ed. Oxford: Clarendon Press, 1770.

————. *A Reply to Dr. Priestley's Remarks.* London: C. Bathurst, 1769.

Boscovich, Roger Joseph. *A Theory of Natural Philosophy.* Cambridge: MIT Press, 1966.

Carlisle, Nicholas. *A Concise Description of the Endowed Grammar Schools in England and Wales.* London: Baldwin, Cradock, and Joy, 1818.

Carruthers, William, ed. *The Grounds and Principles of Religion contained in a Shorter Catechism according to the Advice of the Assembly of Divines at Westminster.* London: for the Lord Wharton Trustees, 1907.

Cook, James. "The Method taken for preserving the Health of the Crew of his Majesty's Ship the Resolution during her late voyage round the World." *Philosophical Transactions* 56 (for 1776): 402–6.

Croft, Herbert. *An Unfinished Letter to the Right Honourable William Pitt.* Facsimile Reprints in English Linguistics, 1500–1800, no. 71. Menston, Eng.: Scholar Press, [1788] reprint, 1968.

[Doddridge, Philip]. *The Correspondence and Diary of Philip Doddridge, D.D.* Edited by John Doddridge Humphreys. London: Henry Coburn and Richard Bentley, 1830.

————. *A Course of Lectures on the Principal Subjects in Pneumatology, Ethics, and Divinity: with References to the most considerable Authors on each Subject.* Edited by Samuel Clark. London: J. W. Clarke and R. Collins, W. Johnston, J. Richardson, S. Crowder and Co., T. Longman, B. Law, T. Field, and H. Payne and W. Cropley, 1763.

———. *The Rise and Progress of Religion in the Soul.* 17th ed. London: W. Baynes, 1808.

Enfield, William. *Institutes of Natural Philosophy.* London: J. Johnson, 1785.

———. Review of "A Course of Lectures on Oratory and Criticism by Joseph Priestley." *Monthly Review* 57 (for August 1777): 89–98.

[Franklin, Benjamin]. *The Writings of Benjamin Franklin.* Edited by Albert H. Smyth. Vol. 8. New York: Macmillan Co., 1907.

'sGravesande, W. James. *Mathematical Elements of Natural Philosophy, confirm'd by Experiments: Or, an Introduction to Sir Isaac Newton's Philosophy.* 2 vols. London: W. Innys, T. Longmon and T. Shewell, C. Hitch, and M. Senex, 1747.

Gentlemen's Magazine. Notice of publication: *New English Grammar. Gentleman's Magazine* 16 (1746): 112.

[Gibbons, Thomas]. "Dr. Thomas Gibbons' Diary." Edited by W. H. Summers. *Transactions of the Congregational Historical Society* 1 (1901–4): 380, 384; 2 (1905–6): 22–38.

Hartley, David. *Observations on Man, his Frame, his Duty, and his Expectations.* 2 vols. Hildesheim: [London, 1749], Georg Olms, 1967.

Hey, William. *Observations on the Blood.* London: J. Wallis, 1779.

Houghton, John. *A New Introduction to English Grammar: In the simplest and easiest Method possible.* Salop: for the Author, by J. Cotton and J. Eddowes, 1766.

Hutcheson, Francis. *An Inquiry into the Original of our Ideas of Beauty and Virtue; in Two Treatises.* London: J. Darby, Wil. and John Smith, W. and J. Innys, J. Osborn and T. Longman, and S. Chandler, 1725.

Kenrick, Timothy. *An Exposition of the Historical Writings of the New Testament, with Reflections subjoined to each Section.* Boston: Monroe and Francis, 1828, from 2d London ed. of 1824 [1807].

[Lamb, Charles]. *The Letters of Charles Lamb, to which are added those of his sister, Mary Lamb.* Edited by E. V. Lucas. London: J. M. Dent and Sons, Methuen and Co., 1935.

Lardner, Nathaniel. *A Letter to Lord Viscount Barrington (written in the year 1730) concerning the Question, Whether the Logos supplied the Place of a Human Soul in the Person of Jesus Christ.* London: for the Unitarian Association, and R. Hunter, 1833.

[Lavoisier, Antoine L.]. *Oeuvres de Lavoisier. Correspondence.* Fasc. 2. Paris: Alben Michel, 1964.

Leeds Intelligencer. "Extracts from the Leeds Intelligencer, 1763–1767; 1768." *Publications of the Thoresby Society, Miscellanea* 33 (1930–32), 209–27.

———. 1767–73; Microfilm, Leeds City Libraries.

Leeds Mercury. 1769–70; Microfilm, Leeds City Libraries. [Nos. between January 1767 and January 1769, January 1771 to January 1773 missing.]

Leeds Intelligencer and Leeds Mercury. "Extracts from the Leeds Intelligencer and the Leeds Mercury 1777–1782 . . ." *Publications of the Thoresby Society* 40 (1955): 1–247, edited by G. D. Lamb and J. B. Place.

[Lenglet du Fresnoy, Pierre Nicolas]. *A Chart of Universal History (Done from the French, with considerable improvements).* London: Thomas Jeffreys, 1750.

Locke, John. *An Essay concerning Human Understanding, &c.* London: A. Church-ill and A. Manship, 1733.

————. *An Essay concerning Human Understanding.* 20th ed. London: T. Long-man, B. Law and Son, J. Johnson, et al., 1796.

Miller, Samuel. *A Brief Retrospect of the Eighteenth Century.* New York: reprinted, London for J. Johnson, 1805.

Monthly Review. Review of "Rudiments of English Grammar." *Monthly Review* 26 (1762): 27–31.

————. Review of "Chart of Biography, Description of a Chart of Biography." *Monthly Review* 32 (1765, misprinted 1764): 160.

[Neumann, Caspar]. *The Chemical Works of Caspar Neumann.* Edited by William Lewis. London: W. Johnston, G. Keith, A. Linde, P. Davey and B. Law, T. Field, T. Caslon, and E. Dilly, 1759.

[Percival, Thomas]. *The Works, Literary, Moral, and Medical of Thomas Percival, M.D.* London: J. Johnson, 1807.

[Price, Richard]. *The Correspondence of Richard Price.* Edited by D. O. Thomas and Bernard Peach. Vol. 1. Durham: Duke University Press, 1983).

Priestley, Timothy. *A Funeral Sermon occasioned by the Death of the late Rev. Joseph Priestley . . . To which is added, a True Statement of many important Circumstances relative to those Differences of Opinion which existed between the Two Brothers . . .* London: for Alex. Hogg, 1805.

Protestant Society for the Protection of Religious Liberty. "Report of the Annual Meeting of the Protestant Society for the Protection of Religious Liberty." *Monthly Repository* 14 (1819): 330–36, 388–94.

[Rogers, Samuel]. *Recollections of the Table-Talk of Samuel Rogers.* Edited by A. Dyce. New York: D. Appleton and Co., 1856.

Rowning, John. *A Compendious System of Natural Philosophy.* London: Sam. Har-ding, 1737–43; published and republished in parts, separately paginated. Consulted for this volume: preface, appendix, index, 1743; part 1, 3d ed., 1738; part 2, 3d ed., 1737; part 2 cont., 1736; part 3 and part 3 cont., 2d ed. 1743; part 4, 1743; part 4 cont., 1743.

Sale, George, John Campbell, Archibald Bower, John Swinton, George Psalma-nazer, et al., comp. *An Universal History, From the Earliest Account of Time: complied from original Authors.* 20 vols. London: T. Osborne, A. Millar, J. Osborn, 1747, 1748.

Sheard, Michael. *Records of the Parish of Batley in the County of York, Historical, Topographical, Ecclesiastical, Testamentary, and Genealogical.* Worksop: Robert White, 1894.

Warrington Trustees. *A Report of the State of Warrington Academy, By the Trustees at their Annual Meeting June 25th. MDCCLXI.* [Copies in Warrington Li-brary, Warrington Cheshire County.]

————. *A Report of the State of Warrington Academy, By the Trustees at their Annual Meeting July 1st-MDCCLXII.*

————. *A Report of the State of Warrington Academy, By the Trustees at their Annual Meeting June 30th MDCCLXIII.*

————. *A Report of the State of Warrington Academy, By the Trustees at their Annual Meeting June 26th MDCCLXVI.*

Watts, Isaac. *Logic: or, the Right Use of Reason in the Enquiry after Truth. With a Variety of Rules to guard against Error, in the Affairs of Religion and Human Life, as well as in the Sciences.* London: T. Longman, T. Sewell, J. Brackstone, 1745.

————. *Philosophical Essays on Various Subjects.* 2d ed., corr. London: Richard Ford and Richard Hett, 1734.

Monographs and Papers

Aaron, Richard I. *John Locke.* Oxford: Oxford University Press, 1955.

Aarsleff, Hans. *The Study of Language in England, 1780–1860.* Princeton: Princeton University Press, 1967.

Adamson, John W. "Education." Chap. 15 in *From Steele and Addison to Pope and Swift.* Vol. 9 of *The Cambridge History of English Literature,* edited by A. W. Ward and A. R. Waller. New York: Macmillan Co.; Cambridge: Cambridge University Press, 1933.

Alexander, H. G., ed. *Leibniz-Clarke Correspondence.* Manchester: Manchester University Press, 1956.

Alison, Archibald. *Essays on the Nature and Principles of Taste.* New York: G. and C. Carvill, 1830.

Allan, D[avid] G. C., and R[obert] E. Schofield. *Stephen Hales: Scientist and Philanthropist.* London: Scolar Press, 1980.

Anning, Steven J. *The General Infirmary at Leeds.* Vol. 1, *The First Hundred Years, 1767–1869.* Edinburgh: E. and S. Livingstone, 1963.

[Aspland, Robert B.]. "Brief Memoir of Rev. John Seddon, of Warrington, with Selections from His Letters and Papers." *Christian Reformer* 10, n.s. (1854): 224–40, 358–68, 618, 629; 11, n.s. (1855): 365–74.

————. "Some Account of Edward Elwall and his Writings." *Christian Reformer* 11, n.s. (1855): 329–45.

Axon, Ernest. "Yorkshire Nonconformity in 1743." *Transactions of the Unitarian Historical Society* 5 (1931–34): 244–61.

[Barbauld, Anna Laetitia]. *The Works of Anna Laetitia Barbauld.* New York: Carvill, Bliss and White, et al., 1826.

Barlow, Richard Burgess. *Citizenship and Conscience: A Study in the Theory and Practice of Religious Toleration in England During the Eighteenth Century.* Philadelphia: University of Pennsylvania Press, 1962.

Barnes, Harry Elmer. *A History of Historical Writing.* Norman: University of Oklahoma Press, 1937.

Baugh, Albert C. *A History of the English Language.* New York: Appleton-Century-Crofts, 1935.

Beaglehole, John C., ed. *The Endeavour Journal of Joseph Banks, 1768–1771.* [Sydney], Trustees of the Mitchell Library, State Library of New South Wales, Angus and Robertson, [1962].

———. *The Voyage of the Resolution and Adventure: 1771–1775.* Cambridge: Cambridge University Press, for the Hakluyt Society, 1961.

Beckwith, Frank. "The Beginning of the Leeds Library." *Miscellanea of the Thoresby Society* 37 (1941).

Bennett, Arthur. "Glimpses of Bygone Warrington." *Proceedings of the Warrington Literary and Philosophical Society* (1898–99).

[Bentham, Jeremy]. *The Collected Works of Jeremy Bentham.* Edited by John Bowring. London: Simpkin, Marshall and Co., 1838–43.

———. *The Correspondence of Jeremy Bentham.* Edited by Timothy L.S. Sprigge. London: University of London, Athlone Press, 1968.

Beresford, N.W., and G. R. J. Jones, eds. *Leeds and its Region.* Leeds, for the British Association for the Advancement of Science, 1967.

Besterman, Theodore, ed. *Publishing Firm of Cadell & Davies: Select Correspondence and Accounts, 1793–1836.* London: Oxford University Press, 1938.

Bonser, K. J., and H. Nicholls, eds. "Printed Maps and Plans of Leeds, 1711–1900." *Publications of the Thoresby Society* 47 (1958): 1–148.

Brewer, John. *Party Ideology and Party Politics at the Accession of George III.* New York: Cambridge University Press, 1976.

Brewster, David, ed. *The Edinburgh Encyclopaedia.* American ed. Philadelphia: Joseph and Edward Parker, 1832, viii.

Bridenbaugh, Carl. *Mitre and Sceptre: Transatlantic Faiths, Ideas, Personalities, and Politics, 1689–1775.* New York: Oxford University Press, 1962.

Brown, Goold. *The Grammar of English Grammars, with an Introduction Historical and Critical.* New York: Samuel S. and William Wood; London: Samson Low, Son and Co., 1857.

Browne, John. *History of Congregationalism and Memorials of the Churches in Norfolk and Suffolk.* London: Harrold and Sons, 1877.

Bryan, W. F. "A Late Eighteenth-Century Purist." *Studies in Philology* 24 (1927): 368–69.

[Burney, Charles, Jr.]. "Character of the Philosopher of Massingham." *London Magazine*, 1, n.s. (1783): 258–59.

Bury, J.B. *The Idea of Progress: An Inquiry into its Origin and Growth.* New York: [Macmillan, 1932], Dover, 1955.

Cameron, H.C. "The Failure of the Philosophers to Sail with Cook in the Resolution." *Geographical Journal* 116 (1950): 49–54.

Carter, Harold B. *Sir Joseph Banks.* London: British Museum (Natural History), 1988.

Cassidy, F. G. "Case in Modern English." *Language* 13 (1937): 240–45.

Cassirer, Ernst. *The Platonic Renaissance in England.* Translated by James P. Pettengrove. Austin: University of Texas Press, 1953.

[Channing, William Ellery]. *Correspondence of William Ellery Channing, D.D. and Lucy Aikin, from 1826 to 1842.* Edited by Anna Letitia LeBreton. Boston: Roberts Brothers, 1879.

Christian Reformer. "An Interesting Revival of Unitarianism in Cheshire." *Christian Reformer* 6, n.s. (1850): 55–60.

Clarke, Henry. *Practical Perspective. Being a Course of Lessons exhibiting Easy and Concise Rules for drawing justly all Sorts of Objects. Adapted to the Use of Schools.* London: for the author, sold by J. Nourse, 1776.

Cohen, I. Bernard. *Franklin and Newton.* Memoirs 43. Philadelphia: American Philosophical Society, 1956.

Cohen, Murray. *Sensible Words: Linguistic Practice in England, 1640–1785.* Baltimore: Johns Hopkins University Press, 1977.

Colie, Rosalie. *Light and Enlightenment: A Study of the Cambridge Platonists and the Dutch Arminians.* Cambridge: Cambridge University Press, 1957.

Cranston, Maurice. *John Locke: A Biography.* London: Longmans, Green and Co., 1957.

[Crippen, T. G., ed.]. "Congregational Fund Board." *Transactions of the Congregational Historical Society* 6 (1913–15): 209–13.

———. "Early Nonconformist Academies: Heckmondwike and Northowram." *Transactions of the Congregational Historical Society* 6 (1913–15): 291–96.

———. "Protestant Society for the Protection of Religious Liberty." *Transactions of the Congregational Historical Society* 6 (1913–15): 364–76.

———. "A View of English Nonconformity in 1773." *Transactions of the Congregational Historical Society* 5 (1911–12): 205–22, 261–77, 372–85.

Dale, Bryan. "Early Congregationalism in Leeds." *Transactions of the Congregational Historical Society* 2 (1905–6): 247–61, 311–25.

Dale, R. W. *History of English Congregationalism.* New York: A. C. Armstrong and Son, 1907.

Davis, Arthur P. *Isaac Watts: His Life and Works.* New York: Dryden, 1943.

Dixon of Thearne, Ronald A.M. "James Priestley and Cumberland University, U.S.A." *Transactions of the Unitarian Historical Society* 4 (1927–30): 412–16.

Elby, Frederick, and Charles Flinn Arrowood. *The Development of Modern Education.* New York: Prentice Hall, 1934.

Elledge, Scott. "The Naked Science of Language." In *Studies in Criticism and Aesthetics, 1660–1800: Essays in Honor of Samuel Holt Monk,* edited by Howard Anderson and John S. Shea, 266–95. Minneapolis: University of Minnesota Press, 1967.

Ellis, Grace A. *A Memoir of Mrs. Anna Laetitia Barbauld, with Many of her Letters.* Boston: James R. Osgood and Co., 1874.

Everett, Charles Warren. *The Education of Jeremy Bentham.* New York: Columbia University Press, 1931.

[Faraday, Michael]. *Faraday's Diary: Being the Various Philosophical Notes of Experimental Investigation.* London: G. Bell and Sons, 1932.

———. *Selected Correspondence of Michael Faraday.* Edited by L. Pearce Williams. Cambridge: Cambridge University Press, 1971.

Faraday, Michael. "A Speculation touching Electric Conduction and the Nature of Matter." *Philosophical Magazine* 24, 3d ser. (1844): 136–44.

Firth, C. H. "Modern History in Oxford, 1724–1841." *English Historical Review* 32 (1917): 1–21.

Fischer, E. S. *Elements of Natural Philosophy*. Translated by John Ferrar. Boston: Hillard, Gray, Little and Wilkens, 1827.

Frazer, N. L. "The Claim of Batley Grammar School to be the Alma Mater of the Rev. Joseph Priestley, F.R.S." *Transactions of the Unitarian Historical Society* 5 (1931–34): 133–44.

Fries, Charles C. "The Rule of Common School Grammars." *PMLA* 42 (1927): 221–37.

———. *The Structure of English: An Introduction to the Construction of English Sentences*. New York: Harcourt, Brace and Co., 1952.

Gentleman's Magazine. "Henry Clarke." Obituary. *Gentleman's Magazine* 88 (1818): 465–66.

Gibbs, F. W. *Joseph Priestley: Adventurer in Science and Champion of Truth*. London: Thomas Nelson and Sons, 1965.

Gilbert, Alan D. *Religion and Society in Industrial England: Church, Chapel and Social Change, 1740–1914*. London: Longman, 1976.

Godley, A. D. *Oxford in the Eighteenth Century*. New York: G.P. Putnam's Sons; London: Methuen and Co., 1908.

Good, H. G. "The Sources of Spencer's Education." *Journal of Educational Research* 13 (1926): 325–35.

Gordon, Alexander. *Cheshire Classis Minutes, 1691–1745*. London: Chiswick Press, for the Provincial Assembly of Lancashire and Cheshire, 1919.

———. *Heads of English Unitarian History, with appended Lectures on Baxter and Priestley*. Bath: Cedric Chivers, 1970, [1895].

———. "John Horsley (1685–1732)." *Dictionary of National Biography* 9 : 1276–77.

Guerlac, Henry. "Joseph Priestley's first papers on Gases and their Reception in France." *Journal of the History of Medicine* 12 (1957): 1–12.

Guthrie, Warren. "The Development of Rhetorical Theory in America, 1635–1850." *Speech Monographs* 14 (1947): 38–54.

Hall, James. *A History of the Town and Parish of Nantich, or Wich-Malbank, in the County Palatine of Chester*. Nantwich: for the Author, 1883.

Hans, Nicholas. *New Trends in Education in the Eighteenth Century*. London: Routledge and Kegan Paul, 1961.

Hazlitt, William. "The Late Dr. Priestley." *The Atlas*, 14 June 1829. Reprint, *Complete Works of William Hazlitt*, edited by P. P. Howe, 20 : 237–38. London: J. M. Dent and Sons, 1934.

Heilbron, John L. "Aepinus." In *Dictionary of Scientific Biography*, edited by Charles C. Gillispie, 1 : 61–68. New York: Charles Scribner's Sons, 1970.

———. *Electricity in the 17th and 18th Centuries: A Study of Early Modern Physics*. Berkeley and Los Angeles: University of California Press, 1979.

Hengel, Martin. *The Atonement: The Origins of the Doctrine in the New Testament*. London: SCM Press, 1981.

Hertling, Georg Freiherrn von. *John Locke und die Schule von Cambridge*. Freiburg im Breisgau: Herder, 1892.

Hill, Andrew M. "The Death of Ordination in the Unitarian Tradition." *Transactions of the Unitarian Historical Society* 14 (1967–70): 190–208.

Holland, T. C. "Brief History of the Dissenters from the Revolution." *Monthly Repository* 12 (1817): 201–3, 384-87, 454–60.

Howell, Wilbur Samuel. *Eighteenth-Century British Logic and Rhetoric.* Princeton: Princeton University Press, 1971.

Hutton, A. W. "Divines." Chap. 15 in *The Age of Johnson.* Vol. 10 of *The Cambridge History of English Literature,* edited by A. W. Ward and A. R. Waller. Cambridge: Cambridge University Press, 1933.

Jeffrey, Francis. Review of "Essays on the Nature and Principles of Taste by Archibald Alison." *Edinburgh Review* 18 (1811): 1–46.

Johansen-Berg, J. "Arian or Arminian," *Journal of the Presbyterian Historical Society* 14 (1968–72): 33–58.

Kallich, Martin. "The Association of Ideas and Critical Theory in XVIII-century England." Ph.D. diss., Johns Hopkins University, 1945.

Kittridge, George L. *Some Landmarks in the History of English Grammars.* Boston: Athanaeum Press, 1906).

Knox, H. M. "Joseph Priestley's Contribution to Educational Thought." *Studies in Education: The Journal of the Institute of Education* 1 (1949): 82–89.

Langley, Arthur S. "Baptist Ministers in England about 1750." *Transactions of the Baptist Historical Society* 6 (1918–19): 138–62.

Leonard, Stirling. *The Doctrine of Correctness in English Usage.* Madison: University of Wisconsin Press, 1929.

Lester, Derek N.R. *The History of Batley Grammar School, 1612–1962.* Batley: J. S. Newsome and Sons, [1962].

Levine, Joseph M. "Ancients and Moderns Reconsidered." *Eighteenth-Century Studies* 15 (1981): 72–89.

Lodge, Oliver J. *Lightning Conductors and Lightning Guards.* London: Whittaker and Co., 1892.

———. *Modern Views of Electricity.* London: Macmillan and Co., 1892.

Lonsdale, Roger. *Dr. Charles Burney: A Literary Biography.* Oxford: Clarendon Press, 1965.

———. "William Bewley and The Monthly Review: A Problem of Attribution." *Papers of the Bibliographical Society* 55 (1961): 309–18.

Mack, Mary P. *Jeremy Bentham: An Odyssey of Ideas, 1748–1792.* New York: Columbia University Press, 1963.

Manuel, Frank E. *Isaac Newton, Historian.* Cambridge: Belknap Press of Harvard University Press, 1963.

Marsh, Robert. "The Second Part of Hartley's System." *Journal of the History of Ideas* 20 (1959): 264–73.

McCormmach, Russell. "Henry Cavendish: A Study of Rational Empiricism in 18th-Century Natural Philosophy." *Isis* 60 (1969): 293–306.

McEvoy, John G. "Joseph Priestley, 'Aerial Philosopher': Metaphysics and Methodology in Priestley's Chemical Thought, from 1762 to 1781." *Ambix* 25 (1978): 1–55, 93–116, 153–75; 26 (1979): 16–38.

McLachlan, Herbert. *English Education under the Test Acts.* Manchester: University of Manchester Press, 1931.

———. *The Unitarian Movement in the Religious Life of England: Its Contributions to Thought and Learning 1700–1900.* London: George Allen and Unwin, [1934].

———. "Warrington Academy: Its History and Influence." *Remains: Historical and Literary connected with the Palatine Counties of Lancaster and Chester, Chetham Society* 107, n.s. (1943).

McLachlan, H. John. "Mill Hill, Leeds and a Ministerial Appointment 1817." *Transactions of the Unitarian Historical Society* 10 (1951–54): 26–28.

Michael, Ian. *English Grammatical Categories and the Tradition to 1800.* Cambridge: Cambridge University Press, 1970.

[Mill, John Stuart]. *The Collected Works of John Stuart Mill.* Edited by J. M. Robson. Toronto: University of Toronto Press, 1977.

Monk, Samuel H. *The Sublime: A Study of Critical Theories in XVIIIth-Century England.* New York: Modern Language Association, 1935.

Mottelay, Paul Fleury. *Bibliographical History of Electricity and Magnetism chronologically arranged.* London: C. Griffin and Co., 1922.

Mullett, Charles F. "The Legal Position of English Protestant Dissenters, 1660–1689." *Virginia Law Review* 22 (1936): 495–526.

———. "The Legal Position of English Protestant Dissenters, 1689–1767." *Virginia Law Review* 23 (1937): 389–418.

———. "The Legal Position of English Protestant Dissenters, 1767–1812." *Virginia Law Review* 25 (1939): 671–97.

Murray, V. Victor. "Doddridge and Education." In *Philip Doddridge, 1702–1751: His Contribution to English Religion.* Edited by Geoffrey F. Nuttall. London: Independent Press, 1951.

N., J.N. *O_2: The Bicentenary of the Discovery of Oxygen by Joseph Priestley.* Leeds: Leisure Services Department, Leeds City Museums, 11 September–26 October 1974, Exhibition Hall, City Square.

Nadel, George H. "Philosophy of History before Historicism." *History and Theory* 3 (1963–64): 291–315.

[Neville, Sylas]. *The Diary of Sylas Neville, 1767–1788.* Edited by Basil Cozens-Hardy. London: Oxford University Press, 1950.

Nicholson, William. *An Introduction to Natural Philosophy.* London: J. Johnson, 1790.

Orton, Job. *Memoirs of the Life and Writings of the Rev. Philip Doddridge.* Salop: J. Cotton and J. Eddowes, 1766.

Owen, W. J. B. *Wordsworth's Preface to Lyrical Ballads.* Vol. 9 of *Anglistica.* Copenhagen: Rosenkilde and Bagger, 1957.

Palmer, Alfred Neobard. *A History of the Town and Parish of Wrexham.* Part 4. Wrexham: Woodall and Thomas, 1893.

Palmer, S[amuel]. "Memoir of Dr. Caleb Ashworth." *Monthly Repository* 8 (1813): 693–96.

Parker, Irene. *Dissenting Academies in England.* New York: [1914], Octagon Books, 1969.

Parkes, Samuel. "Mr. Parkes's Account of a Visit to Birstall, Dr. Priestley's Native Place." *Monthly Repository* 11 (1816): 274–76.

Parsons, Edward. *Civil, Ecclesiastical, Literary, Commercial, and Miscellaneous History of Leeds, Bradford, Wakefield, Dewsbury, Otley, and the District within Ten Miles of Leeds.* Leeds: Frederick Hobson; London: Simkin and Marshall, 1834.

Peardon, Thomas P. *Transition in English Historical Writing, 1760–1830.* Studies in History, Economics and Public Law, no. 390. New York: Columbia University Press, 1933.

Peel, Frank. *Nonconformity in Spen Valley.* Heckmondwike: Senior and Co., 1891.

Poggendorff, J. C. *Lebenslinien zur Geschichte der Exacten Wissenschaften seit Wiederherstellung Derselben.* Berlin: Alexander Duncker, 1855.

Poldauf, Ivan. *On the History of some Problems of English Grammar before 1800.* LV. Prispevkky k Dejinan Reci a Literatury Anglike (Prague Studies in English). Prague: Philosophical Faculty, Karlovy University, 1948.

"Priestley." *Encyclopaedia Britannica.* 11th ed. 22:322b.

"R.,P." [Palmer, Samuel]. "Brief Memoirs of the Rev. Mr. Samuel Clark . . ." *Monthly Repository* 1 (1806): 617–22.

Rees, Abraham, ed. *Cyclopaedia: or Universal Dictionary of Arts, Sciences and Literature.* Vol. 2. London: Longman, Hurst, Rees, Orme and Brown, et al., 1820.

Richey, Russell E. "Joseph Priestley: Worship and Theology." *Transactions of the Unitarian Historical Society* 15 (1972): 41–53, 98–103.

Rimmer, W. G. "William Hey of Leeds, surgeon (1736–1819): A reappraisal." *Proceedings of the Leeds Philosophical and Literary Society* 9, part 8 (1961): 187–217.

Robbins, Caroline. *The Eighteenth-Century Commonwealthman.* Cambridge: Harvard University Press, 1959.

Robinson, Myron. "A History of the Electric Wind." *American Journal of Physics* 30 (1962): 366–72.

Robison, John. *System of Mechanical Philosophy.* Edinburgh: J. Murray, 1822.

Robson, Derek. *Some Aspects of Education in Cheshire in the Eighteenth Century.* Manchester: for the Chetham Society, 1966.

Roddis, Louis H. *James Lind, Founder of Nautical Medicine.* New York: Henry Schuman, 1950.

Rude, George. *Paris and London in the Eighteenth Century.* New York: Viking Press, 1971.

Russell, W. Vine. *Looking back over three centuries: 1662–1962.* Needham Market: Needham Market Congregational Church, 1962.

———. Personal communication, 1962.

Sandford, William P. *English Theories of Public Address, 1530–1828.* Columbus, Ohio: H. L. Hedrick, 1931.

Schofield, Robert E. "Electrical Researches of Joseph Priestley." *Archives Internationale d'Histoire des Sciences* 64 (1963): 277–86.

———. "Joseph Priestley, eighteenth-century British neo-Platonism, and S. T. Coleridge." In *Transformation and Tradition in the Sciences: Essays in Honor of I. Bernard Cohen,* edited by Everett Mendelsohn. Cambridge: Cambridge University Press, 1984.

————. *The Lunar Society of Birmingham: A Social History of Science and Industry in Eighteenth-Century England.* Oxford: Clarendon Press, 1963.

————. *Mechanism and Materialism: British Natural Philosophy in an Age of Reason.* Princeton: Princeton University Press, 1970.

————, ed. *A Scientific Autobiography of Joseph Priestley (1733–1804): Selected Scientific Correspondence.* Cambridge: MIT Press, 1966.

Schonland, B. F. J. *The Flight of Thunderbolts.* Oxford: Clarendon Press, 1950.

Schroeder, W[illiam] Lawrence. *Mill Hill Chapel Leeds, 1764–1924.* [Leeds, Mill Hill Chapel, 1924].

Smith, Andrew C. *Theories of the Nature and Standard of Taste in England 1700–1790.* Chicago: University of Chicago Libraries, 1933.

Smith, Edgar F. *Priestley in America, 1794–1804.* Philadelphia: P. Blakiston's Son and Co., 1920.

Smith, J. W. Ashley. *The Birth of Modern Education: The Contribution of the Dissenting Academies, 1660–1800.* London: Independent Press, 1954; United Reform Church in the United Kingdom, 1972).

Sorby, W. R. "Philosophers." Chap. 14 in *The Age of Johnson.* Vol. 10 of *The Cambridge History of English Literature,* edited by A. W. Ward and A. R. Waller. Cambridge: Cambridge University Press, 1933.

Spalding, James C. "The Demise of English Presbyterianism: 1660–1760." *Church History* 28 (1959): 63–83.

Sperry, Willard L. *Wordsworth's Anti-Climax.* Cambridge: Harvard University Press, 1935.

Stamp, Gavin. *The Great Perspectivists.* London: Trefoil Books, 1982.

Stedman, Thomas. "A List of the Pupils educated by P. Doddridge, D.D." *Monthly Repository* 10 (1815): 686–88.

Stewardson, Thomas. "The Destruction of Priestley's Library." *Notes and Queries,* 34th ser. (1969): 64–65.

Taylor, E. G. R. *Mathematical Practitioners of Hanovarian England.* Cambridge: Institute of Navigation, at the University Press, 1966.

Thomas, T. "Supplementary Hints to the Memoir of Dr. Ashworth." *Monthly Repository* 9 (1814): 10–12, 78–80.

Thomson, John. *Account of the Life, Lectures, and Writings of William Cullen, M.D.* London and Edinburgh: Wm. Blackwood and Sons, 1859.

Thorpe, T. E. *Joseph Priestley.* London: J. M. Dent, 1906.

Turner, A. J., ed. *Science and Music in Eighteenth-Century Bath.* Bath: University of Bath, Exhibition Pamphlet for Holburne of Menstrie Museum, 1977.

[Turner, William] "V.F." "Historical Account of Warrington Academy." *Monthly Repository* 8 (1813)–10 (1815). Republished under his name as *The Warrington Academy,* introd. G. A. Carter. Warrington, Library and Museum Committee, 1957.

Tyerman, L. *The Life and Times of the Rev. John Wesley, M.A.,* New York: [1872], Burt Franklin, 1973.

Urwick, William. "Deanery of Nantwich." in *Historical Sketches of Nonconformity in the County Palatine of Chester,* edited by William Urwick. London: Kent and Co.; Manchester: Septemius Fletcher, 1864.

Waddington, John. *Congregational History, 1700–1800.* London: Longmans, Green and Co., 1876.

Walker, George. *Essays on Various Subjects, to which is prefixed, a Life of the Author.* London: J. Johnson, 1809.

Warboise, Emma Jane. *The Life of Dr. Arnold (Late Head Master of Rugby School)* (London, Burnet and Isbister, 1897).

Ward, A. W., and A. R. Waller, eds. *The Cambridge History of English Literature.* Vol. 9, *From Steele and Addison to Pope and Swift;* vol. 10, *The Age of Johnson.* Cambridge: Cambridge University Press, 1933.

Weaver, William D., ed. *Catalogue of the Wheeler Gift of Books, Pamphlets and Periodicals in the Library of the American Institute of Electrical Engineers.* New York: American Institute of Electrical Engineers, 1909.

Webb, Thomas E. *The Intellectualism of Locke: An Essay.* Dublin: William McGee & Co., 1857.

Weld, Charles R. *History of the Royal Society.* London: J. W. Parker, 1848.

Wellbeloved, Charles. "Presbyterian Nonconformity at Leeds." *Christian Reformer* 3, n.s. (1847): 392–400, 477–87, 522–37.

Westfall, Richard S. *Never at Rest: A Biography of Isaac Newton.* Cambridge: Cambridge University Press, 1980.

Whitaker, Sir Edmund. *A History of the Theories of Aether and Electricity.* New York: Philosophical Library, 1951.

Whitney, Lois. *Primitivism and the Idea of Progress.* Baltimore: Johns Hopkins University Press, 1934.

Wilbur, Earl M. *A History of Unitarianism: In Transylvania, England, and America.* Cambridge: Harvard University Press, 1952.

Williams, John. *Memoirs of the Late Reverend Thomas Belsham.* London: for the Author, 1833.

Wilson, R. G. *Gentlemen Merchants: The Merchant Community in Leeds, 1700–1830.* Manchester: Manchester University Press, 1971.

Winstanley, D. A. *Unreformed Cambridge.* Cambridge: Cambridge University Press, 1935.

Wordsworth, Christopher. *Scholae Academicae: Some Account of the Studies at English Universities in the Eighteenth Century.* Cambridge: Cambridge University Press, 1877.

Wright, Lawrence. *Perspective in Perspective.* London: Routledge and Kegan Paul, 1983.

Yates, James. "Ecclesiastical Proceedings against Dr. Doddridge." *Christian Reformer* 1, n.s. (1861): 552–57.

Young, Thomas. *A Course of Lectures on Natural Philosophy and the Mechanical Arts.* London: J. Johnson, 1807.

Index